D0203430

BODYSPACE

Anthropometry, Ergonomics and the Design of Work

Third Edition

Stephen Pheasant

---

30 March 1949–30 March 1996
(Reprinted from the Second Edition of Bodyspace)

Stephen, who died at the tragically early age of 47, will be remembered by a large and diverse group of friends, colleagues, students, courtroom colleagues and musicians. This alone is testimony to a man whose undoubted intellectual, creative and communicative skills were matched only by his verve and energy in a wealth of areas.

Stephen was raised in Islington before going up to Gonville and Caius College, Cambridge to read Medical Science in 1968. His contemporaries will perhaps remember him best for his passion for free jazz and his role in taking the musically based shows 'Stony Ground' and 'Make Me, Make You' to the Edinburgh Fringe in consecutive years. His earlier experience with the National Youth Jazz Orchestra, and the inspiration of his hero Charlie Parker, no doubt influenced him to form the Steve Pheasant Quintet, which played at the White Hart Inn, Drury Lane from the mid-1970s to the early 1980s. A close friend and band member, Iain Cameron, recalls Steve's versatility and be-bop creativity on sax, his occasional vocal rendering of 'Let the good times roll' and the band's 'sit in' style, in a manner which reflects the enthusiasm and participative spirit of the man. This, coupled with a burning commitment, are instantly recognized in his professional career.

Stephen's students at the Royal Free Hospital and University College, where he lectured for many years in anatomy, biomechanics and ergonomics, could rarely have encountered a more exceptional communicator. His ability to conceptualize and then project complex biomechanical functions in a suitable mode for student learning were testimony to his instinct for education and scholarship. He followed with keen interest the progress of the ergonomists he helped train. His academic and textbook publications were recognized for their application and clarity, a talent acknowledged through the 1985 award, sponsored by the *New Scientist*, for writing about science in plain English. Such skills were inevitably sought by other academic institutions and learned societies; thus he was always high on the invited speaker lists of conference organisers. Professional societies, including the Royal Society of Medicine and the Royal College of Nursing recognized his abilities, as did the British School of Osteopathy, where he held an honorary chair.

Stephen's written output was prolific, and his textbooks, including the first edition of *Bodyspace* (1986) and *Ergonomics, Work and Health* (1991) have become standards on reading lists around the world. His research output was recognized by the Ergonomics Society with the award of the Sir Frederick Bartlett Medal in 1982, jointly with his close colleague Professor Donald Grieve. His published data of human dimensions have been cited in more ergonomic designs than perhaps any other, and we are grateful too for his contribution to improved design of equipment, tools and many other artefacts of work and leisure use.

When he moved from the academic world, he chose to enter the field of personal injury litigation. In particular, Stephen specialized in work-related musculoskeletal damage, including back pain and repetitive strain injury. As an expert witness, most frequently acting on behalf of the injured party, he was perhaps at his most fulfilled. His desire to challenge orthodoxy, his intellectual skills, his ability to communicate, his love of fierce debate and his instinct for 'telling a good story' were all given full rein in such an arena. I have rarely seen him happier than when we developed litigious arguments or exchanged courtroom anecdotes with the help of a good Bordeaux. I am sure that adversaries and colleagues alike will sorely miss his presence and his skills.

Stephen knew of his failing health but never slowed in his endeavours; his output was prodigious. His mother and his partner, Sheila Lee, have much to bear. Family, colleagues, students and friends will remain indebted to Stephen, each in our own way. He will be remembered with affection, respect and regard. I know I speak for many when I say I have lost an inspiring friend.

Dr. Peter Buckle
April 1996

BODYSPACE

Anthropometry, Ergonomics and the Design of Work

Third Edition

Stephen Pheasant
Christine M. Haslegrave

Boca Raton London New York Singapore

A CRC title, part of the Taylor & Francis imprint, a member of the Taylor & Francis Group, the academic division of T&F Informa plc.

Published in 2006 by
CRC Press
Taylor & Francis Group
6000 Broken Sound Parkway NW, Suite 300
Boca Raton, FL 33487-2742

No claim to original U.S. Government works
Printed in the United States of America on acid-free paper
10 9 8 7 6 5 4 3 2 1

International Standard Book Number-10: 0-415-28520-8 (Hardcover)
International Standard Book Number-13: 978-0-415-28520-9 (Hardcover)
Library of Congress Card Number 2005002712

Library of Congress Cataloging-in-Publication Data

Pheasant, Stephen.
Bodyspace : anthropometry, ergonomics, and the design of work / Stephen Pheasant and Christine Haslegrave. -- 3rd ed.
p. cm.
Includes bibliographical references and index.
ISBN 0-415-28520-8 (alk. paper)
1. Human engineering. 2. Anthropometry. 3. Engineering design. I. Haslegrave, C. M. II. Title.

TA166.P49 2005
620.8'2--dc22 2005002712

Taylor & Francis Group
is the Academic Division of T&F Informa plc.

Visit the Taylor & Francis Web site at
http://www.taylorandfrancis.com

and the CRC Press Web site at
http://www.crcpress.com

He who would do good to another must do it in Minute Particulars:
General Good is the plea of the scoundrel, hypocrite and flatterer,
For Art & Science cannot exist but in minutely organized Particulars.
And not in generalising Demonstrations of the Rational Power.

William Blake, *Jerusalem*, 1815, pl. 55, 11.60–64

I design plain truth for plain people.

John Wesley, Sermon, 1746.

Foreword

It has been almost 20 years since the first edition of *Bodyspace* appeared. Over this time it has become clear that the science of ergonomics and its application to modern work systems has never been needed more. The benefits of good ergonomics accrue to individuals, organisations and society alike. Sadly, the catalogue of disasters such as Chernobyl, Bhopal, Piper Alpha and a series of high profile rail accidents provide graphic examples of why ergonomists are needed. The need for an understanding of behaviour, capacities and needs of humans prior to the implementation of a complex system has been identified over and over again. Tragically, the professionals with the required knowledge and skills are consulted, too frequently, only after the event. I'm sure that many of my ergonomics colleagues would agree that the call to action rarely comes during the design process but rather as a desperate plea following an acute or chronic system failure.

If the major acute complex system failure is the focus of public and media attention, then chronic system failure is the silent enemy. The increase in sickness days and work output lost to musculoskeletal problems such as back pain and so called RSI (repetitive strain injury), the rapidly escalating problems of stress-related disorders and their increasing cost to our economies are all testament to poorly designed work systems.

As ergonomists struggle to communicate to others the need for system design, we continue to see chronic disasters developing around us. One such problem is that of the healthcare system which frequently fails the end user, that is the patient, when safety is compromised.

At a more readily appreciable level, the continuous rise of technology and communication systems has often failed to consider the wider needs of the intended user group. For example, how many senior members of the population do we now hear complaining of the inaccessibility of modern technological gadgets, for example mobile phones, whose text and keypad size is starting to exclude all but the most sharp-eyed youngsters with nimble fingers?

At another level, we are frequently asked to assist companies at an organisational level where there remains a reluctance to understand and implement basic concepts based on ergonomics principles that could have lasting impact on their efficiency. It is, of course, of concern that the business case for even simple, user-focused design is still under-documented. It is perhaps too obvious that a well-designed tool will perform better in the hands of the skilled operator than a poorly designed one. The failure to record and cost this adequately leads, too frequently, to the good design being replaced by cheaper, less effective substitutes.

The knowledge base on which ergonomics rests grows significantly year by year. The need for authoritative, contemporary and above all usable reference sources is therefore great. *Bodyspace* is an example of that rare breed of text that, upon publication of the first edition, found favour with both academics and practitioners. Such publications do not happen by chance, and it is undoubtedly a lasting testimony to Stephen that his writing is as accessible and entertaining now, some 20 years after the original edition, as when it was first put into print.

When Stephen died in March 1996, it was difficult to see how such an important text could subsequently be updated to reflect the inevitable new knowledge base that would develop. The update in this third edition has been provided by Christine Haslegrave. Christine's task should not be underestimated, as not only has she skillfully integrated new knowledge into the existing text but she has achieved this without losing the unique, idiosyncratic, English style of writing that has proved so immensely popular with students and others. Whilst the underlying concepts of *Bodyspace* have remained constant, the book now reflects contemporary knowledge in areas such as office ergonomics, the design of hand tools, the development of standards and new developments in methods, for example three-dimensional anthropometry. Importantly, these developments lead us to see *Bodyspace* as not solely a reference text, but also as a thought-provoking and challenging document that enables us to think more clearly about where and how ergonomics impacts on the world today.

As a director of a large and successful postgraduate degree course in ergonomics, I know what a valuable text this is not just because of its content, but in how it engages students and practitioners alike and, like all great educational texts, how it encourages its readers to think beyond the written page.

Professor Peter Buckle
University of Surrey

October 2004

Acknowledgments

Various figures, diagrams and graphs are reproduced from work published by other researchers, and we are grateful for permission to include these. Our thanks also go to Johan Molenbroek and Bill Evans for allowing us to use their unpublished anthropometric datasets. We should particularly like to thank Keith Morton who, as Stephen Pheasant said, "drew all the remaining figures which show any signs of artistic talent (those which show no such talent are my own responsibility)". The Department of Education and Science gave permission to publish the data in Tables 10.23 to 10.38.

Editor

Christine M. Haslegrave is a senior lecturer at the Institute for Occupational Ergonomics at the University of Nottingham and an editor of the journal *Ergonomics*. She is a chartered engineer as well as a fellow of the Ergonomics Society. In 1995, she received the Otto Edholm Award of the Ergonomics Society for significant contributions to applied research in ergonomics. Her research at the Institute for Occupational Ergonomics includes investigation of workplace design, working postures, the biomechanical demands of manual materials handling tasks, work redesign in relation to health and safety problems in industry, and vehicle ergonomics in design and manufacture. She is involved in the training of groups of industrial engineers and health and safety professionals, and has served on various standards committees. She was previously head of the ergonomics section at the Motor Industry Research Association, Nuneaton for several years, with interests in vehicle safety and ergonomic design and evaluation. Her research there included interior packaging for vehicles, fit of safety belts and other restraint systems, and design of impact test dummies, as well as organising a large scale anthropometric survey of U.K. vehicle occupants.

CONTENTS

PART I Ergonomics, Design and Anthropometry

PART II Application of Anthropometry in Design

PART III The Bodyspace Tables – Anthropometric Database

Part I

Ergonomics, Design and Anthropometry

1 Introduction to Ergonomic Design

1.1 INTRODUCTION

> Several similar contests with the petty tyrants and marauders of the country followed, in all of which Theseus was victorious. One of these was called Procrustes or the stretcher. He had an iron bedstead on which he used to tie all travellers who fell into his hands. If they were shorter than the bed he stretched their limbs to make them fit; if they were longer than the bed he lopped off a portion. Theseus served him as he had served others.
>
> **From *The Age of Fable* by Thomas Bulfinch (1796–1867)**

Prior to her injury, 'Janice' worked as a word processor operator for a medium-sized firm of management consultants just outside London. She worked in a typing pool with three other girls. One day, one of the partners in the firm needed to get a lot of information entered onto a database in a hurry — and it occurred to him that Janice might work faster if she was in a room on her own where she could not waste time chattering with her friends. So he had a computer terminal set up for her in the firm's library. It was placed on an antique wooden desk. This was somewhat higher than the standard office desk (antiques often are). It had two plinths and a 'kneehole drawer' in the space between them where the user sits. Janice found that however she sat at this desk she could not get into a comfortable working position. She noticed in particular that her wrists were not at their normal angle to the keyboard. It was during the early part of the afternoon that she first began to be aware of a dull ache at the backs of her wrists. This rapidly became worse until she was in considerable discomfort. So she told her boss about it. His response (as it was subsequently alleged) was to say: "Stop whingeing and get on with your work!" So Janice did. As a result, she developed an acute tenosynovitis affecting the extensor tendons of both wrists. Her condition subsequently became chronic, and she was no longer able to type. She lost her job and was forced to take up less well-paid employment as a traffic warden. She took legal action against her employers, who eventually settled 'on the courtroom steps' for a substantial sum of money.

What lessons may we learn from the story of 'Janice', over and above the more obvious ones concerned with management style and so on? Janice's injury was the result of a *mismatch* between the *demands* of her *working task* and the *capacity* of the muscles and tendons of her forearms to meet those demands. To put it another way, the excessive stresses to which these body structures were exposed stemmed from her being forced to *adapt* to an unsatisfactory working position, which was in turn the result of a *mismatch* between the dimensions and characteristics of her *workstation* and those of its *user*.

Injuries of this sort are common enough (although in Janice's case the causative factors in question are perhaps unusually clear-cut ones). Indeed in many parts of the world the incidence of such injuries is said to be reaching epidemic proportions. The problem of musculoskeletal injury at work — important as it may be in both economic and human terms — is but one small facet of a much larger class of issues involving the interactions between human beings and the objects and environments they design and use.

To say that we live in an artificial world is something of a truism. Look around you. It is unlikely that you are reading this in a desert wilderness. More probably you are indoors in a furnished room, or in a moving vehicle, or at least in a cultivated garden. It is all too easy to ignore the simple fact that most of the visible and tangible characteristics of the artificial environments in which we spend the greater part of our lives are the consequences of design decisions. By no means are all of the decisions that lead to the creation of these artificial environments made by professional designers. They may be the results of extensive planning or of momentary whims. They represent choices that have been made, which could have been made differently but were by no means inevitable.

All too often, however, the artefacts that we encounter in our human-made environment are like so many Procrustean beds to which we must adapt. Why should this be so? There is a science that deals with such matters. It is called *ergonomics*.

1.2 WHAT IS ERGONOMICS?

Ergonomics is the science of work: of the people who do it and the ways it is done, of the tools and equipment they use, the places they work in, and the psychosocial aspects of the working situation.

The word *ergonomics* comes from the Greek *ergos*, work, and *nomos*, natural law. The word was coined by the late Professor Hywell Murrell, as a result of a meeting of a working party which was held in Room 1101 of the Admiralty building at Queen Anne's Mansions on 8 July 1949, at which it was resolved to form a society for 'the study of human beings in their working environment'. The members of this working party came from backgrounds in engineering, medicine and the human sciences. During the course of the war, which had just ended, they had all been involved with research of one sort or another into the efficiency of the fighting man, and they took the view that the sort of research they had been doing could have important applications under peacetime conditions. There did not seem to be a name for what they had been doing, however, so they had to invent one and finally settled on ergonomics.

The word *work* admits a number of meanings. In a narrow sense it is what we do for a living. Used in this way, the activity in question is defined by the context in which it is performed rather than by its content. Unless we have some special reason for being interested in the socioeconomic aspects of work, however, this usage is arbitrary. Some people play the violin, keep bees or bake cakes to make a living; others do these things solely for pleasure or for some combination of the two. The content of the activity remains the same.

There is a broader sense, however, in which the term *work* may be applied to almost any planned or purposeful human activity, particularly if it involves a degree of skill or effort of some sort. In defining ergonomics as a science concerned with human work, we will in general be using the word in this latter and broader sense. Having said this, it would also be true that throughout its 50 years of history, the principal focus of the science of ergonomics has tended to be upon work in the occupational sense of the word.

Work involves the use of tools. Ergonomics is concerned with the design of these — and by extension with the design of artifacts and environments for human use in general. If an object is to be used by human beings, it is presumably to be used in the performance of some purposeful task or activity. Such a task may be regarded as work in the broader sense. Thus to define ergonomics as a science concerned with work or as a science concerned with design means much the same thing at the end of the day.

The ergonomic approach to design may be summarised in the *principle of user-centred design:*

> If an object, a system or an environment is intended for human use, then its design should be based upon the physical and mental characteristics of its human users (insomuch as these may be determined by the investigative methods of the empirical sciences).

The object is to achieve the best possible match between the *product* (object, system or environment) being designed and its *users*, in the context of the (working) *task* that is to be performed (Figure 1.1). In other words, ergonomics is the science of fitting the job to the worker and the product to the user.

1.2.1 What Criteria Define a Successful Match?

The answer to this question will depend upon the circumstances. Criteria that are commonly important in achieving a successful match include the following:

- Functional efficiency (as measured by productivity, task performance, etc.)
- Ease of use
- Comfort
- Health and safety
- Quality of working life

The ergonomic approach is to consider all relevant criteria, not simply to design for one criterion at the expense of others. Fitting the job to the worker involves consideration of health and quality of working life just as much as of productivity, and efficiency and quality of performance are influenced by all three (see Figure 1.1).

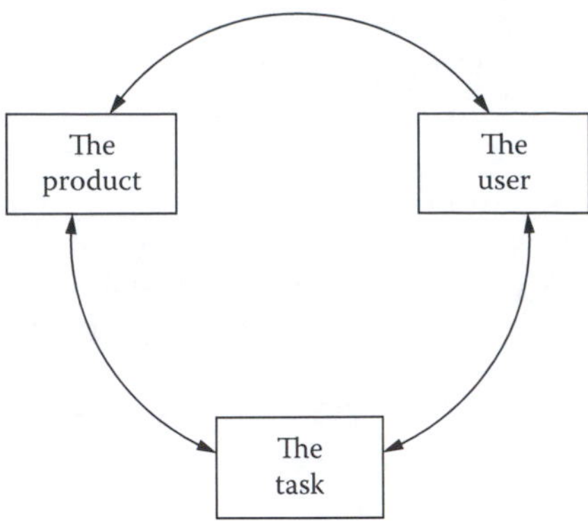

FIGURE 1.1 User-centred design: the product, the user and the task.

1.2.2 What If These Criteria Prove Incompatible?

Ergonomists often argue that this problem is not as big as it seems. There is some truth in this. There are without doubt circumstances in which ergonomic improvements introduced in the interests of health and safety have a positive pay-off in terms of productivity — and vice versa. Likewise, the product that is easy to use will probably, for that very reason, be both safe and efficient in its operation. It is the difficult-to-use products that are, in general, unsafe and inefficient. It would be naïve to pretend, however, that these sorts of basic criteria that we have invoked to define a good fit are *never* in conflict, and the deeper we fish in these waters, the more difficult the problem becomes.

The celebrated American product liability case of *Grimshaw* v. *Ford Motor Company* in 1981 (Jones, 1986) is illustrative — notwithstanding that it does not deal with ergonomic issues as such. Briefly, the facts were these. The defendants discovered a fault in the design of the petrol tank of one of their models, which meant that it was likely to explode in rear-end collisions. On the basis of certain alleged cost-benefit analyses, they decided that it would be cheaper in the long run to pay damages for the fatalities and injuries that resulted than to redesign the car and opted to take no further action. Outraged by this cynical view of the economic value of human life and limb, an American jury awarded punitive damages of $125 million against the defendants — much greater than any economic benefits that might have accrued to the defendants from failing to take proper steps to contain the hazard. Regrettably perhaps, this was reduced upon appeal to $3.5 million.

This is a somewhat gross example. Cost-benefit trade-offs with implications for health and safety are a fact of everyday industrial life (as any personal injury lawyer will tell you). The fragmented and repetitive short-cycle time tasks of industrial assembly remain an efficient enough way of producing many of the manufactured goods demanded by the consumer economy, but does this production process not have hidden costs? The physical injuries that result are easy enough to recognise, and we could in principle (if we chose) compute the costs of such injuries and incorporate them into some overall system of cost-benefit analysis or social audit. But does it stop there? Does work of this sort result in more subtle sorts of personal injury?

By its very nature, the applied discipline that we call ergonomics sits on the boundary between the domain of empirical science and the domain of ethical values. That is one very good reason that it is important.

1.3 ANTHROPOMETRICS

Anthropometry is the branch of the human sciences that deals with body measurements, particularly with measurements of body size, shape, strength, mobility and flexibility and working capacity. Humans are variable (in dimensions, proportions and shape, as in all other characteristics), and user-centred design requires an understanding of this variability. Anthropometrics is an important branch of ergonomics. It stands alongside (for example) cognitive ergonomics (which deals with information processing), environmental ergonomics and a variety of other identifiable subdisciplines which progress (in parallel, as it were) towards the same overall end of fitting the job to the worker and the product to the user.

This book is principally concerned with the anthropometric side of ergonomics, that is, with matching the physical form and dimensions of the product or workplace to those of its user and likewise with matching the physical demands of the working task to the capacities of the workforce. We shall be developing these issues at length in due course — but first a brief digression on human proportion (the relationships between anthropometric dimensions of body regions).

1.4 HUMAN PROPORTION: AN HISTORICAL PERSPECTIVE

In discussing classical styles of architecture, people often use the expression 'designed to the human scale'. The implication is that such buildings are aesthetically well proportioned and convey a certain sense of rightness and harmony. What does this mean? The idea certainly goes back a very long way, and it is closely linked historically to the various canons of human proportion which have been employed by artists and sculptors since ancient times.

The tomb painters of ancient Egypt (who worked in elevation only and knew no perspective) are known to have employed a modular grid for the preparation of their preliminary drawings of the human figure. The standing figure was divided into 14 equal parts, and the grid intersections corresponded to certain predetermined anatomical landmarks.

Modular systems of this sort (and their equivalents in terms of mathematical ratios between the dimensions of body parts) evolved initially as simple aids to drawing, and indeed rules of thumb of this sort are still taught in life classes today.

In classical times, however, the theory of human proportions began to assume a deeper significance, and it came to be thought that certain whole-number ratios between the dimensions of the body and its component parts were inherently 'harmonious' in the sense of being aesthetically pleasing. The argument was probably made, in the first instance, by analogy with musical harmony. The physics of vibrating pipes and stretched strings was known to Pythagoras (c. 582–500 B.C.).

Unfortunately, the systems of human proportions used by the sculptors of classical antiquity are for the most part lost to us. The single remnant of these systems which has been passed down to modern times concerns the female nude, in which the nipples and umbilicus are represented as making an equilateral triangle. (We have to allow for the effects of side bending of the trunk, or *contrapposto*, as it is called by artists.) You can see this relationship clearly in the Venus de Milo for example, as well as in paintings of the Renaissance and Baroque periods, as diverse in the actual physical types they portray as Botticelli and Rubens. It is absent, however, in painters who derive their style from Northern Gothic tradition, for example Cranach.

The most detailed system of human proportions which has come down to us from classical times is that of the Roman architectural theorist Vitruvius, writing some time around the year 15 B.C. Many of Vitruvius' body-part ratios are familiar to us from archaic units of measurement. The stature of a 'well-made man', for example, is held to be equal to his arm span (one fathom or two yards), which in turn is equal to four cubits (from the elbow to the fingertip), six foot lengths and so on. Vitruvius makes it clear that he regards this 'science' of human proportions as being a fundamental principle in building design.

The celebrated drawing of *Vitruvian Man* by Leonardo da Vinci, in which a male figure is drawn circumscribed within a square and a circle, must be one of the most overworked visual images around. By Leonardo's day, the theory of human proportions had become bound up with that of the 'golden proportion' or 'golden ratio'. It became accepted as fact that the umbilicus divides the stature of the standing (male) person in golden section: that is, such that the ratio of the greater part to the whole is equal to that of the lesser part to the greater part.

By this stage the entire affair was acquiring distinctly metaphysical overtones. It is these overtones that are invoked perhaps in the expression 'designing to the human scale'. If the phrase has any more pragmatic meaning we have been unable to discern it.

We may think of Leonardo (1452–1519) and his younger contemporary Albrecht Dürer (1471–1528) as standing on the watershed between modern empiricism and the earlier classical tradition, with Leonardo looking backwards and Dürer looking forwards. The classical tradition was prescriptive. It dealt with idealised human beings as they ought to be according to some preexisting aesthetic or metaphysical principle, rather than as real human beings as they actually are. Dürer's *Four Books of Human Proportions*, by contrast, may be regarded as the beginnings of modern scientific anthropometry. In them Dürer attempts to categorise and catalogue the diversity of human physical types, and his exquisite illustrations are, by his account of the matter at least, based upon the systematic observation and measurement of large numbers of people.

There is a curious footnote to this history. The classical tradition briefly reasserted itself in the middle years of the twentieth century in the work of the celebrated French architect Le Corbusier (1887–1965). His definitive treatment of the subject, *The Modulator: A Harmonious Measure to the Human Scale Universally Applicable to Architecture and Mechanics*, is an obscure work thought by many to be profound. It was the same Le Corbusier who said, 'A house is a machine for living in', and thus became one of the patron saints of the school of design known as 'functionalism' (of which more anon).

1.5 ERGONOMICS AND DESIGN

What is meant when we hear that a product is 'ergonomically designed'? Regrettably, the short answer to this question is all too often 'not very much'. Nowadays the term is widely used (or misused) in advertising circles. One frequently sees it employed, for example, in the marketing of fancy, overpriced and overdesigned furniture (particularly office furniture) which is supposed to be good for you in terms of some theory or another (which may or may not be correct) concerning how to sit correctly. The worst examples of these are very expensive indeed and ergonomically quite unsatisfactory. We can, of course, choose to shrug this off with the thought that 'if people are daft enough to buy this, it's their own silly fault'. However, to the responsible professional ergonomist this state of affairs is regrettable in the extreme, not least in that it can only bring his or her profession into disrepute. (We shall return to the ergonomics of furniture in general and office furniture in particular in Chapter 7.)

Occasionally, the misuses of the term *ergonomically designed* have an appealing surrealist quality. There was once an account in a Sunday newspaper of 'ergonomically designed pasta', which was (we were told) designed for ease of straining and sauce retention. (This could be called fitting the noodles to the user.)

Here is a good, straightforward, common sense way to recognise an ergonomically designed product, which is quoted from a pamphlet published by the Ergonomics Society (now unavailable) entitled *Ergonomics: Fit for Human Use*.

> Try using it. Think forward to all of the ways and circumstances in which you might use it. Does it fit your body size or could it be better? Can you see and hear all you need to see and hear? Is it hard to make it go wrong? Is it comfortable to use all the time (or only to start with)? Is it easy and convenient to use (or could it be improved)? Is it easy to learn to use? Are the instructions clear? Is it easy to clean and maintain? Do you feel relaxed after a period of use? If the answer to all of these is 'yes' then the product has probably been thought about with the user in mind.

Let us now look a little harder at the issue of functional design — first from the standpoint of design history. The American architect Louis Sullivan is credited with originating the slogan 'form follows function' (c. 1895), his implication being that functional considerations alone are sufficient to determine the form of an object and that ornament is therefore superfluous. According to this theory, functional objects are, of necessity, aesthetically pleasing. This is called 'functionalism'. It was the dominant theory underlying the so-called Modern Movement in design.

When we consider such modern classics as the Marcel Breuer Wassily chair (1925) or the Mies van der Rohe Barcelona chair (1929), we find very little relationship between the form of these seats and that of the human body which it is (presumably) their function to support. The fact that such pieces are commonly referred to as 'occasional chairs' implies that they are without particular function

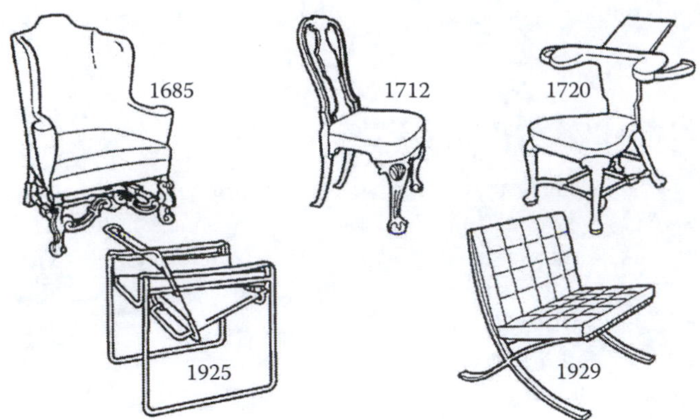

FIGURE 1.2 Form and function: eighteenth-century style and twentieth-century style. (Upper row, left to right) William and Mary winged armchair, Queen Anne dining chair, early Georgian library chair. (Lower row) Wassily chair by Marcel Breuer, Barcelona chair by Mies van der Rohe. For a contrast see Figure 5.10.

— except to be used occasionally. (In fairness we must admit that the Barcelona chair was in fact designed for the King of Spain to sit on at the opening of an exhibition.)

If we look back to earlier periods of furniture design, for example to the early years of the eighteenth century in Britain, we find a very different state of affairs. The William and Mary, Queen Anne and early Georgian periods produced furniture in general, and chairs in particular, that showed a closeness of functional relationship with the human body which has never been excelled (Figure 1.2). Consider the William and Mary winged chair and the variety of ways in which it may provide the postural support necessary for relaxation or the Queen Anne dining chair (sometimes known as the Hogarth chair) with its gently curved back which reflects the form of the human spine. Neither should we ignore those furniture types of the Georgian period designed for various specific functions: the library or 'cock-fighting' chairs which gentlemen would sit astraddle, the feminine equivalent for kneeling upon, the reading stands and even the 'night table' on which to empty the contents of the pockets. All these bespeak a paramount concern for user requirements — a relationship between maker and user which is also apparent in much vernacular design (perhaps most clearly so in the hand tools used by wood-workers and other craftsmen).

At some time around the midpoint of the eighteenth century, we see function gradually playing an increasingly accessory role as design became dominated by a succession of aesthetic theories or styles: neoclassicism, Gothic, etc. Paradoxically, the most recent of these styles is called 'functionalism', but it should be seen as an aesthetic demand for absence of ornament, 'truth to materials', etc., rather than a particular concern with end use. Functionalism is essentially a visual metaphor by which a designed object may acquire certain desirable connotations.

Discussions during a period of time teaching design students were the origin of the *five fundamental fallacies* of design which are set out in Table 1.1. They revolve

TABLE 1.1
The Five Fundamental Fallacies

No. 1	This design is satisfactory for me — it will, therefore, be satisfactory for everybody else.
No. 2	This design is satisfactory for the average person — it will, therefore, be satisfactory for everybody else.
No. 3	The variability of human beings is so great that it cannot possibly be catered for in any design — but since people are wonderfully adaptable, it doesn't matter anyway.
No. 4	Ergonomics is expensive, and since products are actually purchased on appearance and styling, ergonomic considerations may conveniently be ignored.
No. 5	Ergonomics is an excellent idea. I always design things with ergonomics in mind — but I do it intuitively and rely on my common sense so I don't need tables of data or empirical studies.

around two principal themes. The first is the contrast between the investigative methods of the empirical sciences and the creative problem-solving methods of the designer which, for want of a better word, we could call intuitive. The second theme is that of human diversity. This can be thought of as the single most important characteristic of people to be borne in mind in the world of practical affairs in general and of design in particular. To put it plainly, people come in a variety of shapes and sizes — to say nothing of their variability in strength, dexterity, mentality and taste. As we shall see, the five fallacies are increasingly difficult to refute.

Not many people would express the first fallacy in so many words, but in implicit form it is very widespread. How many products are actually tested at the design stage on a representative sample of users? More commonly the evaluation of a design proposal is entirely subjective. The designer considers the matter, tries out the prototype and concludes that it 'feels all right to me', with the clear implication that, if it is satisfactory for me, it will be for other people too. In general, objects designed by the stronger or more able members of the population can create insurmountable difficulties for the weaker and less able. Women frequently say with exasperation, 'You can tell it was designed by a man!'

The first fallacy is closely linked with the last by the concept of empathy, of which more anon; it is also closely linked to the second since most people consider themselves to be more or less average. Suppose we were to determine the dimensions of a door by the average height and breadth of the people who were to pass through it. The 50% of people taller than average would bang their heads; the 50% wider than average would have to turn sideways to squeeze themselves through. Since the taller half of the population are not necessarily the wider half, we would, in fact, satisfy or accommodate less than half of our users. Nobody would make such an elementary mistake in designing a door, but in our experience, the second fallacy turns up quite frequently in the work of students, of both design and ergonomics, who have only partially grasped the principles of anthropometrics. Obviously enough, we must seek to accommodate the largest percentage possible of the user population (see Chapter 2).

The third fallacy really has the ring of truth. Human beings are indeed very adaptable — they will put up with a great deal and might not necessarily complain. In the example we have just quoted, the taller half of the population would presumably learn to duck. This is the Procrustean approach to design. Adaptation to the Procrustean bed commonly has hidden costs in terms of ill health, although only rarely are these as dramatic as an amputated limb. Consider the economic losses occasioned by the extensive range of musculoskeletal disorders which may be attributed to faulty workspace design: back pain, neck pain, repetitive strain injuries, and so on (see Chapter 9). The variability of human beings is great, but it can be accommodated through empirical design (see Chapters 2 and 3).

Part of the refutation of the fourth fallacy rests upon these hidden costs of adaptation. In addition we should consider that the design process not only responds to consumer needs but in some measure creates them as well. We could question the extent to which (a) the public gets what the public wants, (b) the public wants what the public gets, or (c) the public knows perfectly well what it wants but can't get it and puts up with whatever is available. Superimposed over these possibilities are the effects of marketing and advertising on the one hand and consumer pressure groups and legislation on the other. The objects that the designer creates reflect the society in which they are created. In some cases, consumer pressure leads to the introduction of ergonomic features into design. This has happened quite dramatically in recent years in the area of computer technology. The computer workstations of today are much better than those of a decade or more ago — principally because of the effects that consumer pressure has had on market forces. In some areas, consumers are prepared to pay extra for quality. In Chapter 8 we shall consider the desirability of providing kitchen worksurfaces at a range of heights — this is perfectly possible technically but is generally deemed uneconomic. For which quality would the informed consumer rather pay extra: an elegant finish with gleaming worktops and polished brass cupboard door furniture or ease of use and less backache? However, beyond all these considerations is the simple fact that making something the right size is often no more expensive than making it the wrong size. The decision to ignore ergonomics on grounds of economics is often just an excuse.

The fifth and final fallacy involves some rather complex issues. The intuition and common sense of which we speak in this context is sometimes called 'empathy', and if you are a designer you may well have it in abundance. (Whether it is an innate gift or the fruit of experience is another matter.) Empathy is an act of introspection or imagination by which one may place oneself in another person's shoes. It could be argued that, by empathetically casting oneself in the role of the user, the act of designing for others becomes an extension of designing for oneself and the traditional subjective approach becomes valid. In some measure this is probably true, but can these intuitions really circumvent the problems of human diversity? Can we really imagine how somebody quite different from ourselves would experience a certain situation?

This question seems never to have really been put to the test. Psychologically it is a very interesting one. In general we would predict that empathy would increase with things like social and demographic proximity (as measured by age, sex, etc.) or with similarity in physical characteristics such as strength and fitness, attitudinal characteristics and so on. For any given degree of proximity or similarity, we should obviously expect some people to be more empathic than others. Were we able to measure this trait, we might find that it correlates in interesting ways with other personality characteristics. What sorts of people are the most empathic? Regrettably, however, this all remains within the realms of speculation.

The term *common sense* also deserves some scrutiny, not least because you often hear people say (perhaps with a measure of truth), 'Ergonomics — that's just common sense!' As a rule, statements like this should be viewed with circumspection. At one time the term *sensa communis* was used to refer to a (hypothetical) physiological system which integrated the separate functions of the traditional five senses of vision, hearing, touch, taste and smell. *Common sense* underwent a major shift of meaning, however, with its modern usage (as far as we can tell) being established by the eighteenth century or thereabouts. We all think we know what it means because we all have it. At one level, expressions like 'that's just common sense!' can be used as a justification for the blind acceptance of an untested hypothesis. We must also distinguish common sense from common knowledge and conventional wisdom. There are those who think that common sense and the scientific method are much the same thing — the latter being a refined version of the former. There seems to be a measure of truth in this. We would only add that common sense sometimes seems remarkably rare.

1.6 THE USER-CENTRED APPROACH

We have described the ergonomic approach to design as user-centred. How may we characterise this description more fully? One way would be in terms of methodology. The science of ergonomics has built up both a substantial organised body of knowledge about human capacities and limitations and a repertoire of investigative methods for acquiring such knowledge and for practical problem solving. Two particular techniques deserve special attention: *task analysis* and the *user trial*. For a more detailed treatment of these and of ergonomics methodology in general, the reader is recommended to turn to Wilson and Corlett (2005).

Advice to students of both design and ergonomics is that 'every good project starts with a task analysis and ends with a user trial'. All too often this goes unheeded, and the resulting design solution is inadequate, either facilitating only some aspects of the use of the product (just those that occurred to the designer without any systematic analysis) or satisfying only some of the potential users. Task analysis and user trials are both extremely simple in concept — to the point perhaps of being 'just common sense' — but they help the designer avoid falling into the trap of any of the five fallacies.

A *task analysis* is a formal or semiformal attempt to define and state what the user/operator *is actually going to do* with the product/system/environment in question. This is stated in terms of the desired ends of the task, the physical operations the user will perform and the information processing and decision making it entails. Each of these steps in the task is then considered in turn to identify the physical

TABLE 1.2
User-Centred Design

User-centred design is empirical.
It seeks to base the decisions of the design process upon hard data concerning the physical and mental characteristics of human beings, their observed behaviour and their reported experiences. It is distrustful both of grand theories and intuitive judgements — except insomuch as these may be used as the starting points for empirical studies.
User-centred design is iterative.
It is a cyclic process in which a research phase of empirical studies is followed by a design phase, in which solutions are generated which can in turn be evaluated empirically.
User-centred design is participative.
It seeks to enrol the end-user of the product as an active participant in the design process.
User-centred design is non-Procrustean.
It deals with people as they are rather than as they might be; it aims to fit the product to the user rather than vice versa.
User-centred design takes due account of human diversity.
It aims to achieve the best possible match for the greatest possible number of people.
User-centred design takes due account of the user's task.
It recognises that the match between product and user is commonly task-specific.
User-centred design is systems-orientated.
It recognises that the interaction between product and user takes place in the context of a bigger socio-technical system, which in turn operates within the context of economic and political systems, environmental ecosystems and so on.
User-centred design is pragmatic.
It recognises that there may be limits to what is reasonably practicable in any particular case and seeks to reach the best possible outcome within the constraints imposed by these limits.

and cognitive requirements and assess these against the capacities, skills and limitations of the expected users. The environmental constraints that might pertain and any potential hazards are also noted. An effective task analysis will clarify the overall goals of the project, establish the design criteria that need to be met and point out the most likely areas of *mismatch*. This clearly establishes the *user's needs* and expresses them in terms of design criteria against which the final prototype can be evaluated.

A *user trial* is just what its name suggests: an experimental investigation in which a sample of people test a prototype version of the product under controlled conditions. The subjects in the trial must be chosen with care. Ideally they should be a representative sample of the population of users for whom the end product is ultimately intended. There would be little point in trying out some new high-tech product on the technophiles down the corridor if it is ultimately going to be used by the technophobes in the street. Sometimes, as a deliberate strategy, it makes sense to test a product on those sorts of people who are likely to have the most difficulty using it — the technologically naïve, the elderly and infirm, and so on — on the grounds that, if they can cope, then the product will also be acceptable for the more able majority. (This is the equivalent of the *principle of the limiting user* in anthropometrics which we shall encounter in the next chapter.) We must likewise take care to ensure that the circumstances under which the trial is conducted are a reasonably valid approximation to those of real-world use and that the product is tested in a range of scenarios typical of its various uses.

The *user-centred approach* can be characterized in terms of the features in Table 1.2, which will also serve as a summary of much of what has gone before in this chapter. The methodology of task analysis and user trials focuses the design requirements clearly on the users. User trials can collect opinions and suggestions from the representative sample who participate in the trials as well as taking measurements (whether by observation of behaviour, recording physical responses or requesting subjective judgements), and these can then be used to evaluate the success or degree of inadequacy of the prototype. Design is, by its nature, an iterative process, continuously looking for and testing solutions to design problems — or, for the ergonomic aspects of the design, mismatches — until the prototype is judged to meet the design criteria which were defined after the task analysis. This empirical approach has an additional advantage in providing data which can be used in a cost-benefit analysis of alternative design solutions, assisting the ergonomist or designer in overcoming the scepticism of project managers and finance controllers who are susceptible to the third and fifth fallacies. This accompanies the pragmatic approach to give ergonomics requirements equal weight with other design requirements and to achieve the best possible outcome for the user within the constraints imposed by technology, cost and other constraints.

2 Principles and Practice of Anthropometrics

2.1 INTRODUCTION

There are a few situations in which it is possible to design a product or workstation for a single user: bespoke tailoring, *haute couture*, the customised seats used by racing drivers and the workstations of astronauts are examples. These are all essentially *luxury goods*. For a very small number of especially unfortunate individuals, the luxury of custom design becomes a necessity. The physical characteristics of the very severely disabled are so diverse that aids to mobility and independence must often be made for the individual concerned. However, in the great majority of real-world design problems our concern will be with a population of users. The product must therefore be designed to be adequate for all the population, include adjustability or be produced in a range of sizes.

We all acknowledge the necessity of manufacturing garments in a range of sizes, but would it be true to say that chairs and tables, for example, should be supplied in a range of sizes as well? The answer is 'only to a limited extent'. We do not expect adults and children to use the same-sized writing desks in their offices and schools, although they seem to cope perfectly well with the same dining table at home. We commonly supply typists with adjustable chairs, but their desks are usually of fixed height. Obviously, we are prepared to accept a less accurate fit from a table and chair than from a shirt and trousers. What is rather less obvious is how we should choose the best compromise dimensions for equipment to be employed by a range of users, and at what point we should conclude that adjustability is essential. In order to optimise such decisions we require three types of information:

1. The anthropometric characteristics of the user population.
2. The ways in which these characteristics might impose constraints upon the design.
3. The criteria that define an effective match between the product and the user.

Before discussing these matters further we shall need to establish some of the mathematical foundations upon which the applied science of anthropometrics rests. In the section that follows we have endeavoured to do this with the minimum possible recourse to the use of mathematical equations and formulae. The reader who requires a more detailed mathematical treatment of the subject is referred to the Appendix.

2.2 THE STATISTICAL DESCRIPTION OF HUMAN VARIABILITY

2.2.1 FREQUENCY DISTRIBUTION OF A DIMENSION WITHIN A POPULATION

In order to establish the statistical concepts that describe human variability, let us conduct what earlier scientific writers would have called an experiment of the imagination. Supposing you are in a large public building frequented by a fairly typical cross-section of the population. A companion, who is an inveterate gambler, offers to take bets on the stature (standing height) of the next adult man to walk down the corridor. (We could just as well bet on women, children or everyone taken together, but it is a little easier to deal with the problem mathematically if we only consider adults of one sex.) On what height would you be best advised to place your money (assuming of course that you have no prior knowledge of people who happen to be in the area)? You will probably pick a stature that is somewhere near the average, since experience has told you that middling-sized people are relatively common, whilst tall or short people are rare by comparison. You have in essence made a judgement as to the relative probability of people of different statures, or the relative frequency with which such people are encountered by chance. Average people are more probable than extremes, in that you encounter them more frequently.

The statistically minded punter, offered a bet of this kind, could optimise the chance of winning by going out and measuring all the men in the building. With these data we could plot a chart like the one shown in Figure 2.1, in which probability (frequency of encounter) is plotted vertically against stature, which is plotted horizontally. The smooth curve on this chart is known as a probability density function or a *frequency distribution*. The particular curve we have drawn here is symmetrical about its highest point; this is the average stature, otherwise known as the mean, and is also the most probable stature. Since the curve is symmetrical, it follows that 50% of the population are shorter than average and 50% are taller. We would say, therefore, that in this distribution the mean is equal to the fiftieth percentile (commonly abbreviated as 50th %ile). In general, n% of people are shorter that the nth %ile. Hence, somewhere near the left-hand end of the horizontal axis there is a point, known as the fifth percentile (5th %ile), of which we could say 'exactly 5% of men are shorter than this' or 'there is only a one-in-twenty chance of encountering a man shorter than this'. Similarly, an equal distance from the mean towards the right of the chart is a point known as the 95th %ile, of which we could say 'only 5% of men are taller than this'. Ninety percent of the population are between the 5th and 95th %ile in stature — but the same could be said for the 2nd and 92nd %ile or the 3rd and 93rd %ile. It is important to note that, by virtue of their symmetrical positions about the mean, the 5th and 95th %ile define the shortest distance (or range) along the horizontal axis to enclose 90% of the population.

Two further points must be borne in mind when discussing percentiles. Firstly, percentiles are specific to the populations that they describe. Hence, the 95th %ile stature for the general public might only be the 70th %ile for a specially selected occupational group like the police force or perhaps the 5th %ile for a sample made

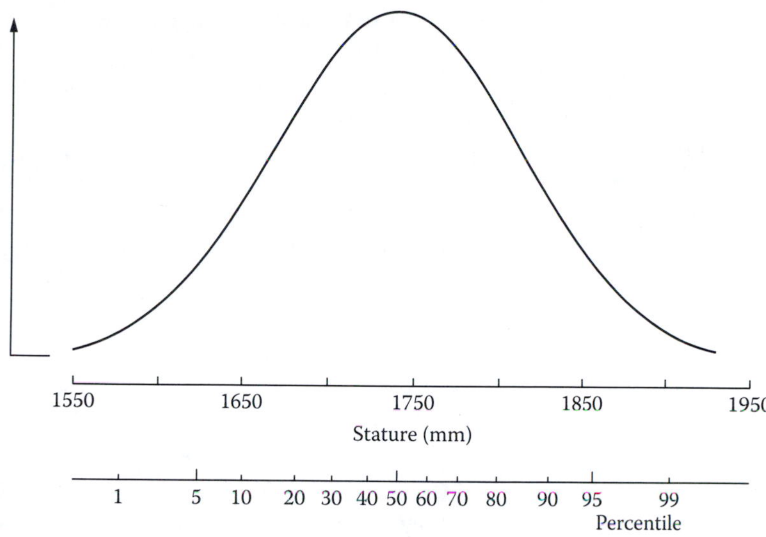

FIGURE 2.1 The frequency distribution (or probability density function) for the stature of adult British men. This is an example of the normal or Gaussian distribution.

up from the Harlem Globetrotters and other professional basketball teams. Secondly, percentiles are specific to the dimension that they describe. Hence, a person who is a particular percentile in stature may or may not be the same percentile in shoulder breadth or waist circumference, since people differ in shape and proportions as well as in size.

The frequency distribution shown in Figure 2.1, with its characteristic symmetrical bell-shaped curve, is very common in biology in general and anthropometry in particular. It is usually known as the normal distribution. We should not, however, infer from this name that the distribution is in some way associated with normal people as against abnormal ones. We might conveniently think of the term as meaning something like the distribution which you will find most useful in practical affairs. To avoid this possibility of confusion, some statisticians prefer to call it the 'Gaussian distribution', after the German mathematician and physicist Johann Gauss (1777–1855), who first described it (in the context of random errors in the measurement of physical quantities). It is possible to predict that a variable such as stature will be normally distributed in the general population if we are prepared to make certain plausible assumptions concerning the way it is inherited from one generation to the next (see any textbook of genetics).

Indeed, it is empirically true that most anthropometric variables conform quite closely to the normal distribution (at least within reasonably homogeneous populations). This is an exceedingly convenient state of affairs since the normal distribution may be described by a relatively simple mathematical equation. The exact form of this equation need not concern us here since we are unlikely to employ it in practice. The important thing is that it has only two parameters. (In mathematics a parameter is a quantity that is constant in the case considered but variable in different cases.) One of these parameters is the mean, which tells us where the distribution is located on the horizontal axis. The other is a quantity known as the standard deviation (SD), which is an index of the degree of variability in the population concerned, that is the 'width' of the distribution or the extent to which individual values are scattered about or deviate from the mean. If we were to compare, for example, the general male population with the police force, we would find that the latter had a greater mean but a smaller standard deviation; that is, they are on average taller than the rest of us, and they are less variable amongst themselves. The SD of a population is usually estimated from a sample of individuals drawn from the population, when it is given by the equation

$$SD = \sqrt{\frac{\sum (x-m)^2}{n-1}} \tag{2.1}$$

where m is the mean, x is the value of the dimension concerned for any individual in the sample, and n is the number of subjects in the sample. (We use $n - 1$ in the equation in the hope of correcting any bias introduced by the finite size of our sample and making a better prediction of the standard deviation of the population from which it was drawn, since this is what in general concerns us.)

In this book we shall, in the interests of brevity, commonly adopt a convention for describing the parameters of normal distributions. Whenever a figure is followed by another in square brackets [ ] it refers to a mean and standard deviation. Hence, the statement that 'the stature of British men is 1740 [70] mm' should be taken as meaning 'the stature of British men is normally distributed, with a mean of 1740 mm and a standard deviation of 70 mm'. (This is a purely local convention; you will not encounter it outside this book.)

2.2.2 Calculating Percentile Values for a Body Dimension

A normal distribution is fully defined by its mean and standard deviation. If these are known, any percentile may be calculated without further reference to the raw data (i.e., the original measurements of individual people). The pth %ile of a variable X is given by

$$X_p = m + z\text{SD} \tag{2.2}$$

where z is a constant for the percentile concerned, which we look up in a statistical table. A selection of z values for some important percentiles is given in Table 2.1. For a more detailed table of p and z, turn to the beginning of the Appendix.

Suppose we wish to calculate the 90th %ile of stature for the adult male population of Britain. It happens that British men have a mean stature of 1740 mm with a standard deviation of 70 mm (as shown later in Table 2.5). From Table 2.1 we see that for $p = 90$, $z = 1.28$. This simply means that the 90th %ile is greater than the mean by 1.28 times the standard deviation. Therefore, using Equation 2.2, the 90th %ile value of stature = 1740 + 70 x 1.28 = 1830 mm. Taking another example, if we wish to calculate the 25th %ile male stature, Table 2.1 shows us that for $p = 25$,

TABLE 2.1
Values of *z* for Selected Percentiles (*p*)

p	z	*p*	z
1	−2.33	99	2.33
2.5	−1.96	97.5	1.96
5	−1.64	95	1.64
10	−1.28	90	1.28
25	−0.67	75	0.67
50	0.00		
0.1	−3.09	99.9	3.09
0.01	−3.72	99.99	3.72
0.001	−4.26	99.999	4.26

$z = -0.67$; that is, the 25th %ile is less than the mean by 0.67 times the standard deviation, and the 25th %ile value would be 1693 mm.

Alternatively, we might wish to do the calculation in reverse and determine the percentile value for a person of a particular stature. Hence, a stature of 1625 mm is 115 mm below the mean, for which Equation 2.2 tells us $z = -1.64$. Looking this up in Table 2.1, we find that this is equivalent to the 5th %ile male stature.

2.2.3 Effects of Deviation from a Normal Distribution

Most linear dimensions of the body are normally distributed, and this certainly makes life easier for the user of anthropometric data. There are, however, other kinds of frequency distribution which turn up occasionally in anthropometric practice. Some other possibilities are shown in Figure 2.2. In most populations body weight and

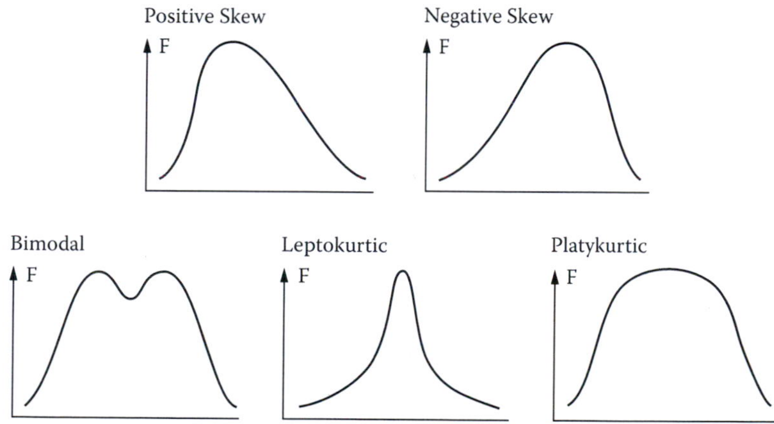

FIGURE 2.2 Deviations from normality in the statistical distributions of anthropometric data.

muscular strength show a modest positive skew; it seems that there are a disproportionate number of heavy, strong people and a dearth of light, weak ones. The combination of two normal distributions, such as a mixed population of men and women or of adults and children, will give us a new distribution that is flat-topped (platykurtic) or even double-peaked (bimodal).

What will happen, in these cases, if we work on the erroneous assumption that such a combined distribution is normal and go ahead and calculate percentiles by the means described above? Errors will accrue, the magnitude of which will be determined by the extent of the deviation from normality in the population distribution. In practice, however, the errors will in many circumstances be negligible. Combining data for adult men and women is a case in point. In theory, the resultant unisex distribution is platykurtic. In practice, the deviations from normality are so small that we can ignore them. The only alternative, which avoids the assumption of normality, is to determine percentiles directly by simply counting heads, but since this requires large numbers of subjects, it is rarely feasible and few datasets in the literature have been established with this degree of certainty. In general, the best practice is to assume normality but to proceed with circumspection in those situations (mentioned above) where we have reason to doubt the assumption. From now on, our discussion will be almost entirely confined to normal distributions.

2.3 DESIGN LIMITS: ACCOMMODATION PROVIDED BY A DESIGN DECISION

For some purposes it may be especially informative to plot out the normal distribution in its cumulative (or integral) form. In this version percentiles are plotted against values of the dimension concerned (or, if we calibrate the horizontal axis in standard deviations, we have in effect a plot of p against z). The curve that we obtain is known as the normal ogive, as in Figure 2.3, which is the cumulative form of the data in Figure 2.1. The advantage of such a plot is that, since we may read off percentiles directly, it enables us to evaluate the consequences of a design decision in terms of the percentage of users accommodated. To take a simplistic example, Figure 2.3 would tell us directly the percentage of British men who could pass beneath an obstruction of a given height without stooping or banging their heads.

The slope of the normal ogive is greatest at the mean value (i.e., the point of maximum probability) and steadily diminishes as we approach the extreme tails of the distribution. The curve is asymptotic to the horizontal axis at 0 and 100% (i.e., in theory meets these axes at infinity). Hence, it is increasingly difficult to accommodate extreme percentiles. (We note in Figure 2.1 that the percentiles are densely packed near the centre and thinly spread at the extremes.) The practical consequence of this is that each successive percentage of the population we wish to accommodate imposes a more severe requirement upon our design. In cost-benefit terms we are in a condition of steadily diminishing returns.

Figure 2.4 illustrates this *design problem* with respect to the case of the adjustability of a seat. The graph shows the benefit to be gained (in terms of percentage of the population accommodated) by providing seat adjustment of a particular range.

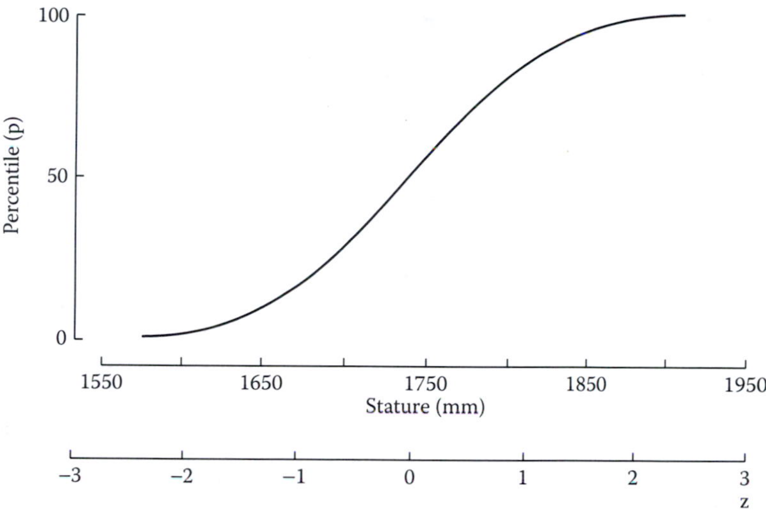

FIGURE 2.3 The frequency distribution of the stature of adult British men, plotted in cumulative form.

The cost in this example is represented by the length of the range of adjustment. Calculations were based on the criterion that seat height should be equal to the vertical distance from the sole of the foot to the crook of the knee (popliteal height), which for the unisex distribution of adult British men and women (shod) is 455 [30] mm.

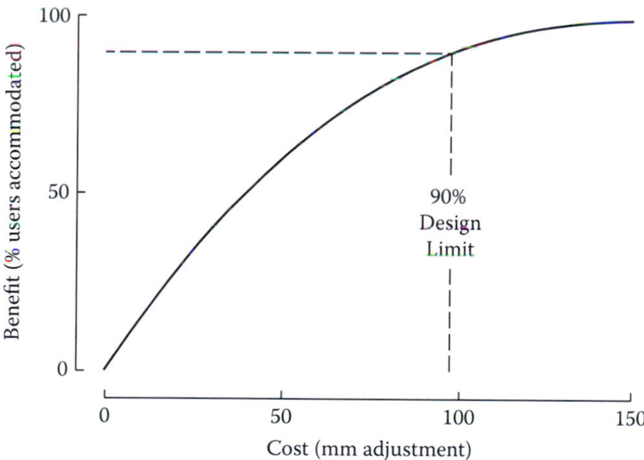

FIGURE 2.4 Anthropometric cost-benefit function showing the percentage of members of a target population accommodated by various ranges of adjustment in the height of a seat. When this information is applied to the seat itself, the adjustment range chosen should be 'centred' at a height of 455 mm.

How then should we draw the line in this increasingly costly and constraining process of accommodating the extreme members of the user population? In other words, where should we set our *design limits*? A purely arbitrary answer to this question, which has been found to work well enough in practice for many purposes, is to design for the 5th to 95th %ile range, that is, for the middle 90% of the user population. When using this rule of thumb, however, we must always bear in mind the consequences of a *mismatch* for those 5% smaller and 5% larger members of the user population who are outside our design limits. Will a mismatch merely cause mild discomfort and inconvenience or might it compromise overall working efficiency? Are there implications for the health and safety of the user, in either the short or long term? A less than 5th %ile person sitting at a dining chair that is too high may be mildly uncomfortable over dinner but if he or, more probably, she has to work at a desk that is too high for seven hours a day, five days a week, the consequences may be much more severe (see Chapters 7 and 9). In the latter case, there is a stronger argument for setting the design limit to accommodate more than 90% of the user population.

In practice, when designing for a mixed user population of adult men and women, it is very common to take the 5th %ile female and 95th %ile male dimensions as the lower and upper design limits. If these were applied to the seat adjustment range in the example above, 95% of the user population would be accommodated. Other design limits may be used under certain circumstances (and some guidance on the selection of representative samples of users is given in ISO 15537 [ISO 2004]). Supposing we were asked to specify dimensions for an escape hatch in a confined working area. A hatch width based upon the appropriate body dimension of a 99th %ile user would mean that one person in 100 would get stuck. This would clearly not be acceptable. (Actually the problem is further compounded by the fact that the distributions of the body bulk dimensions involved will probably be positively skewed, so that accommodating larger percentiles would require increasingly larger increments to the dimension.) In *safety-critical applications* of this sort, each individual case must be judged on its own merits. We might, for example, decide, as a matter of policy, that less than one user in 10,000 should be mismatched, and set our design limits at around four standard deviations from the mean (see Table 2.1).

In a more general sense, it is only possible to specify percentiles at all if we can first define the *user population*. This would be simple enough in the case of a fighter aircraft, for example, where the numbers of pilots are small and they are a well-defined population, but the users of a public transport system would be quite another matter. They are a much more diverse population. Here we must consider children as well as adults, the elderly and infirm, pregnant women and wheelchair users. These people may not fit readily into the percentile tables of the anthropometrist, but can they legitimately be excluded from participation in the system or environment in question? We shall return to the issue of *barrier-free design* in due course, but first we shall deal with the narrower problem of designing for the majority.

2.4 DESIGN CONSTRAINTS AND CRITERIA

The design problem therefore requires us to identify the user population and its characteristics (particularly the anthropometric distributions of the relevant dimensions) and to understand how these characteristics may impose constraints on the design, and then to set criteria which can be used to assess whether a satisfactory match has been achieved in the design. The foregoing discussion of the statistics of anthropometric distributions (defining the variability within the user population) provides us with the tools with which we can calculate values for the various design parameters (such as the range of seat adjustment and the escape hatch width in the earlier examples). The aim is to follow the fourth and fifth principles of user-centred design (Table 1.2) and achieve the best possible match for the greatest possible number of people.

In anthropometrics a *constraint* is an observable, preferably measurable, characteristic of human beings, which has consequences for the design of a particular artefact. A *criterion* is a standard of judgement against which the match between user and artefact may be measured. We may distinguish various hierarchic levels of criteria. Near the top are overall desiderata such as comfort, safety, efficiency, aesthetics, etc., which we may call high-level, general or primary criteria. In order to achieve these goals, numerous low-level, special or secondary criteria must be satisfied. The relationship between these concepts may be illustrated by way of an example. In the design of a chair, comfort would be an obvious primary criterion; the lower leg length of the user imposes a constraint upon the design since, if the chair is too high, pressure on the underside of the thigh will cause discomfort. This leads us to propose a secondary criterion: that the seat height must not be greater than the user's popliteal height (shod). A table of data will tell us the distribution of this dimension. It would seem reasonable to choose the 5th %ile value, since if a person this short in the leg is accommodated, the 95% of the population who are longer-legged will also be accommodated, providing they have room in front to stretch their legs. This leads more or less directly to a *design specification* or tertiary criterion: that the height of the seat shall not be greater than 406 mm. (Note that if we propose an adjustable seat we will use our criterion differently, as in Figure 2.4; see Chapter 5 for a more general discussion of this particular problem.)

Taken in isolation, the primary criterion will usually be what is known, amongst certain ergonomists, as a stunning glimpse of the obvious (SGO). In general, it is necessary to work down through successive levels of the hierarchy of criteria before defining any operationally useful recommendations (which is the top down approach of working from the general to the specific).

However, it is rare that there is only one criterion or one constraint on a design, and the interactions between different criteria need to be considered in arriving at a combined design solution. Thus, at any level in the hierarchy conflicts between criteria may arise, which will necessitate trade-offs. Hence, in the example we took above, our secondary criterion tells us when a seat is too high but not when it is too low. The criteria for this latter case are less well defined — we might call them fuzzy rather than sharp. None the less, it is perfectly possible that a tall man might feel uncomfortably cramped in a seat designed to accommodate the lower leg length of a 5th %ile woman, and some suitable compromise might have to be reached in the interest of the greatest comfort for the greatest number. Similarly, there might be circumstances in which it was necessary to trade off, say, comfort against efficiency or safety. These latter circumstances are probably few, but they raise the interesting point of what superordinate criterion could be used to measure both.

In practical matters, the middle of the hierarchy is often the best place to start (which some call the 'middle-out' approach). We shall therefore consider four sets of constraints which between them account for the vast majority of everyday problems in anthropometrics per se and, hence, a sizeable portion of ergonomics. We shall call them the four cardinal constraints of anthropometrics: clearance, reach, posture and strength.

2.4.1 Clearance

In designing workstations it is necessary to provide adequate head room, elbow room, leg room, etc. Environments must provide adequate access and circulation space. Handles must provide adequate apertures for the fingers or palm. These are all clearance constraints. They are *one-way constraints* and usually determine the minimum acceptable dimension in the object. If such a dimension is chosen to accommodate a bulky member of the user population (e.g., 95th %ile in height, breadth, etc.), the remainder of the population, smaller than this, will necessarily be accommodated.

In a few safety-critical circumstances, a related criterion of maximum acceptable dimension to exclude people (or some part of their body) is needed. An example of this would be gaps between bars in a safety guard on a machine tool, where a maximum gap size would be specified to prevent fingers contacting the moving parts of the machine. Here a 1st %ile criterion (or even smaller) would probably be set.

2.4.2 Reach

The ability to grasp and operate controls is an obvious example of a reach constraint, as is the constraint mentioned above on the height of a seat or the ability to see over a visual obstruction. Another example of a visual reach constraint is the distance at which a display screen should be placed so that text on the screen can be read comfortably. Reach constraints determine the maximum acceptable dimension of the object. They are again usually one-way constraints but this time are determined by a small member of the population, for example, 5th %ile.

2.4.3 Posture

A person's working posture will be determined (at least in part) by the relationship between the dimensions of his or her body and those of the workstation. Postural problems are commonly more complex than problems of clearance and reach, since posture will almost certainly be affected by more than one dimension of the workplace (not to mention the task being performed). Moreover, the needs of one person may conflict with those of another person. For example, a working surface that is too high for a small person is just as undesirable as one that is too low for a tall person (see Sections 2.6.1 and 4.7). Hence we have a *two-way constraint* in which both a maximum and a minimum value of a workstation dimension must be specified. There are several ways of satisfying such design requirements, and these are discussed in Sections 2.5 and 2.6.

2.4.4 Strength

A fourth constraint concerns the application of force in the operation of controls and in other physical tasks. Often, limitations of strength impose a one-way constraint, and it is sufficient to determine the level of force that is acceptable to a weak user. There are cases, however, where this may have undesirable consequences for the heavy-handed (or heavy-footed) user, or in terms of the accidental operation of a control, etc. In these cases a two-way constraint may apply.

2.5 DEFINING DESIGN REQUIREMENTS TO SATISFY THE FOUR CARDINAL CONSTRAINTS

In practice, a design specification must define specific design requirements (the 'tertiary criteria') in terms of dimensions of the design parameters, and the specification should also state ways in which a judgement may be made as to whether or not each criterion has been satisfied. In making this judgement for one-way criteria, it may be sufficient to compare an objective measure of a dimension (clearance, reach or control operating force) with the design requirement, but for two-way criteria further analysis is often required; appropriate methods are introduced in this section.

All four cardinal constraints are influenced by the task being performed as well as by body dimensions. Thus the need for elbow room when sitting still in a bus is quite different from the clearance needs of packers at an assembly line conveyor belt, who not only need adequate elbow room to perform the task but may well on occasions need additional space to work ahead of their station to gain extra time or fall behind their station if they become fatigued and slow down. Posture in particular is influenced by the task being performed at the workstation, and the criteria must be defined with this in mind. For example, if the task involves exerting force, the work surface will need to be lower so that the person can lean forward to use body weight to assist the force exertion.

When choosing the appropriate criterion it is often helpful to think in terms of the *limiting user*. The limiting user is that hypothetical member of the user population who, by virtue of his or her physical (or mental) characteristics, imposes the most severe constraint on the design of the artefact. In clearance problems the bulky person is the limiting user; in reach problems the small person is the limiting user. One of the reasons that postural design problems are commonly more complex than problems of clearance and reach is that we may have limiting users in both tails of the distribution, resulting in a two-way constraint.

In summary, two of the main issues that need to be considered in finding a design solution are how best to achieve a match between the workstation dimensions and the user's anthropometry and whether the task will have a significant influence on this. Common criteria for judging the match (the resulting fit or posture) relate to whether it is comfortable, efficient and safe.

There are three types of design solution (or design strategy):

1. Design for the limiting user.
2. Define an area of common fit.
3. Provide adjustment.

The first strategy can be applied to a one-way constraint, while the other two strategies can be considered for two-way constraints. An area of common fit is a range of a given design dimension within which all users (or the 90% of users between the 5th %ile and 95th %ile) can be satisfied, few having their preferred or optimal solution but all finding it acceptable. The design dimension can then be specified anywhere within the area of common fit. This solution can often be found for objects or workplace parameters which are not used repetitively or for long periods of time. An example might be the diameter of a handle. However, the more frequently this is used or the longer the period of use, the more critical the choice of dimension and the less likely it is that an area of common fit will be found to satisfy the design criterion. Where an area of common fit cannot be found, adjustability should be provided or the object should be produced in a range of sizes.

2.6 METHODS FOR ANALYSIS OF DESIGN PROBLEMS

Some analysis and possibly some experimentation is often needed to find design solutions. Three simple methods — fitting trials, the method of limits and body link diagrams — will be described. Such experimentation can be carried out in trials either with representative samples of users or as virtual trials with digital human models using workspace simulation software.

2.6.1 Fitting Trials

A *fitting trial* is an experimental study in which a sample of subjects use an adjustable mock-up of a workstation in order to make judgements as to whether a particular dimension is 'too big', 'too small' or 'just right'. During the fitting trial they can simulate performing the tasks, or critical aspects of the tasks, to make their judgements more realistic.

Figure 2.5 shows the results of a simple fitting trial, the purpose of which was to determine the optimum height for a lectern in a lecture theatre. Ten people (five male, five female) acted as subjects in the fitting trial. A music stand served as the adjustable mock-up simulating the lectern. Each subject set the music stand to the lowest and highest heights that he or she considered acceptable and then to their own personally preferred optimum height. The means and the standard deviations of the lower and upper limits for all ten subjects were calculated. These were used to plot the smooth curves (cumulative distributions) which define the thresholds of 'too low' and 'too high' shown in Figure 2.5 (by using z and p values in the same

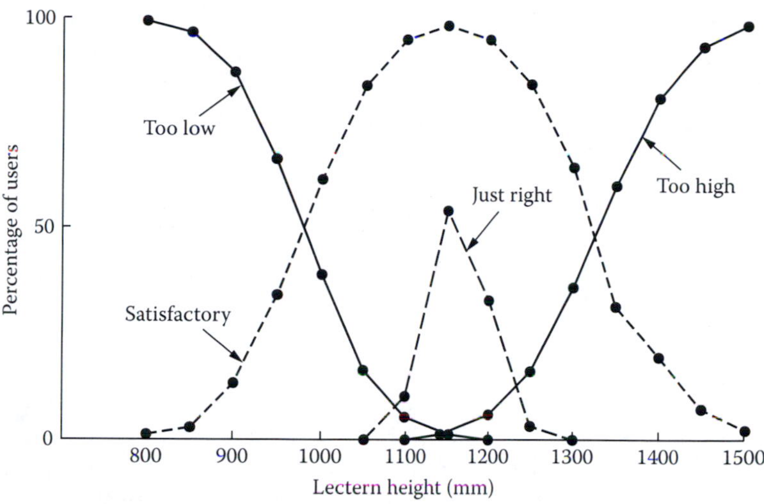

FIGURE 2.5 Results of a fitting trial to determine the optimal height of a lectern. (From A. S. Nicholson and J. E. Ridd, *Health, Safety and Ergonomics*, Butterworth-Heinemann, 1988, Figure 7.2. Reproduced with kind permission.)

way as for the construction of Figure 2.3). These cumulative distributions represent the best estimates for a population represented by the subject sample.

Another smooth curve for 'satisfactory' is plotted in a similar way in Figure 2.5 by calculating the percentage of people for whom each height was neither 'too low' nor 'too high'. (Thus for any given height, the responses 'too low' + 'too high' + 'satisfactory' = 100%.) In can be seen that the majority of subjects in the sample could be 'satisfied' by lectern heights between approximately 1050 mm and 1250 mm. Obviously, few of the subjects found these heights 'just right', but this is naturally the case when trying to find a design solution for a population rather than an individual. The distribution for the subjects' preferred optimum heights ('just right') was derived directly from their responses (i.e., without calculating the cumulative distribution), rounding off each individual subject's response to the nearest 50 mm and plotting these as the dashed curve shown in Figure 2.5. It is interesting to compare the curves for 'just right' and for 'satisfactory'. Subjects could accept a range of lectern heights for the given task even though most heights within the range were not ideal for them. Thus, in this case, there is some flexibility in choosing a design solution which will satisfy the majority of users. Were this not the case, it would be necessary to build adjustability into the lectern. In fact, in this experiment, a clear optimal height of 1150 mm emerged (at least for this group of subjects). At this height, more than 50% of users regarded the lectern as 'just right' (within ± 25 mm), and over 95% considered it satisfactory.

The design of the experiment could perhaps have been improved, although a relatively crude experiment of this nature will suffice for many real-world design problems. We could also question the extent to which the subjects in this experiment (students) were a representative sample of the true population of end-users of the lectern. If the problem were more critical it would merit a more detailed investigation.

A fitting trial is a type of *psychophysical experiment* in which people make subjective (i.e., psychological) judgements concerning the objective properties of physical objects or events. The form that the plotted results of the lectern experiment (Figure 2.5) took is characteristic of psychophysical experiments. It is made up from the ogival curves of two normal distributions, facing each other, which define a third normal distribution by subtraction of their summed values from 100%. Similar results may be found in other areas of ergonomics. Judgements about thermal comfort — for example 'too hot', 'too cold', 'just right' — are distributed in this way (see Grandjean 1988). In principle we might expect to encounter this form in any situation in which people are asked to express a subjective preference on a bipolar continuum: too fat/too thin, too young/too old, and so on.

A more systematic method of obtaining the judgements in a fitting trial was developed by Jones (1963) to take account of human biases and uncertainties in making judgements. (It was adapted from the 'method of limits' which is discussed in more detail, in an analytical rather than an experimental application, in the next section.) Jones' method for conducting the fitting trial to determine the best height for the lectern would be to adjust the lectern to set heights, in steps of perhaps 25 or 50 mm, and at each height to ask the subject to say whether or not it was 'acceptable' (i.e., for performing the given task). Responses of 'not sure' would not be allowed; it can be difficult to make a definite judgement close to a threshold of acceptability (or comfort), but it is better to press the subject for a decision rather than allow hesitation or opting out (and the analysis of the results is also simpler).

The sequence of heights would be presented twice, once in ascending order and once in descending order, these being alternated between subjects to minimise systematic bias. The ascending sequence would start with a very low lectern, and it would then be raised in the steps until a point was reached at which the lectern was definitely too high. This would be repeated in the descending sequence until a point was reached at which it was definitely too low. Curiously enough, the ascending and descending trials tend to give slightly different answers, with the thresholds between unacceptable and acceptable (or vice versa) being lower in the descending trial. This can be seen in the data from the hypothetical fitting trials shown in Figure 2.6. The reason for this is that we all find it difficult to change from a negative judgement to a positive judgement (or vice versa) at a threshold, being influenced by the judgement we made in the previous condition (a carry-over effect). This is the main reason for including both ascending and descending trials. The upper and lower thresholds of acceptability are calculated for each subject by averaging the threshold values found in their ascending and descending trials. The general procedure for conducting a fitting trial is shown in Table 2.2.

The data collected by using Jones' method for fitting trials (as in Figure 2.6) could be processed to present the results in the same way as for the original lectern experiment in Figure 2.5. However, even the raw results in Figure 2.6 allow us to reach some conclusions about the design solution, and specifically to decide that there is an area of common fit, which (for this user population) indicates that a fixed height lectern would be satisfactory and that adjustability is not essential. The form of presentation in Figure 2.5 goes further than this and provides the results from which cost-benefit analyses could be made.

Lectern height* (mm) * as adjusted in the mock-up	Subjects' judgements (for a group of 5 women and 5 men) √ acceptable, x unacceptable										Design options
	F1	F2	F3	F4	F5	M1	M2	M3	M4	M5	
1500	x	x	x	x	x	x	x	x	x	x	
1450	x	x	x	x	x	x	x	x	√	√	
1400	x	x	x	x	x	x	√	x	√	√	
1350	x	x	x	x	x	x	√	x	√	√	
1300	√	x	x	√	x	√	√	√	√	√	
1250	√	x	√	√	√	√	√	√	√	√	
1200	√	√	√	√	√	√	√	√	√	√	area of common fit
1150	√	√	√	√	√	√	√	√	√	√	
1100	√	√	√	x	√	√	√	√	√	√	
1050	√	√	√	x	√	√	√	√	√	x	• adjustability is not essential • lectern height can be between 1125 mm and 1225 mm
1000	√	√	√	x	x	√	√	√	√	x	
950	√	√	√	x	x	x	x	√	x	x	
900	√	x	x	x	x	x	x	x	x	x	
850	x	x	x	x	x	x	x	x	x	x	
800	x	x	x	x	x	x	x	x	x	x	

FIGURE 2.6 Hypothetical responses which might be obtained from subjects in a fitting trial using Jones' method. The arrows indicate the upper and lower thresholds of the range of lectern heights judged acceptable by each subject (where each threshold shown is the average value for the subject's ascending and descending trials).

TABLE 2.2
Procedure for Conducting a Fitting Trial

Construct a mock-up of the workplace (with adjustability for all components representing the features being evaluated).

Include a simulation of the task(s) which will be performed in the workplace.

Select a sample of subjects to represent the relevant characteristics of the eventual user population.

Decide the order of presentation of the workplace features being evaluated (an important decision when assessing whether adjustability will be essential for more than one of the features).

Determine the range which will be tested for each component, which should extend beyond the expected minimum and maximum thresholds of acceptability.

Determine the interval between settings which will be tested over this range, depending on the sensitivity required of the evaluation.

For each subject:

- Decide the order of ascending and descending presentation for each component (which should vary between subjects to minimise bias due to any systematic effects); usually there will be several repetitions.
- Set the component at the specified intervals throughout its range.
- At each setting:
  - Where appropriate, ask the subject to perform the real or simulated task(s).
  - Ask the subject to judge whether the setting is acceptable.

Plot the results for all subjects.

Determine which features need to be adjustable and which can be fixed.

Specify the best design compromise — dimension (or range of adjustability) of each feature of the workplace.

Evaluate the final design solution, initially perhaps with the mock-up but ultimately in the workplace itself, and with the actual user population.

2.6.2 Analytical Application of the Method of Limits

Consider now the problem of seat height which we discussed in Section 2.4. We could in principle have solved this empirically by conducting a fitting trial, but instead we solved it analytically by the application of anthropometric data. The line of reasoning that we adopted could be written out in a formal way as follows:

- Criterion: seat height ≤ popliteal height
- Limiting user: 5th %ile woman, shod popliteal height = 406 mm
- Design specification: maximum seat height = 406 mm

Let us now apply a similar line of reasoning to the analysis of a more complex problem involving a two-way postural constraint. The technique we are going to use is called the *method of limits*. (The name is borrowed from that of a technique in psychophysics which is equivalent in form.) In essence this technique is a model or analogue of the fitting trial, in which anthropometric criteria and data are used as

TABLE 2.3
Calculation of Percentage of Men Accommodated by a Workbench that is 1000 mm in Height

Criterion	Distribution	Percentile	Conclusion
EH – 150	965 [52]	75	25% — much too low
EH – 100	1015 [52]	39	61% — too low
EH – 50	1065 [52]	11	11% — too high
EH	1115 [52]	1	1% — much too high
			28% — just right

EH, elbow height

substitutes for the subjective judgements of real people. The problem is to determine the optimum height for a workbench to be used in a certain industrial task which involves a moderate degree of both force and precision. (It is assumed in this example that the object being handled is small, so that the workbench height and the working height are essentially the same; this is not always the case, and analyses should always be made for the working height — the height of the hands when performing the task.) To simplify the calculation we shall also assume that the task will be performed by male workers. The workers will be standing.

According to Grandjean (1988), the optimum working height for a task involving moderate force and precision is between 50 and 100 mm below the person's elbow height. We note that it is a two-way criterion since there may be a mismatch in either direction (with the workbench height too low or too high). The elbow height (EH) of British men is 1090 [52] mm (as shown later in Table 2.5). To this we must add a 25 mm correction for shoes, giving 1115 [52] mm (see Section 2.7.4). Combining these data with the above criterion from Grandjean (1988) gives us the upper and lower limits of optimal working level: EH – 50 = 1065 [52]; EH – 100 = 1015 [52]. We can treat these just as if they were new normally distributed anthropometric dimensions and calculate the percentile in these distributions to which any particular workbench height corresponds (as in Table 2.3 for a workbench height of 1000 mm). However, we should bear in mind that the criterion refers to a zone of 'optimal' bench heights. Since we may reasonably assume that users may be prepared to accept less than absolute perfection, we may find it useful to consider two further zones extending 50 mm above and 50 mm below the optimum, which we would characterize as 'acceptable' but not perfect (see Figure 2.7). We choose 50 mm pragmatically because it seems reasonable rather than on the basis of any particular scientific evidence.

Table 2.3 shows a set of calculations performed for a workbench height of 1000 mm. We find that this workbench height corresponds to the 75th %ile in the lowest criterion (EH-150) distribution, from which we infer that a workbench of 1000 mm would be 'much too low', or 'unsatisfactory', for the 25% of men who are larger than this. Similarly, the central criteria (EH-100 and EH-50, bounding the optimal zone) correspond to the 39th and llth %iles, respectively, from which we infer that

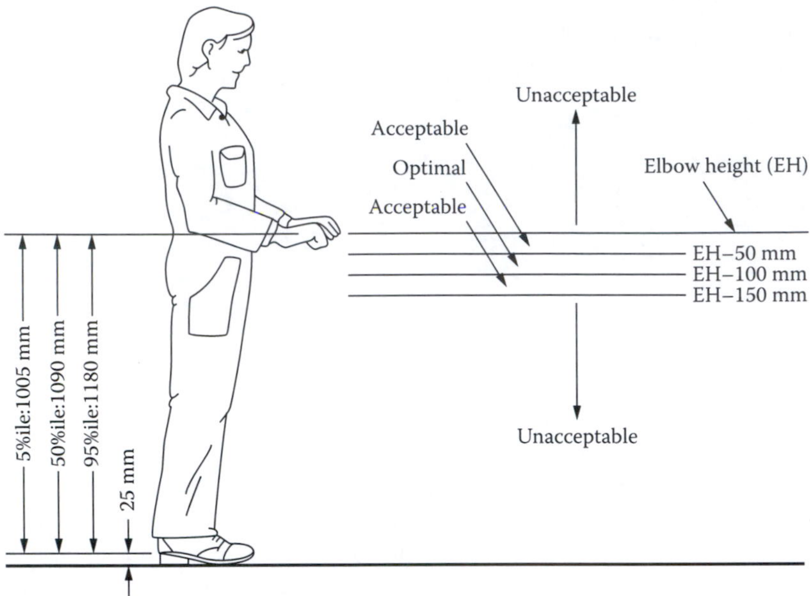

FIGURE 2.7 Criteria for optimal and satisfactory working heights in an industrial assembly task.

the 28% of men between these heights would find the workbench 'just right' or 'optimal'.

We could keep on performing such calculations for different workbench heights until we homed in on a value that maximised the percentage optimally matched and minimised the unsatisfactory matches. (Here, of course, the computer would help.) At this point we are like the statistically minded punter searching for the best bet. The full results of a series of such calculations are plotted in Figure 2.8. It comes as no surprise to discover that the 'optimal' figures describe a normal curve (e), whereas the 'too high' and 'too low' figures yield normal ogives facing in opposite directions (a, b, c, d). We might also lump together those who were optimally matched with those who were a little too high and a little too low into an 'acceptable' category (f), leaving a residual 'unacceptable' category (g) outside these limits (which would be 26% unacceptable and 74% acceptable for a workbench height of 1000 mm). The statistically minded punter looking at Figure 2.8 should settle for a working height of a little less than 1050 mm.

This is not quite the end of the process, since at the best compromise height 15% of users will have an 'unsatisfactory' match. Is this an acceptable (or tolerable) situation or will they be severely uncomfortable or suffer long-term damage? Is it better to have a bench that is too high or one that is too low? Do we require an adjustable workbench or some similarly varied solution? In order to make such judgements, more information would be needed about the demands of the tasks performed and about the duration and frequency of use of the workbench. In general, neither a work surface that is too low nor one that is too high can be said to be the

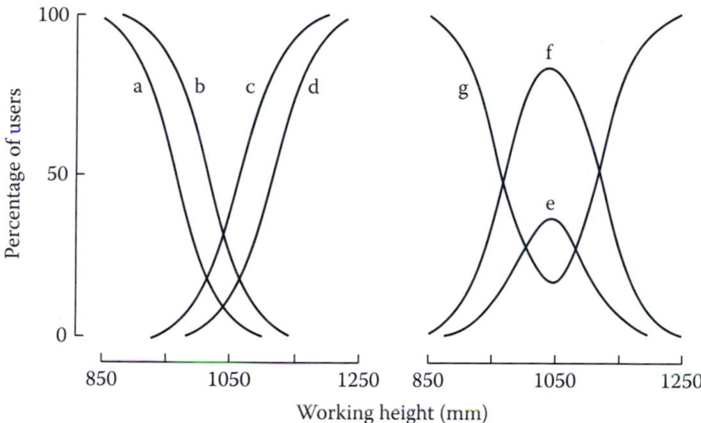

FIGURE 2.8 The anthropometric method of limits, applied to the determination of the optimal working height for an industrial assembly task. Curves show the percentage of users accommodated or otherwise: (a) much too low; (b) too low; (c) too high; (d) much too high; (e) just right; (f) acceptable; (g) unacceptable. See text for definition of categories and discussion of technique.

better compromise — the effects will differ, but both can be uncomfortable or potentially harmful under some working conditions.

If a fixed height solution is chosen, the best possible compromise height is 75 mm below the elbow height of the average user (i.e., at the mid-point of the optimum range). With the wisdom of hindsight we can see that this follows necessarily from the shape of the normal distribution. Having laboriously analysed the problem, we find that it could have been solved by inspection. We could write out our reasoning as follows:

- Criterion: elbow height – 100 mm ≤ bench height ≤ elbow height – 50 mm
- Best possible compromise: bench height = average elbow height – 75 mm
- Design specification: bench height = 1040 mm

Again, a solution can be found through simple reasoning, but the full analysis of Figure 2.8 is needed to establish the extent of mismatches and the severity of the consequences and to provide a cost-benefit justification to argue for the design solution.

2.6.3 Body Link Diagram

Another way of analysing a workplace layout to accommodate a range of people of different sizes is by using a *body link diagram*. An example of this approach is shown in the layout of a driver's workstation in Figure 2.9. The analysis, based on an original study by Rebiffé et al. (1969), shows the zones within which the grip point on a steering wheel should be located for a small woman and a tall man, and similarly the eye levels which result when they have adopted the posture considered

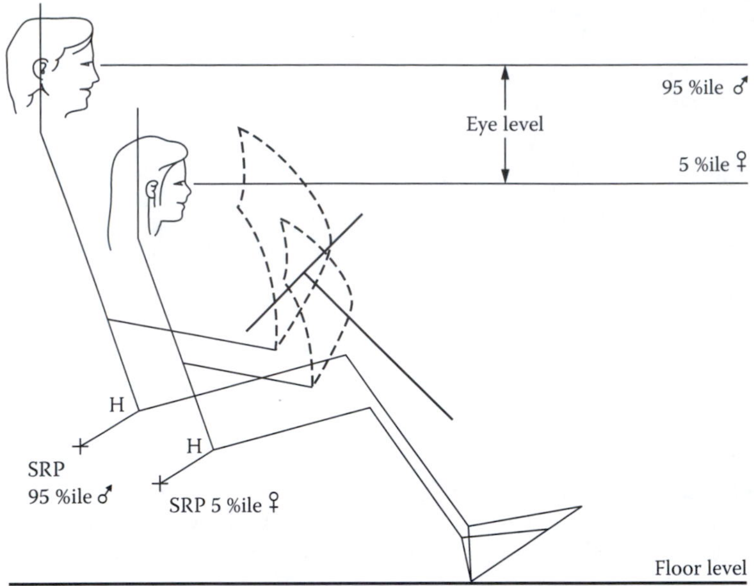

FIGURE 2.9 Body link diagram used to analyse the layout of a driver's workstation, based on the following postural criteria: thigh angle (i.e., seat angle) to horizontal = 15°; knee angle = 110°; ankle-foot angle = 90°; seat back reclined at 20° to the vertical. The hand position zones are defined by the ranges of shoulder and elbow postures: shoulder flexion 0 to 45°; elbow flexion 60 to 110°.

to be comfortable (as defined by the postural criteria — Rebiffé's (1966) ranges of comfortable angles of joints [see Section 4.5] — used in constructing the link diagram). From this analysis, the car designer has information related to vision, helping to set limits on the upper and lower boundaries for the windscreen and location of the rear view mirror, an area of common fit for locating the steering wheel and the ranges of adjustment necessary for horizontal and vertical positioning of the seat. In the figure, a relatively small 360 mm steering wheel set at 45° has been drawn, centred on the area of common fit for the hand grips. In this position it appears possible that it only just provides sufficient abdominal and knee clearance. Hence there is a strong argument for making the steering column adjustable. Another example of the use of a link diagram is the definition of the zones of reach on a work surface which is illustrated in Figure 4.6 and Figure 4.7.

In a link diagram, the human body is represented by links (bony segments) articulating (pivoting) about joint centres. For many purposes, the lengths of the links can be approximated from known anthropometric data, such as knee height, buttock–knee length or seated shoulder height, with corrections estimated for the location of the joint within the skin surface dimensions. Tabulations of link lengths can be found in biomechanics textbooks, often based on regression analysis of data on internal and external skeletal dimensions, and other research has defined locations of joint centres from studies of motion (Seitz and Bubb, 1999). While there is obviously a degree of error in any of these estimations, a link diagram is an effective means of helping a designer to visualise the design problem in its entirety and to appreciate the interactions between the various design criteria.

More accurate data on joint centres, link lengths and locations of centres of gravity of the body segments represented by the links has been collected. Historically, the parameters of body segments have been determined by the careful and systematic dissection of cadavers. The collection of data by this means commenced with the work of German anatomists such as Harless, Braune and Fischer at the end of the nineteenth century. The definitive study of the subject was that of Dempster (1955), whose data have been quoted and requoted and applied in applications ranging from astronautics to industrial safety, automobile crash test dummies and sport. It is sobering to record that Dempster's work was based upon only eight cadavers. Subsequent studies increased this number to 65 (Reynolds, 1978). More recently, data has been collected from x-rays and other medical scans. Where accuracy is important for a body link diagram, a compilation of this data can be found in Chaffin et al. (1999).

Information is also needed on ranges of joint movements and, for setting the posture criteria, on the comfortable range within the maximum limits of movement at a particular joint. These data are discussed in Chapter 4.

2.6.4 Workspace Simulation and Digital Human Models

The analysis of design problems can now made through simulation with digital human models in virtual environments as well as by the empirical and analytical methods just described. Anthropometrics is developing very rapidly with advances in computer simulation, automated measurement technology and image processing techniques. The effects of this on applications of anthropometrics will be important and are briefly discussed below. For a more comprehensive overview of these, and of digital human models in general, the reader is recommended to turn to Chaffin (2001, 2004) and Robinette et al. (2004). The underlying principles of anthropometrics, as given in *Bodyspace*, are fundamental and will continue to apply even as the practice changes and the power of analysis increases.

Anthropometric data collection surveys have been greatly enhanced in the types and quantity of anthropometric dimensions which can be measured and in the speed with which data can be collected (which in turn has improved accuracy by reducing the extent of postural changes made by subjects while they are being measured). Whole body scanners (reviewed together with other technologies such as landmark capture and tracking devices by Rioux and Bruckart 1997) provide three-dimensional (3-D) anthropometric data which capture the contours and shape of the body surface. This opens up possibilities for entirely new types of measurements (termed 3-D anthopometry) and ones which are being collected for very large population surveys. In the international CAESAR study, the 3-D surface anthropometry of diverse populations from North America and Europe have been surveyed, and data from approximately 4400 individuals have been collected (Robinette, 2000; Robinette et al., 2002; Blackwell et al., 2002). Human motion can also be captured and analysed to investigate both task behaviour and workspace parameters (Chaffin et al., 2000).

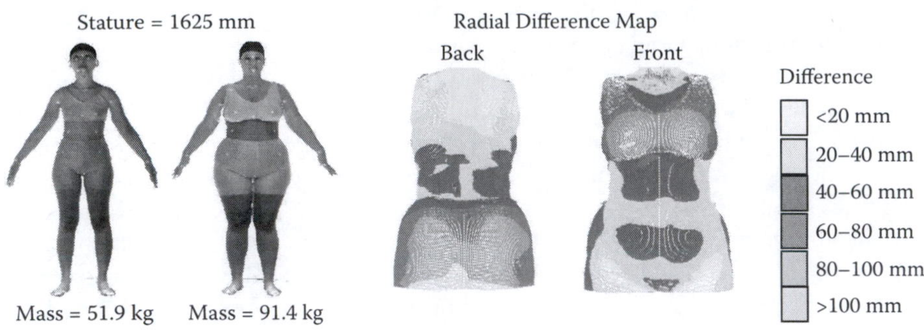

FIGURE 2.10 Radial differences between the torsos of two women with the same torso height but different masses (Robinette et al., 2004).

Figure 2.10 shows one example of the potential applications for 3-D anthropometric data, taken from the work of Robinette et al. (2004), in which the 3-D differences in shape between people of different sizes and body proportions were mapped and defined in quantitative terms and the consequences for design evaluated. The superposition of scans can similarly be used to assess the fit of apparel or of personal protective equipment. 3-D scans have even been used to estimate functional lung capacity and to show that the radial differences at shoulder level and in the chest exceed 1 cm in different phases of the breathing cycle (Kovatz et al., 1988). Jones and Rioux (1997) have described other applications, including human recognition and design of prostheses. The increased power of such analyses is apparent when it is further realised that the variation within a whole population sample can be mapped in this way, no longer requiring the artificial representation by percentiles of dimensions or manikins. The measurement of changes over time is also facilitated, whether for the tracking of movements while performing tasks (from which dynamic reach or space envelopes can be derived) or longer-term bodily changes such as those occurring during pregnancy. Volume scanners, using ultrasound imaging, x-ray computed tomography and magnetic resonance imaging offer further potential for recording dimensions and movements of the internal structures of the body. These latter applications are largely beyond the scope of the design applications of anthropometry, but some of the possibilities can be imagined from the anatomical cross-sectional scanned images of the male and female bodies recorded in the National Library of Medicine's Visible Human Project® (Patrias, 2000).

Computer aided design (CAD) has long been used to simulate workspaces and equipment, but realistic digital human models (also known as avatars) have now been incorporated in these computer simulations (Chaffin, 2004), and animation techniques allow the analysis and evaluation of movements performed during work tasks, as well as consideration of the interactions between pairs and teams of workers. The human models extend the body link diagram approach described earlier by displaying the body form (or 'skin') within which there is a representation of the skeleton or musculoskeletal system. The construction of the digital human models relies on knowledge (and representation through algorithms) of body linkages, joint motion centres and joint ranges of motion, but they are subject to similar errors of approximation as in the more traditional approaches (and perhaps less obviously so where algorithms simplify the reality that they represent or are based on incomplete knowledge of human behaviour). Effects of skin compression artefacts and identification of comfort thresholds pose much the same difficulties.

As Chaffin (2001) notes, an important attribute of these human models will be their ability to use either inverse kinematics or motion capture data to predict the postures that people use when performing a task. At present these methods can be used to create realistic static postures and some simple motions for a variety of common tasks, (e.g., walking, kneeling, stooping, etc.). However, knowledge of postural behaviour while performing work tasks needs much further study, but predictive modeling of some dynamic actions is underway (Zhang et al., 1997; Chaffin, 2002). Few validation studies have yet been reported of the algorithms controlling posture adjustment and animation of actions of digital human models. Their users should be aware of the possibility of such limitations and of the effect of user's background and experience on their interpretation of the results (Che Doi and Haslegrave, 2003; Dukic et al., 2002). User trials with real human subjects, performing their real tasks, are just as necessary to evaluate and confirm design solutions following computer simulation as when following the more traditional anthropometric analysis techniques.

The opportunities offered by 3-D anthropometry and digital human modeling are intensely exciting. The former is already introducing possibilities of mass customisation to replace the mass production of consumer goods, with its disadvantages of ill-fitting clothes, inappropriate work surface heights, inefficient tools, etc. Bespoke tailoring or footwear, for example, could become available for everyone. Perhaps, too, workstations and conveyors on industrial production lines could be designed to be reconfigured to fit the incoming operators at shift changes. Already the digital human models and workspace simulations (virtual mock-ups) provide designers with tools to evaluate alternative design solutions both rapidly and at an early stage in the design process (exploring a wide variety of task scenarios and performing 'what if' exercises) (Robertson and Bedford, 1999; Chaffin, 2001). Currently, designers are mainly using these to investigate fit, clearance and line of sight issues and for solving problems related to strength (Chaffin, 2001), but, as more knowledge of task behaviour is developed, it will be possible to use them to assess comfort or endurance and to model motion in simulated activities.

2.7 USING ANTHROPOMETRIC DATA

Whether for traditional or 3-D anthropometry, the basis of the foregoing discussion and analysis is that the user population of interest is known (or that a similar surrogate population can be found) and that its anthropometric characteristics have been measured. Apart from a few special small populations who are completely known (such as astronauts or Formula One racing drivers), it is rare to have anthropometric data directly applicable to a target population (such as production line operators within a particular factory or purchasers of a particular product). Usually the general (civilian) population anthropometric statistics for the appropriate nationality will be used (although the choice of data source is obviously more complicated for design of products that are to be marketed internationally). The following sections give guidance on the points to consider when applying anthropometric data for a specific design question (and user population).

2.7.1 Sources of Anthropometric Data

The anthropometric data available for different populations vary greatly, in both quantity and quality. Students in ergonomics and design courses are often critical and surprised by this, but the reality is that the collection of anthropometric data is extremely time-consuming and expensive, particularly when surveys attempt to be truly representative. Few countries have yet attempted to collect representative and comprehensive anthropometric data, although many countries are currently undertaking surveys. Moreover, we know that secular changes can occur quite rapidly in populations (see Chapter 3), but surveys are much too expensive to repeat at frequent intervals. As a result, we are normally using anthropometric data that is 10, 20 or more years old and which may only be the best approximation available for the actual user population. Corrections may need to be made for this, as well as predictions of future secular trends when products are expected to remain in use for many years into the future (e.g., when designing buses or aircraft). Few organizations outside the military have the resources to mount a full-scale anthropometric survey. As a consequence, we have extensive and detailed anthropometric data for many of the world's armed services but relatively few data for the civilian populations from whom they were recruited and of whom they may or may not be representative samples. Military data should be used with caution when applied to a civilian target population — the military populations surveyed tend to be young, fit and subject to selection biases.

Despite these cautions, there are many sources of anthropometric data for national populations, which can be found in papers in scientific journals, in compilations such as the *Anthropometric Source Book* (Webb Associates, 1978), Jurgens et al. (1990), *ADULTDATA* (and its companion volumes *CHILDATA* and *OLDER ADULTDATA* [Peebles and Norris, 1998; Norris and Wilson 1995; Smith et al. 2000]) and in the anthropometric database PeopleSize 2000 (Open Ergonomics Ltd., 2000). However, few of the surveys for either national or special populations provide a comprehensive set of anthropometric dimensions (if such a set were possible), and designers are often faced with the need to estimate dimensions for some design parameters even when using the most comprehensive data set. Approaches to dealing with this are discussed in Section 2.8.

At the end of this chapter you will find a table of best estimate figures for the body dimensions of the adult population of the United Kingdom aged 19 to 65 years (Table 2.5). In the chapters that follow we shall treat this as the *standard reference population* on which we shall base our design recommendations and other anthropometric calculations. Data for other target populations and details of sources, etc., can be found in Chapter 10.

2.7.2 Defining the Target User Population

The principal factors to take into account when defining a target population of users, for the purpose of selecting an appropriate source of anthropometric data, will in general be sex, age, nationality (or ethnicity) and occupation (or social class), generally in that order of importance. Where the target population includes children, then age will take first place. Taking account of the presence of ethnic minorities within a national population tends to be more of a problem in theory than in practice. As a general guideline, percentile values are unlikely to be affected to any significant extent until the minority group reaches 30% or more of the total population. Again, however, there may be exceptions for certain safety-critical applications (e.g., guarding of machinery; see Thompson and Booth, 1982) and for very ethnically diverse populations (Al-Haboubi, 1992).

If no adequate data are available for the user population, anthropometric data may need to be collected from a sample representative of the user population (or even from all users if dealing with a workspace for a small number of people). This book does not set out to explain how to conduct an anthropometric survey, but practical guidance on measurement can be found in Roebuck (1995), Bradtmiller et al. (2004) and ISO standards ISO 7250 Basic Human Body Measurements for Technological Design and ISO 15535 General Requirements for Establishing Anthropometric Databases (ISO, 1996, 2003b). ISO 15535 includes a method of estimating the sample size which will be adequate for different purposes.

2.7.3 Accuracy of Anthropometric Data

Without attempting to discuss all of the many considerations necessary when designing an anthropometric survey, some important issues related to accuracy, repeatability and reproducibility can be highlighted. There is always some variation between repetitions of measurements (usually termed measurement error [ISO, 1994]). Within measurement error, there are at least four components: error from the measuring equipment itself (whether systematic or random), error from locating the landmark, error from standardising the posture of the subject and error from the subject's understanding of or response to instructions for adopting the required posture. Variability may also occur through natural biological fluctuations, two of the most common being the diurnal (circadian) change in stature due to the response to loading of the complex structures in the spine and changes in chest dimensions through the breathing cycle. Stature varies typically by about 15 mm over 24 hours, being greatest first thing in the morning after the spine has been relieved of supporting body weight during the night lying in bed. The rate of shrinkage is also greatest in the first three hours after rising, and recovery fastest in the early part of the night's rest (Reilly et al., 1984).

Reproducibility, too, the repeatability between measurers and measuring environments, is important when undertaking anthropometric surveys. The reader is referred to Kanis and Steenbekkers (1995) and Kanis (1997) for discussion of some of these issues.

What accuracy is actually required in anthropometric data? This is a very difficult question which must be studied at several different levels. In the purely formalised statistical sense we may consider what a given percentile of a dimension that is erroneously quoted actually represents. For example, if a 95th %ile value were in error, the figure quoted might in truth represent the 93rd or 98th %ile, with a consequence of mismatching a greater or lesser percentage of the target population in the design. In a validation study of the ratio-scaling estimation technique (discussed later in Section 2.8; Pheasant, 1982), the estimates of the 1st and 99th %iles were checked in this way. On average the estimates would have included 96% of the population as against 98% for perfect data.

It is more informative, however, to consider the likely errors of prediction alongside those that might arise in other ways. The human body has very few sharp edges — its contours are rounded, and it is generally squashy and unstable. The consequent difficulty in identifying landmarks and controlling posture makes it virtually impossible to achieve an accuracy of better than 5 mm in most anthropometric measures, and for some dimensions the errors may be much worse (and sitting elbow height is a notorious example). An accuracy of ±5 mm is more than adequate for most design applications.

In applying anthropometric data we commonly need to make corrections for shoes, clothing, postural variation and so on (see below), since the data are collected in standardised conditions and postures during surveys (for reasons of ensuring repeatability and reproducibility). The necessary corrections, although they are better than arbitrary, will usually be inexact. However (and perhaps more importantly), we also have to consider the accuracy of the anthropometric criteria we apply to define a match. Take the case of seat height that we discussed earlier. The sensations of discomfort which the user experiences will become progressively more pronounced as the height of the seat exceeds his or her popliteal height. However, there is no obvious and clearly defined cut-off point at which we should say 'thus far and no further'. In practice, anthropometric criteria are almost always 'fuzzy' in this way. At the very least, a design study should specify criteria clearly and in quantitative terms as far as possible.

There are doubtless certain safety-critical applications in which accuracy would be at a premium, but experience indicates that these are the exception rather than the rule. In practice there would be few everyday workspace problems requiring an ergonomic specification to an accuracy of better than ±10 mm.

2.7.4 Clothing Corrections

Most anthropometric measurements are made on lightly clothed (or sometimes on unclothed) people, whereas of course most products and environments are used by clothed people. The data tabulated in Table 2.5 are for unclothed and unshod people, so before applying these data to any particular problem it will in general be necessary to add an appropriate correction for clothing. It makes sense to do it this way, rather than to quote figures with clothing corrections already added, because the original measurements will be more accurate and because the magnitude of the necessary correction may vary greatly depending upon the circumstances.

The most important of these corrections is an increment for the heels of shoes which must be added to all vertical dimensions measured from the floor. The thinnest pair of carpet slippers has a heel height of only 10 mm. The most outrageous pair of high heels may add 150 mm to stature. A typical heel height for men's everyday shoes and for women's flat shoes is around 25 ± 5 mm.

Women's shoes (and men's shoes to a lesser extent) are subject to periodic changes in fashion. The products and spaces we design will presumably remain in use over several of these fashion cycles. In theory therefore we should base our heel height correction on the midpoint about which these cycles oscillate. In theory also we should add an increment to the standard deviation of our dimension as well as the mean, to allow for variability in heel height. In practice, however, variability in heel height is small compared with anthropometric variability, and a uniform increment to all percentiles will be adequate. Taking one consideration with another, the following corrections would seem appropriate for shoes worn in public places on formal and semiformal occasions:

- For men, add 25 mm to all dimensions
- For women, add 45 mm to all dimensions

Corrections for situations where other types of footwear are the norm should be made on an ad hoc basis.

Other clothing corrections are in general likely to be small, except for very heavy outdoor clothing, for safety helmets or other specialised protective gear or when people are wearing equipment (e.g., tool belts or backpacks). Some examples are given in Section 2.8 when discussing individual body dimensions.

2.7.5 Standard Anthropometric Postures

Most of the measurements in anthropometric surveys (and likewise, those given in Table 2.5 and Chapter 10) were made in one of two standard postures, which can be seen in Figure 2.11.

In the *standard standing posture* the subject stands erect, pulling himself or herself up to full height and looking straight ahead, with shoulders relaxed and arms hanging loosely by the sides. Head posture in "looking straight ahead" is standardised by aligning the upper edge of the external opening of the ear (auditory meatus) and the lower edge of the eye socket (orbital margin) horizontally, in the so-called Frankfurt plane.

In the *standard sitting posture* the subject sits erect on a horizontal, flat surface, pulled up to full height and looking straight ahead. The shoulders are relaxed, with the upper arms hanging freely by the sides and the forearms horizontal (i.e., the elbows are flexed to a right angle). The height of the seat is adjusted (or blocks are placed under the feet) until the thighs are horizontal and the lower legs are vertical (i.e., the knees are flexed to a right angle). Measurements are made perpendicular to two reference planes. The *horizontal reference plane* is that of the seat surface. The *vertical reference plane* is a real or imaginary plane which touches the back of the uncompressed buttocks and shoulder blades of the subject. The *seat reference*

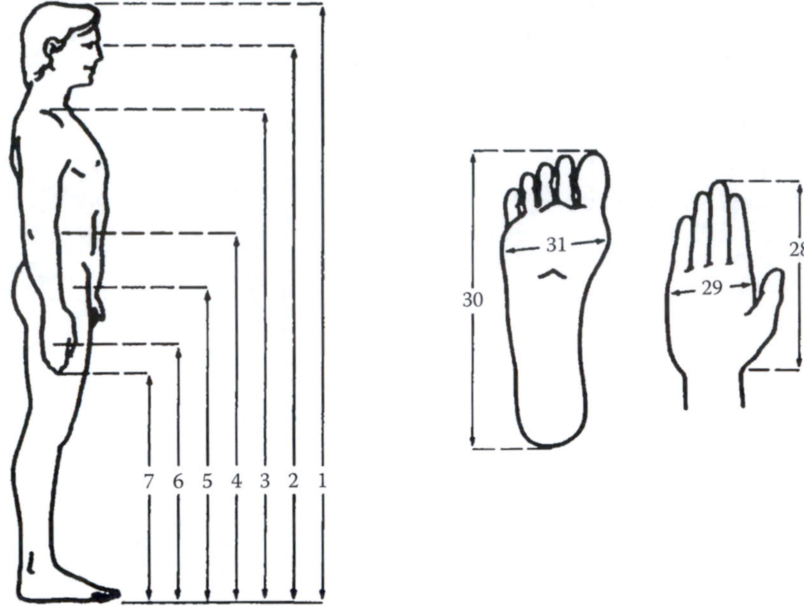

FIGURE 2.11 Body dimensions (continued on the following pages).

point (SRP) lies at the point of intersection of these two planes and the *mid-sagittal plane* of the body (i.e., the plane that divides it equally into its right and left halves). (It should be noted that the seating reference point can be defined differently in some studies and applications, particularly when applied to upholstered (compressible) seats, and the definition in the data source should be consulted before using dimensions related to this reference point.)

People, of course, rarely use these upright positions in everyday life. In practice this may not be so much of a problem as it seems, since we shall commonly set our criteria in such a way as to take this into account. There are circumstances, however, where it may be appropriate to make a nominal correction for normal slump in a relaxed sitting posture. Where this is the case, as a rough approximation for adult populations, subtract 40 mm from all percentiles of relevant vertical sitting dimensions; this is based on the findings of a large-scale anthropometric survey of the U.S. civilian population (HES, 1962). The degree of slump will increase when sitting for a long period of time, which could be important in applications such as vehicle interior design.

2.7.6 Body Proportions

A mistake that is sometimes made is to assume that a person having one dimension of, say, 95th %ile, will have all other dimensions 95th %ile. This is far from the case — we vary in body proportions just as we do in individual segment dimensions. One person of 95th %ile stature can be long-legged and short in the torso, while another can be just the opposite. Compare the lengths of fingers on your right and

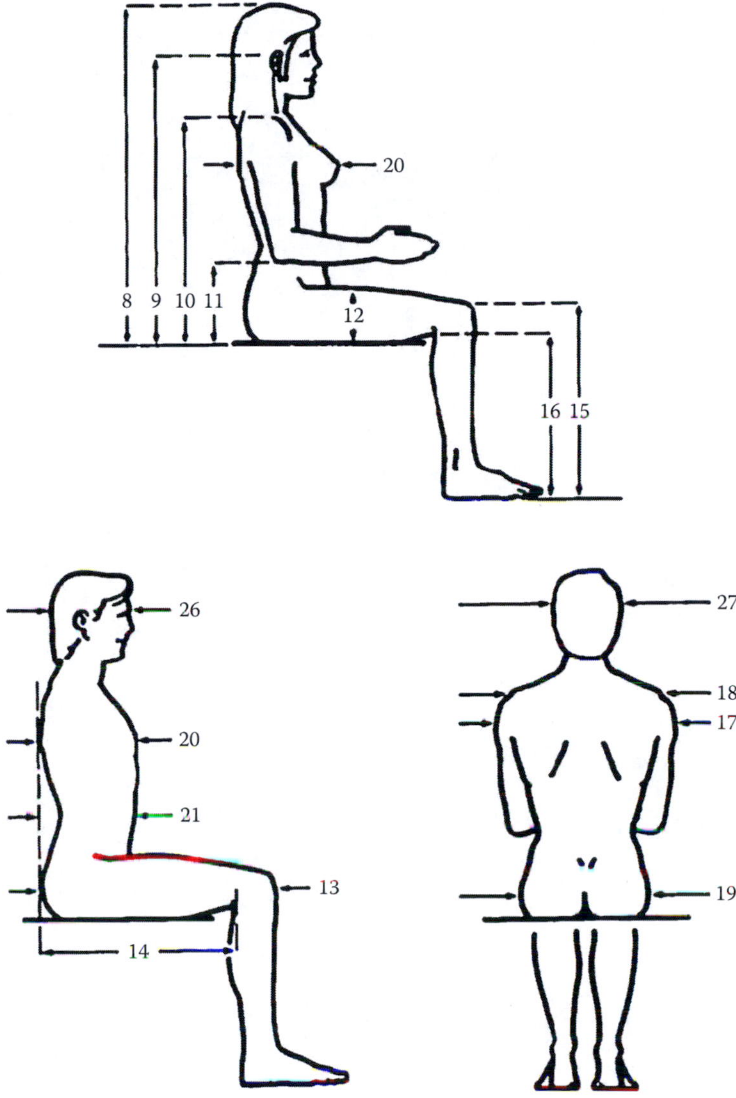

FIGURE 2.11 (Continued).

left hands. Daniels (1952) studied dimensions relevant to clothing sizes for a sample of over 4000 individuals, categorising them as 'small', 'average' or 'large' in each dimension (the average group being those who fell within ±0.3 standard deviations of the mean, rounded to the nearest whole centimetre, on the particular dimension). He found that, while 26% of the sample fell into his average group, by the time three dimensions were considered (stature, chest circumference and sleeve length) only 3.5% of the original sample were average in all three dimensions. When seven dimensions were considered, only 0.2% of the original sample were average in all

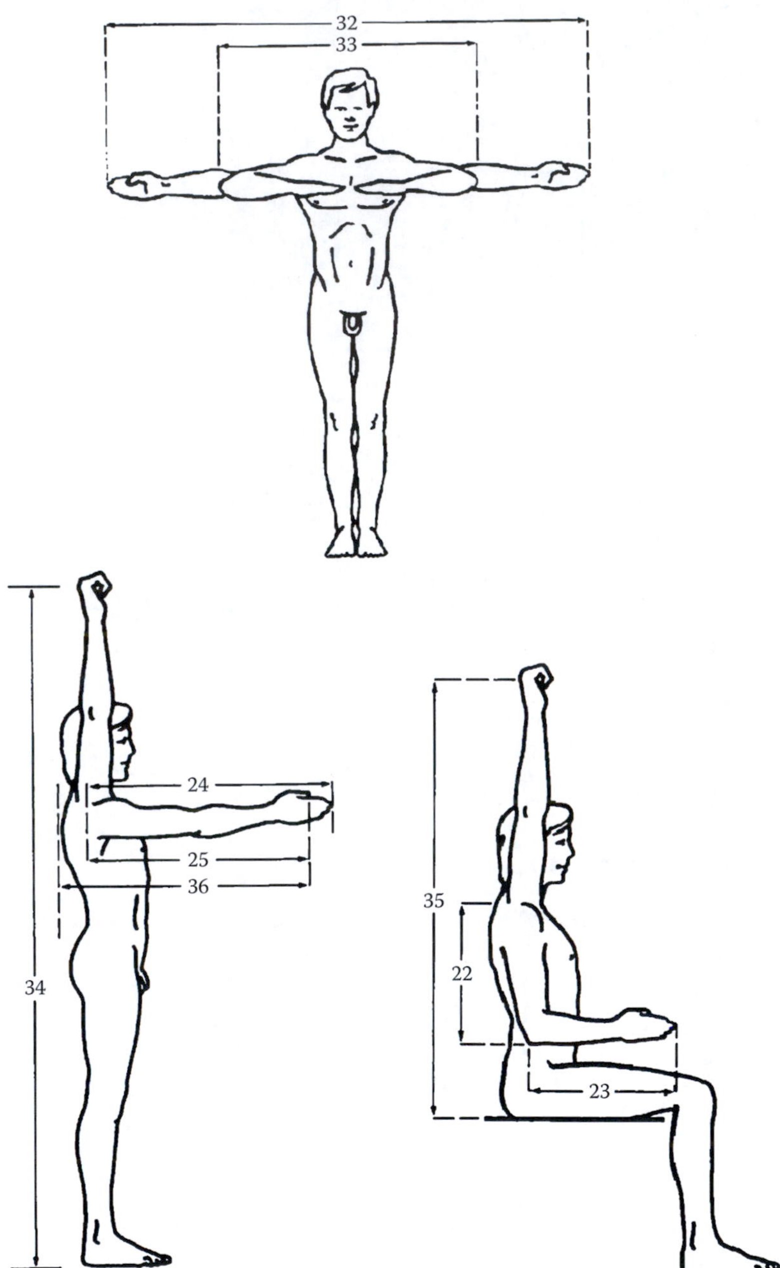

FIGURE 2.11 (Continued).

seven dimensions, and by ten dimensions none were average in all dimensions. The overall 'average man' really does seem to be mythical.

Thus, although a statistical model of an average person would be 50th %ile in all dimensions, in reality such a person might not even exist. Certainly it is incorrect, both statistically and in reality, to combine 95th (or 5th or nth) %ile dimensions to create a human model. If we tried to combine 95th %ile leg segments, 95th %ile trunk length, and 95th %ile head/neck length in such a model, the result would be very much taller than 95th %ile stature. McConville and Churchill (1976) performed this exercise using anthropometric data for 14 body segments and found that it would produce an overall stature of 2022 mm, overestimating the real 95th %ile stature by 300 mm and exceeding the tallest person within their population (of nearly 2000 people) by 192 mm. We cannot therefore refer to a 'person of a particular percentile', only to individual dimensions of the given percentile. In other words, percentiles are specific to the dimension, not the person. It is important to remember this when designing a workplace with several interacting anthropometric constraints.

This problem also arises when creating manikins to represent humans of different sizes (Robinette and McConville, 1981; Annis and McConville, 1990). Manikins are used, for example, in crash testing vehicles to understand the interaction between the occupant and the vehicle interior, and families of virtual manikins are still needed for the digital human models in workplace simulation. Given the problems described, one approach to constructing a manikin could be to represent the required percentile in one key dimension (usually stature) and to chose all other dimensions as typical for a person of that stature. Typical dimensions for British women of 5th %ile stature and men of 95th %ile stature were calculated by Haslegrave (1986).

It is worth finishing this section on a positive note and quoting Annis and McConville (1990), who conclude that

> The cautionary tone…should in no way be construed to mean that designing workspaces and equipment for use by both men and women of all ages and races presents insoluble problems. Rather, we want to emphasise that the simplistic plucking of a value or values from a convenient handbook may be of little more value than the rules of thumb that are stated in various folk norms and in some instances will be down right wrong. It is therefore, incumbent on the designer to ascertain who is being designed for and then locate the data base that best characterizes the body size variability of that population. With this in hand the appropriate design value necessary to accommodate the body size variability of the population can usually be determined.

Various multivariate techniques are mentioned in the Appendix for dealing with the inherent variability in and between human populations, while the potential for accessing comprehensive pouplation data through 3-D anthropometry is illustrated in Section 2.6.4.

2.8 BODY DIMENSIONS

Bodyspace could never set out to provide comprehensive data for international use (or even for national use in the longer term), and useful sources of such data have already been mentioned. The main aim is to present the principles of anthropometry and workspace design, providing examples to illustrate their application. In order to do this, the *Bodyspace* tables in Chapter 10 were compiled to give a coherent anthropometric dataset for the U.K. population (based on estimates calculated from a compilation of the best data sources available). This data from the *Bodyspace* tables has been used throughout the book.

The dimensions included in this dataset are illustrated in Figure 2.11 and defined in Table 2.4, which also contains notes on applications for the data and corrections which should be applied when they are used. The composition and limitations of this dataset are explained in detail in Chapter 10. The principles of anthropometry can be applied to other populations by substituting the corresponding data (or data for the nearest equivalent population) in the examples provided throughout this book.

Data on civilian populations are relatively sparse, few surveys covering large numbers of body dimensions. For example, the only currently available dataset that may be reasonably assumed to be representative of the adult population of the United Kingdom was collected by the Office of Population Censuses and Surveys in 1981 (OPCS, 1981; Knight, 1984), and in this survey the only body dimensions measured were height and weight.

Confronted with this situation, we may take the purist approach and only quote sources of unimpeachable accuracy; or we may take the pragmatic approach and fill in the gaps as best we can by using various rule-of-thumb methods of estimation (and a certain amount of informed guesswork). *Bodyspace* has adopted the latter approach.

You can use a number of techniques to fill the gaps in this way when hard data are not available. These are described in detail in the Appendix. The technique used most extensively in this book is the method of *ratio scaling*. This technique assumes that although different samples drawn from a particular 'parent' population (as defined in terms of age, sex and ethnic origin) may vary greatly in size, they are likely to be relatively similar in shape. (Note that, while this is a reasonable pragmatic assumption for the population parameters of mean and standard deviation, it is not true and should not be used to represent an individual person or construct a human model, as discussed in Section 2.7.6.) Thus, if we have stature data only for the target population that interests us, but we have detailed body-part measurements for an equivalent population or population sample, we may use the former to 'scale up' the latter.

A detailed validation study of this technique is described in Pheasant (1982). The 1st and 99th %ile values of 136 dimensions drawn from six surveys were estimated from a knowledge of only the parameters of stature in the survey concerned. (This is a more rigorous test than the 5th and 95th %iles commonly used in design work.) The errors that accrued were random and conformed approximately to a normal distribution with parameters of –3 [13] mm. Ninety-three percent of the errors fell within the range of ±25 mm. In many cases the estimates were within the confidence limits of the original survey.

The resulting anthropometric estimates for the British adult population are given in Table 2.5, and this is used as the *'standard reference population'* throughout this book. The table includes the dimensions that are probably those most frequently used by designers. When using the data, shoe and clothing corrections should be applied (where relevant). Recommended values for these and other corrections can be found in Table 2.4.

TABLE 2.4
Annotated List of Body Dimensions and Their Definitions

Definition of Dimension	Applications	Corrections
1. **Stature:** The vertical distance from the floor to the vertex (i.e., crown of head) Notes: 1. In children who cannot stand unaided, crown–heel length is the nearest equivalent dimension. The child must be stretched out in a supine position and prevented from wriggling. 2. If you ask adults to tell you their height, you must expect them to overestimate by an average of about 25mm.	As a cross-referencing dimension for comparing populations and estimating data; defines the vertical clearance required in the standing workspace; minimal acceptable height of overhead obstructions such as lintels, roofbeams, light fittings, etc.	Shoes as in Section 2.7.4 25 mm for a hat 35 mm for a protective helmet A few design applications call for supine or prone body length (lying on back or front repectively). Such a position lengthens the adult body by approximately 15 mm.
2. **Eye height:** Vertical distance from the floor to the inner canthus (corner) of the eye	Centre of the visual field; reference datum for location of visual displays as in Section 4.6, ‘reach’ dimensions for sight lines, defining maximum acceptable height of visual obstructions; optical sighting devices for prolonged use should be adjustable for the range of users	Shoes as in Section 2.7.4
3. **Shoulder height:** Vertical distance from the floor to the acromion (i.e., the bony tip of the shoulder)	The approximate centre of rotation of the upper limb and, hence, of use in determining zones of comfortable reach; reference datum for location of fixtures, fittings, controls, etc.	Shoes as in Section 2.7.4
4. **Elbow height:** Vertical distance from the floor to the radiale (the bony landmark formed by the upper end of the radius bone which is palpable on the outer surface of the elbow) Note: Some surveys measure to the underside of the elbow when it is flexed to a right angle. This gives a value approximately 15 mm less than the standard measurement.	An important reference datum for the determination of work-surface heights, etc. (see Section 4.7)	Shoes as in Section 2.7.4

(continued)

TABLE 2.4 (CONTINUED)
Annotated List of Body Dimensions and Their Definitions

Definition of Dimension	Applications	Corrections
5. **Hip height:** Vertical distance from the floor to the greater trochanter (a bony prominence at the upper end of the thigh bone, palpable on the lateral surface of the hip)	Centre of rotation of the hip joint, hence the functional length of the lower limb	Shoes as in Section 2.7.4
6. **Knuckle height:** Vertical distance from the floor to metacarpal III (i.e., the knuckle of the middle finger)	Reference level for handgrips; for support (handrails, etc.) approximately 100 mm above knuckle height is desirable. Handgrips on portable objects should be at less than knuckle height. Optimal height for exertion of lifting force (see Section 9.5)	Shoes as in Section 2.7.4
7. **Fingertip height:** Vertical distance from the floor to the dactylion (i.e., the tip of the middle finger)	Lowest acceptable level for finger-operated controls	Shoes as in Section 2.7.4
8. **Sitting height:** Vertical distance from the sitting surface to the vertex (i.e., crown of the head)	Clearance required between seat and overhead obstacles.	10 mm for heavy outdoor clothing beneath the buttocks; variable amount for seat compression; 25 mm for a hat; 35 mm for a safety helmet
9. **Sitting eye height:** Vertical distance from the sitting surface to the inner canthus (corner) of the eye	See dimension 2	10 mm for heavy outdoor clothing beneath the buttocks; up to 40 mm reduction for 'sitting slump'; variable amount for seat compression
10. **Sitting shoulder height:** Vertical distance from the seat surface to the acromion (i.e., the bony point of the shoulder)	Approximate centre of rotation of the upper limb	As for dimension 9
11. **Sitting elbow height (also known as elbow rest height):** Vertical distance from the seat surface to the underside of the elbow	Height of armrests; important reference datum for the heights of desk tops, keyboards, etc., with respect to the seat (Chapter 7)	

12. **Thigh thickness (also known as thigh clearance):** Vertical distance from the seat surface to the top of the uncompressed soft tissue of the thigh at its thickest point, generally where it meets the abdomen	Clearance required between the seat and the underside of tables or other obstacles (see Section 7.2)	35 mm for heavy outdoor clothing
13. **Buttock–knee length:** Horizontal distance from the back of the uncompressed buttock to the front of the kneecap	Clearance between seat back and obstacles in front of the knee (see Section 7.2)	20 mm for heavy outdoor clothing
14. **Buttock–popliteal height:** Horizontal distance from the back of the uncompressed buttocks to the popliteal angle, at the back of the knee, where the back of the lower leg meets the underside of the thigh	Reach dimension, defines maximum acceptable seat depth (see Section 5.4)	
15. **Knee height:** Vertical distance from the floor to the upper surface of the knee (usually measured to the quadriceps muscle rather than the kneecap)	Clearance required beneath the underside of tables, etc. (see Section 7.2)	Shoes as in Section 2.7.4
16. **Popliteal height:** Vertical distance from the floor to the popliteal angle at the underside of the knee where the tendon of the biceps femoris muscle inserts into the lower leg	Reach dimension defining the maximum acceptable height of a seat (see Section 5.4)	Shoes as in Section 2.7.4
17. **Shoulder breadth (bideltoid):** Maximum horizontal breadth across the shoulders, measured to the protrusions of the deltoid muscles	Clearance at shoulder level	10 mm for indoor clothing; 40 mm for heavy outdoor clothing
18. **Shoulder breadth (biacromial):** Horizontal distance across the shoulders measured between the acromia (bony points)	Lateral separation of the centres of rotation of the upper limb	
19. **Hip breadth:** Maximum horizontal distance across the hips in the sitting position Note: This is a dimension with a substantial soft tissue component. In studies of physique, etc., the bony dimension bicristal breadth is generally used (measured between the lateral edges of the crests of the hip bones)	Clearance at seat level; the width of a seat should be not much less than this (see Section 5.4)	10 mm for light clothing; 25 mm for medium clothing; 50 mm for heavy outdoor clothing
20. **Chest (bust) depth:** Maximum horizontal distance from the vertical reference plane to the front of the chest in men or breast in women	Clearance between seat back and obstructions	Up to 40 mm for outdoor clothing

(continued)

TABLE 2.4 (CONTINUED)
Annotated List of Body Dimensions and Their Definitions

Definition of Dimension	Applications	Corrections
21. **Abdominal depth:** Maximum horizontal distance from the vertical reference plane to the front of the abdomen in the standard sitting position	Clearance between seat back and obstructions	Up to 40 mm for outdoor clothing
22. **Shoulder–elbow length:** Distance from the acromion to the underside of the elbow in a standard sitting position		
23. **Elbow–fingertip length:** Distance from the back of the elbow to the tip of the middle finger in a standard sitting position	Forearm reach; used in defining normal working area (see Section 4.3)	For general reach corrections, see dimension 34
24. **Upper limb length:** Distance from the acromion to the fingertip with the elbow and wrist straight (extended)		
25. **Shoulder–grip length:** Distance from the acromion to the centre of an object gripped in the hand, with the elbow and wrist straight	Functional length of upper limb; used in defining zone of convenient reach (see Section 4.3)	Reach corrections as for dimension 34
26. **Head length:** Distance between the glabella (the most anterior point on the forehead between the row ridges) and the occiput (back of head) in the midline	Reference datum for location of eyes, approximately 20 mm behind glabella (see Section 4.6)	
27. **Head breadth:** Maximum breadth of the head above the level of the ears	Clearance	Add 35 mm for clearance across the ears; up to 90 mm for protective helmets
28. **Hand length:** Distance from the crease of the wrist to the tip of the middle finger with the hand held straight and stiff	See dimension 34 and Section 6.2	
29. **Hand breadth:** Maximum breadth across the palm of the hand (at the distal ends of the metacarpal bones)	Clearance required for hand access, e.g., grips, handles, etc. (see Sections 6.2 and 6.4)	As much as 25 mm for some protective gloves

30. **Foot length:** Distance, parallel to the long axis of the foot, from the back of the heel to the tip of the longest toe	Clearance for foot; design of pedals	In many respects surveys of shoes would be more relevant than surveys of feet since their sizes and shapes are often unrelated. For the purposes of argument, we could add 30 mm for men's street shoes and 40 mm for protective boots
31. **Foot breadth:** Maximum horizontal breadth, wherever found, across the foot perpendicular to the long axis	Clearance for foot; spacing of pedals, etc.	See dimension 30; 10 mm for men's street shoes; 30 mm for heavy boots
32. **Span:** Maximum horizontal distance between the fingertips when both arms are stretched out sideways	Lateral reach (see Section 4.3)	See dimension 34
33. **Elbow span:** Distance between the tips of the elbows when both upper limbs are stretched out sideways and the elbows are fully flexed so that the fingertips touch the chest	A useful guideline when considering 'elbow room' in the workspace	
34. **Vertical grip reach (standing):** Distance from the floor to the centre of a cylindrical rod grasped in the palm of the hand, with the arm raised vertically above the head in an easy reach (without excessive stretch)	Functional reach (see Section 4.3 for a discussion of reach envelopes in general)	Some surveys measure reach to the tip of the outstretched middle finger or to the tip of the thumb when it forms a 'pinch' with the index finger. Approximately, fingertip reach = grip reach + 60% hand length; thumbtip reach = grip reach + 20% hand length
35. **Vertical grip reach (sitting):** As for dimension 34 except that the measurement is made from the seat surface	Functional reach (see Section 4.3 for a discussion of reach envelopes in general)	See dimension 34
36. **Forward grip reach:** The arm is raised horizontally forward at shoulder level with a cylindrical rod grasped in the palm of the hand, in an easy reach (without excessive stretch). The measurement is made from the back of the shoulder blades to the centre of the rod	Functional reach (see Section 4.3 for a discussion of reach envelopes in general)	See dimension 34

TABLE 2.5
Anthropometric Estimates for British Adults Aged 19 to 65 Years (all dimensions in millimetres, except for body weight, given in kilograms)

	Men				Women			
Dimension[a]	5th %ile	50th %ile	95th %ile	SD	5th %ile	50th %ile	95th %ile	SD
1. Stature	1625	1740	1855	70	1505	1610	1710	62
2. Eye height	1515	1630	1745	69	1405	1505	1610	61
3. Shoulder height	1315	1425	1535	66	1215	1310	1405	58
4. Elbow height	1005	1090	1180	52	930	1005	1085	46
5. Hip height	840	920	1000	50	740	810	885	43
6. Knuckle height	690	755	825	41	660	720	780	36
7. Fingertip height	590	655	720	38	560	625	685	38
8. Sitting height	850	910	965	36	795	850	910	35
9. Sitting eye height	735	790	845	35	685	740	795	33
10. Sitting shoulder height	540	595	645	32	505	555	610	31
11. Sitting elbow height	195	245	295	31	185	235	280	29
12. Thigh thickness	135	160	185	15	125	155	180	17
13. Buttock–knee length	540	595	645	31	520	570	620	30
14. Buttock–popliteal length	440	495	550	32	435	480	530	30
15. Knee height	490	545	595	32	455	500	540	27
16. Popliteal height	395	440	490	29	355	400	445	27
17. Shoulder breadth (bideltoid)	420	465	510	28	355	395	435	24
18. Shoulder breadth (biacromial)	365	400	430	20	325	355	385	18
19. Hip breadth	310	360	405	29	310	370	435	38
20. Chest (bust) depth	215	250	285	22	210	250	295	27
21. Abdominal depth	220	270	325	32	205	255	305	30
22. Shoulder–elbow length	330	365	395	20	300	330	360	17
23. Elbow–fingertip length	440	475	510	21	400	430	460	19
24. Upper limb length	720	780	840	36	655	705	760	32
25. Shoulder–grip length	610	665	715	32	555	600	650	29
26. Head length	180	195	205	8	165	180	190	7
27. Head breadth	145	155	165	6	135	145	150	6
28. Hand length	175	190	205	10	160	175	190	9
29. Hand breadth	80	85	95	5	70	75	85	4
30. Foot length	240	265	285	14	215	235	255	12
31. Foot breadth	85	95	110	6	80	90	100	6
32. Span	1655	1790	1925	83	1490	1605	1725	71
33. Elbow span	865	945	1020	47	780	850	920	43
34. Vertical grip reach (standing)	1925	2060	2190	80	1790	1905	2020	71
35. Vertical grip reach (sitting)	1145	1245	1340	60	1060	1150	1235	53
36. Forward grip reach	720	780	835	34	650	705	755	31
Body weight	55	75	94	12	44	63	81	11

[a] Definitions of the dimensions are given in Table 2.4 and Figure 2.11.

3 Human Diversity

3.1 INTRODUCTION

In this chapter we shall consider the principal ways in which *samples* and *populations* of human beings differ in their anthropometric characteristics — and the *biosocial* factors that underlie these differences.

The sizes, shapes and strengths of human beings are very often broken down by age and sex when they are tabulated in anthropometric databases. In defining a target user population for anthropometric purposes, we must also take into account ethnicity (and sometimes regional differences within the relatively homogeneous population of a country), social class and occupation. Superimposed over these differences are changes occurring within populations over a period of time. Some of these are attributable to the migration and genetic admixture of hitherto distinct ethnic groups and others to more complex historical processes, which over the past century or so have led to an almost worldwide increase in stature, which is referred to as the '*secular trend*'.

The extent to which these measurable differences between populations of human beings are determined by biological (or genetic) factors, as against social (or environmental) ones, poses a difficult set of questions. This nature/nurture controversy has ramifications in many branches of the human sciences. In reality, asking whether a given characteristic is determined by inheritance or by upbringing and lifestyle is probably a bit like asking whether the area of a field is determined by its length or its breadth. Where anthropometric dimensions are concerned, the environment — health and conditions — of mothers (and perhaps even of preceding generations) is also likely to have had a strong influence.

One further point must be stressed. When comparing and contrasting the measurable characteristics of different groups of people, we will always be dealing with *within-group* variability as well as *between-group* variability. The greater the former compared with the latter, the less significant will be the difference between the groups (both in terms of statistical theory and in terms of ergonomics practice).

Consider humankind as a whole. It is debatable whether anthropometric data available at the present time, even if they could be assembled in one place, would constitute a representative sample of all human beings living at the present time. Anthropometric databases come from surveys which have been carried out at different times, and secular trends vary greatly between countries (as discussed later). Such indications as we have, however (see Tildesley, 1950), suggest that, in round figures, the stature of all adult living male adults has a mean value of about 1650 mm with a standard deviation of 80 mm, which, as the reader will recall from Chapter 2, we write as 1650 [80] mm. Assuming a sex difference in average stature of 7% and an equal coefficient of variation, the stature of living female adults will have a distribution of about 1535 [75] mm. (These figures could doubtless be improved by anyone with the patience to do so but will serve as a starting point.)

The adult population of Great Britain (represented in Table 2.5 as the standard population throughout this book) is well into the taller half of the human race, with stature 1740 [70] mm for men and 1610 [62] mm for women. Hence the average British adult male is about 87th %ile for the human race as a whole.

According to Roberts (1975), the shortest people in the world are the Efe and Basua 'pygmies' of Central Africa, whose average stature is 1438 [70] mm for men and 1372 [78] mm for women. The tallest are the Dinka Nilotes of the southern Sudan: 1829 [61] mm for men and 1689 [58] mm for women. However, differences almost as great as these may be found between particular samples drawn from the British population (representing occupational or age groups). Guardsmen (Gooderson and Beebee, 1977) stand at some 1803 [63] mm, whereas a sample of elderly women measured by Harris and her colleagues (Institute for Consumer Ergonomics, 1983) had a stature of 1515 [70] mm (deducting a modest 20 mm for shoes since the subjects were measured shod).

The human race is more varied still. The limits of what is normally considered to be clinical normality are set at an adult stature of something in the order of between 1370 and 2010 mm. (The exact figures quoted vary somewhat, being essentially arbitrary.) According to the best information available at the time of writing (supported by evidence and as set out in Guinness World Records 2004), the shortest living adult man in the twentieth century was 570 mm tall, and the tallest was 2720 mm.

3.2 SEX DIFFERENCES

It has become fashionable of late to refer to the differences between men and women as ones of 'gender' rather than 'sex'. This is incorrect. The word *gender* applies to the distinction that exists in most European languages (other than English) between masculine and feminine nouns, rather than the differences between male and female living organisms. The human species, like all the higher animals, is sexually dimorphous.

Are the anthropometric differences between men and women attributable to underlying biological (i.e., genetic and physiological) differences or to cultural differences in upbringing and lifestyle? We can be fairly sure that sex differences in stature and related body dimensions and most differences in bodily proportions are almost entirely biological in their origin, although there may be a small overlay of differences attributable to lifestyle, etc. In the case of muscular strength, however, the position is more evenly balanced, and although the male of the species has the greater physiological propensity to the acquisition of muscle strength, the overlay of differences associated with physical training and lifestyle is considerable.

What is the best way of describing sex differences statistically? The most frequent way found in the literature is a straightforward comparison of means. Hence, we read statements like 'on average women are 7% shorter than men' or 'on average women are 66% as strong as men'. Let us call the average female dimension (or strength) divided by the average male dimension (or strength) the F/M ratio.

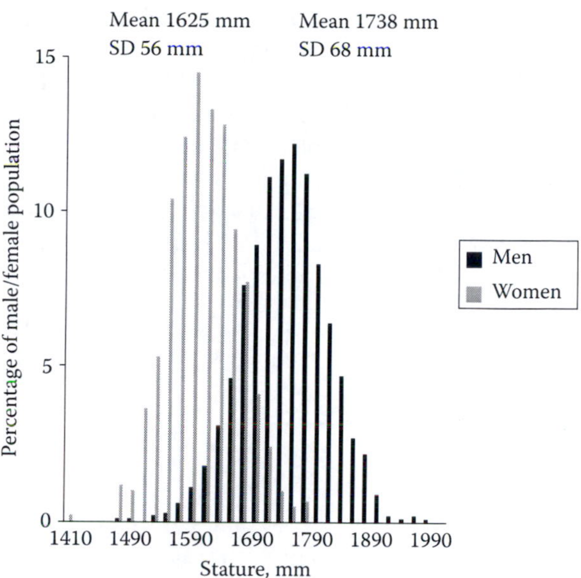

FIGURE 3.1 Comparison of the stature distributions of U.K. female and male car drivers. (Data from Haslegrave, C. M. (1986). *Ergonomics,* 29(2), 281–301.)

However, for all the variables we are likely to consider in this text, there is considerable overlap between the male and female distributions, as can be seen in the stature distributions shown in Figure 3.1. The F/M ratio of means tells us very little about this combined distribution. (Among many other equally interesting descriptions we might include the ratio of the 95th %ile female to the 5th %ile male values or the 5th %ile female to the 95th %ile male, the percentage of women taller than the 5th %ile man or men shorter than the 95th %ile woman, etc.)

At the very least, a descriptive index should reflect both the difference between the means and the magnitude of the variances of the male and female distributions under consideration. It would be useful and informative to know the proportion in the total variance in stature (i.e., in the combined unisex distribution) which is attributable to sex. Afficionados of the one-way analysis of variances will understand that this index is given by the equation for the coefficient of determination R^2 :

$$R^2 = \frac{\text{between sex sum of squares}}{\text{total sum of squares}} \tag{3.1}$$

(If this equation is absolute gibberish to you, don't worry too much. Alternatively turn to any textbook of statistics.)

What does the layperson have in mind when he (or she) asks a question such as 'How true is it that men are stronger than women?' Consider a population of men and a population of women. Suppose we select a man at random followed by a woman at random and compare their strengths. We will call such a comparison a chance encounter. If we perform an infinite number of such comparisons we may generate a statistical distribution of chance encounters. The F/M ratio is equivalent to an encounter between an average woman and an average man. Both the layperson and the human scientist wish to know about the remainder of the distribution. For reasons that would only be comprehensible to a competent mathematical statistician, the distribution of the ratios of two normal distributions is not itself normally distributed. If, however, we forget about ratios and consider absolute differences, the problem becomes much more tractable. If differences are used, the distribution of chance encounters M_e [s_e] is normal, and its parameters are given by

$$M_e = M_m - M_f \quad (3.2)$$

$$s_e^2 = s_m^2 + s_f^2 \quad (3.3)$$

where M and s are the mean and standard deviation; the subscripts m and f refer to men and women, respectively. The value of zero in this distribution represents a chance encounter between a man and a woman of equal strength. It is simple to calculate the proportions of the distribution lying on either side of this point (by calculating z and looking up a table as described in Chapter 2). We therefore know the percentage of chance encounters in which the female is stronger or, in the more general case, the percentage of chance encounters in which the female exceeds the male (% CEFEM). This index is as close as we can reasonably get to the layperson's conception of the question posed.

Obviously, any investigation of sex differences will founder if the samples of men and women who are studied are not truly comparable. Thus a comparison of male navvies with female secretaries or of male secretaries with female navvies, is not solely an investigation of sex differences per se.

In general, men will exceed women in all linear body dimensions except hip breadth (as shown in the data in the tables in Chapter 10). There are ethnic differences in the magnitude of these sex differences — at least for stature. Eveleth (1975) found greater differences between the sexes in Amerindians than in Europeans, who in turn showed greater differences than black Africans.

3.2.1 Variation in Body Proportions

Many sex differences in body proportion are too well known to require further comment. In general, the lengths of the upper and lower limbs are proportionally as well as absolutely greater in men. Thus the ratio of sitting height to stature (sometimes called 'sitting height index' and used as an index of relative trunk length) will be greater in women than in men. The only limb dimension that is proportionally greater in women is buttock–knee length, due to differences in the forms of the male and female buttock. There is no difference between men and women in the proportional values of either head length or head breadth.

In addition to the dimensional anthropometrics described above, men and women differ in their bodily composition. In general, fat represents a greater proportion of body weight in the adult female than in the male. (Subcutaneous fat is also distributed differently, women having a propensity to accumulate fat in the breasts, hips, thighs and upper arms. Abdominal fat accumulates above the umbilicus in men and below the umbilicus in women.) The most direct way of measuring body fat is by densitometry. Fat is a good deal less dense than lean tissue, so if the density of the body is determined (usually by underwater weighing) it is possible to calculate the percentage that fat contributes to the weight of the body. Durnin and Rahaman (1967) found this percentage to be 13.5 [5.8] for adult men and 24.2 [6.5] for adult women (F/M ratio = 179%, R^2 = 43%; % CEFEM = 89).

Data from three-dimensional (3-D) body scanning surveys have now facilitated more detailed study of body shape variation and its relationship with size variation (*allometry*, relative growth). Recent research has shown how the variation arises from size differences, shape differences and allometry differences between the sexes (Cerney and Adams, 2004). For example, the difference in allometry between males and females was found to be characterised as a lengthening of the torso, an increase in the proximal positioning of the sellion (top of the nose), and the locations of the elbow, wrist and hand being further from the body. Thus body shape changes in a different way with respect to body size for the two sexes. As Cerney and Adams point out, this needs to be considered when characterising the extremes of a mixed population.

Friess and Corner (2004) have studied allometeric change in body shape and proportions under the influence of weight. They found that, as body mass index (or relative body weight) increases, there is more pronounced abdominal and gluteal protrusion and an increase in cross-sectional dimensions throughout the leg, although the strongest effect is in the upper thigh. However, women show an increase in the waist and hip regions that is not seen in men and a greater increase in shoulder breadth. Thus the male body becomes more "barrel-shaped" in front view, producing a more even distribution of the excess weight than in women. In side view men carry some excess weight in the lower chest in contrast to in the abdomen for women.

3.2.2 Variation in Strength

Pheasant (1983) published a detailed analysis of sex differences in strength. A survey of the literature located a total of 112 datasets in which a direct and, presumably, valid comparison of the performances of men and women in some test of static strength could be made. Indices of sex differences were calculated for each of these datasets (see Table 3.1). Although the average value of the F/M ratio is 61% — very close to commonly quoted figures of women being two-thirds as strong as men — the ratios found in the whole series range from 37 to 90%. The other indices tell a similar story: sex can account for a major (85%) or a negligible (3%) proportion of the total variance in strength.

An interesting pattern emerges if we divide the datasets into groups according to the part of the body tested. Upper limb tests show greater sex differences than lower limb tests or tests of pushing, pulling and lifting actions, with tests of trunk

TABLE 3.1
Sex Differences in Strength (Classified by Part of Body or Type of Exertion Tested)

Part of Body	Number of Tests	F/M ratio (%)			R^2 (%)			% CEFEM		
		Minimum	Average	Maximum	Minimum	Average	Maximum	Minimum	Average	Maximum
Lower limb	17	50	66	81	11	43	62	4	12	28
Push/pull/lift	41	38	65	90	3	33	72	1	14	37
Trunk	11	37	62	68	35	54	61	1	6	13
Upper limb	29	44	58	86	7	53	77	1	8	35
Miscellaneous	14	43	53	61	35	63	85	0	5	16
All tests	112	37	61	90	3	45	85	0	10	37

Source: Data from Pheasant, S. T. (1983). *Applied Ergonomics,* 14, 205–11.

F/M ratio, female strength divided by male strength; % CEFEM, the percentage of chance encounters in which the female exceeds the male strength being somewhere between the two. Subdivision of the upper limb category reveals that tests of hand and forearm muscles give lesser sex differences than tests of upper arm and shoulder muscles. Taking all three indices together, the least sex differences were found in the push/pull/lift category. (The factors that determine performance in a task of this kind are numerous and complex. The coordinated activity of many muscles may be involved, and in some cases the limiting factor may be body weight and its leverage.) Hettinger (1961), considering the magnitude of the F/M ratio in various muscle groups, suggested that the sex difference is small in those muscle groups that are under-used in everyday life. In the present authors' view it would be just as reasonable to propose the opposite hypothesis on the basis of the available data — and the opposite would be just as biologically plausible!

What is the underlying physiology of sex differences in strength? The strength of a muscle is directly proportional to the effective cross-sectional area of its contractile tissue. The cross-sectional area of a muscle must be closely related to the bulk that is visible to the casual observer or can be measured with a tape. Ikai and Fukunaga (1968), using a sophisticated ultrasound measurement of a cross-sectional area, found strength to be approximately 64 N per cm^2 of muscle tissue, which was independent both of sex and of age from 12 years upwards. Trained judo men had the same strength per unit area as untrained people. In general then, it is the quantity not the quality of muscle that counts, at least in strength measurements of short duration.

It is widely accepted that the secondary sex differences of fat distribution and muscle bulk result from the relative concentrations of the sex hormones — androgens in the male and oestrogens and progestogens in the female. Testosterone, the most important of the androgens, is produced in large quantities in the testes but also in very small quantities in the ovaries. (The level of testosterone in the blood plasma of men is 20 to 30 times that of women.) In response to a given training programme, men show a faster and greater increase in strength than women; this difference is generally attributed to the effects of testosterone (Hettinger, 1961; Klafs and Lyon, 1978). Brown and Wilmore (1974) monitored a small group of female throwing-event athletes over a six months' maximal resistance training programme. Strength increased by 15 to 53%, but there was little evidence of muscular hypertrophy (increase in bulk). How this latter finding is to be reconciled with the results of Ikai and Fukunaga (1968) is not yet clear.

Klafs and Lyon (1978) speculate that women who have a relatively high level of plasma testosterone will 'bulk up' like men in response to intense weight training, and, indeed, the current vogue for female body building shows that a striking degree of muscular hypertrophy may occur in some individuals. It has often been suggested that female athletes are in one respect or another more masculine than their more sedentary sisters. Malina and Zavaleta (1976) calculated an 'androgyny index' as follows:

$$\text{Androgeny Index} = [3 \times \text{biacromial (shoulder) breadth}] - \text{bicristal (pelvic) breadth} \qquad (3.4)$$

Hence a high score on this index is indicative of a masculine skeletal frame and a low score of a feminine frame. Runners (both long and short distance) were not found to differ from nonathletes in androgyny scores, but jumpers and throwers were found to be significantly more masculine according to this criterion. Is this a training effect or does it represent self-selection? The latter is generally believed to be the case amongst physical educationalists (Klafs and Lyon, 1978). Adams (1961) compared young black women who had been engaged in heavy farm labour all their lives with ones who had not. Although the labourers were larger in overall size and muscular development than the controls, the androgyny index was similar for the two groups.

Our discussion of these matters would not be complete without some passing reference to 'norms', 'ideals', 'cultural expectations' and the elusive phenomenon of preference. These can exert extraordinarily strong influences on some members of any particular population to control or change their anthropometric characteristics, as witnessed by the incidence of disorders such as anorexia, discrimination experienced by middle-aged workers in some service industry jobs or the vogue for body building. The cultural expectations do, however, vary between countries and over time. The history of European art reveals considerable diversity in the ideal female form. Consider, for example, the way that Venus was depicted by Rubens, Titian, Botticelli and Cranach, to name but four in order of decreasing radius of her curvature. The ideal male form (Mars, Adam, etc.) seems to have remained remarkably constant by comparison. One of the relatively few empirical studies of such ideals and fashions is that of Garner et al. (1980), who took the strikingly original approach of analysing the recorded heights, weights and bodily circumferences of all Playboy magazine centrefolds between 1959 and 1978. The trend over the period was for an increase in height, reduction in weight for height, bust circumference, and hip circumference and increase in waist circumference, indicating a move towards a body form that the authors characterized (somewhat oddly) as 'tubular'.

3.3 ETHNIC DIFFERENCES

An ethnic group is a population of individuals who inhabit a specified geographical distribution and who have certain physical characteristics in common which serve, in statistical terms, to distinguish them from other such groups of people. These characteristics may be presumed to be predominantly hereditary, although the extent to which this is the case is sometimes contentious.

Ethnic groups may or may not be co-extensive with national, linguistic or other boundaries — hence the various ethnic types to be found within the population of Europe are distributed across national (and linguistic) boundaries and emigrants from an ethnic group may be resident in a different part of the world — and the frequency with which a given ethnic type is encountered will vary from place to place. To some extent, ethnic groups fall into more or less natural clusters, which may be referred to as the Negroid, Caucasoid and Mongoloid divisions or major groups of humankind. The term *race* has tended to disappear from the scientific literature, due, one might suppose, to a collective embarrassment occasioned by its misuse for dogmatic and propagandist purposes. As Gould (1984) has clearly shown, supposedly objective scientific writers have colluded in this misuse.

The Negroid division includes most of the dark-skinned peoples of Africa, together with certain minor ethnic groups of Asia and the Pacific islands. The Caucasoid division includes both light- and dark-skinned peoples resident in Europe, North Africa, Asia Minor, the Middle East, India and Polynesia (together with the indigenous population of Australia and some other ethnic groups who form a subdivision of their own). The Mongoloid division comprises a large number of ethnic groups distributed across central, eastern and south-eastern Asia, together with the indigenous populations of the Americas.

Samples of adults may vary from each other either in overall size (as measured by stature or weight) or in bodily proportions. The most characteristic ethnic differences are of the latter kind since the major divisions of humankind include both tall and short populations. Figure 3.2 illustrates some salient features. Average sitting height (measured from the seat surface) has been plotted against average stature. The ratio of the two (relative sitting height) is plotted as oblique lines on the chart. When relative sitting height is large, the sample is 'short legged' and vice versa. The data points are all male samples taken from Eveleth and Tanner (1976) and Webb Associates (1978). Samples drawn from the civilian or military populations of the United States (of which there are a considerable number in the literature) have been included as 'of predominantly European descent', notwithstanding that around 10% of the membership of such samples are of identifiably different ethnic origins (see Section 2.7.2).

Black Africans have proportionally longer lower limbs than Europeans; Far Eastern samples have proportionally shorter lower limbs, the difference being most

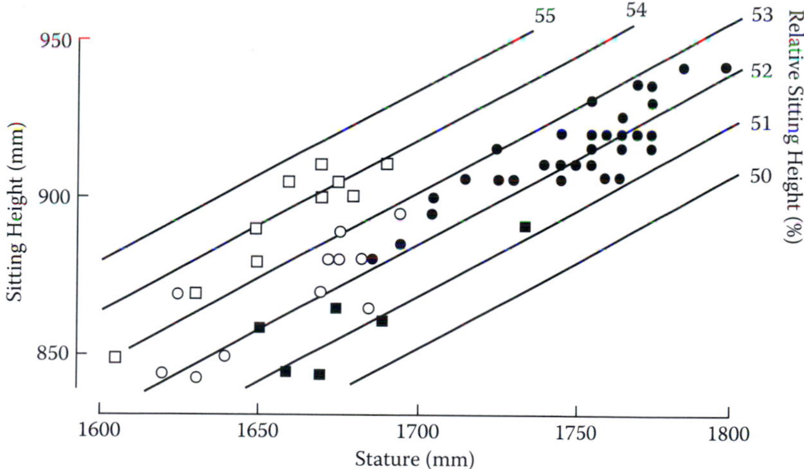

FIGURE 3.2 Ethnic differences in the relationship between average sitting height and average stature in samples of adult men. ● = European (including samples of predominantly European descent); ○ = Indo-Mediterranean; □ = African; ■ = far Eastern.

marked in the Japanese, less in the Chinese and Koreans and least in the Thai and Vietnamese. These differences of proportion occur throughout the stature range. If we consider the European data only, there is a tendency for the ratio of sitting height to stature to be slightly greater for short samples than tall ones, suggesting the interesting hypothesis that the lower limbs contribute more to differences in stature than the trunk. The populations of Turkey, the Middle East and India (labelled 'Indo-Mediterranean'), have proportions similar to Europeans but, typically, a lesser overall stature.

Do these ethnic differences in size and proportion have any evolutionary significance? Zoologists have identified two rules concerning morphological variations of warm-blooded polytypic species, of which humankind is an example. Bergman's rule states that the body size of varieties increases with decreasing mean temperature of the habitat. Alien's rule states that the relative size of exposed portions of the body decreases with decreasing temperature. Roberts (1973), in an extensive survey of the anthropometric literature concerning the world's indigenous populations, showed that these rules are in general applicable to humankind. Body weight is negatively correlated with mean annual temperature. Samples with the lowest body weights are not found outside the tropics, and the highest body weights are not found at latitudes lower than 30°. Furthermore, linearity of bodily form (as indicated by high values for relative limb lengths) shows a strong positive correlation with mean annual temperature. Taken together, these findings indicate that ethnic groups inhabiting hot climates tend to have a high ratio of surface area to body mass, which is advantageous for the loss of heat. Similarly, the inhabitants of cold regions are adapted for heat retention. Roberts concluded, however, that there were differences in form between the major ethnic divisions of humanity even when the effects of temperature had been taken into account.

The relative lengths of the upper limbs show a similar pattern of ethnic differences to the lower limbs, and there is some evidence to suggest that the differences are more due to a lengthening or shortening of the distal segment of the limb (i.e., the forearm or shank) than of the proximal segment (i.e., the upper arm or thigh). The shoulders are a little narrower relative to stature in Africans than Europeans and the hips are considerably narrower in both sexes. In general, African bodily proportions are best described as 'linear'.

It would be a mistake to consider these differences in bodily size or shape to be fixed and immutable characteristics of ethnic groups. Several studies of migrant samples have shown significant differences between the growth patterns or adult dimensions of individuals born in the new environment and equivalent samples in the 'old country'. Boas (1912) and Shapiro (1939) are classic studies of this kind, and subsequent investigations include Kaplan (1954), Greulich (1957) and, more recently, Koblianski and Arensburg (1977). Shapiro (1939) studied Japanese immigrants to Hawaii. He showed that although the Hawaiian-born generation are taller than the immigrants and larger in most other dimensions, the ratios of the major bodily dimensions (i.e., relative sitting height and relative biacromial breadth) are not very different. This relative constancy of proportion has been confirmed by Miller (1961).

This led Roberts (1975) to conclude that 'the data suggest a strong genetic component to body proportion, and a more labile overall size'. Things are not quite this simple, however. There is evidence, for example, that the Japanese are becoming more like Europeans in terms of their relative limb lengths (Tanner et al., 1982, see below) but less like Europeans in terms of their head shapes (Yanagisawa and Kondo, 1973). This seems very curious indeed, but leads into the consideration of the effects of secular and social (environmental) changes in Sections 3.5 and 3.6.

In practical terms, it has to be emphasised that the anthropometric differences between many ethnic groups are sufficiently great that a product or item of equipment designed for one group will be unsuitable for another (and the same caution must be taken with the use of standards and guidelines). Although within-group variability may be large, between–ethnic group variation is generally significantly larger. The types of problems encountered by different populations attempting to use the same equipment are exemplified by two contrasting situations. In the first, when some (second-hand) European buses were first exported to a Southeast Asian country, many drivers found it very difficult even to reach the pedals. The second example relates to the consequences of investment from one country in the industry of another country (in this case from Japan to the United Kingdom). This can lead to importation of organisational systems and designs for facilities as well as of equipment. When Japanese companies set up their first plants in the United Kingdom, some installed identical equipment and facilities to those used in the home country and failed to appreciate the anthropometric (as well as the many other) differences. As a result, some British workers found themselves using machinery with work surfaces that were much too low and started to experience musculoskeletal problems.

Both examples show that managers and policy makers, as well as designers, must consider the effects of anthropometric differences. Sen (1984) gives other examples of similar problems experienced by workers in industrially developing countries and argues cogently for designers and manufacturers of products and equipment, and planners and purchasing managers of international companies, to take responsibility for identifying the relevant characteristics of the user/worker population and their task demands. Often, the local managers and purchasing departments in industrially developing countries do not have the necessary knowledge and expertise to do this.

In the case of ethnic minorities within a working population which is predominantly from another ethnic group, the situation is less clear; however given the relative magnitudes of the within-group and between-group variation concerned, we should not in general expect their presence to be significant in ergonomic and design terms unless they make up more than about one-third of the total. This is only a rough rule of thumb, however, and there may be circumstances in which the presence of ethnic minorities in a working population is more critical. Thompson and Booth (1982), for example, show that there are circumstances in which people from certain ethnic groups may be more at risk if industrial safety guarding standards are not modified to take into account their particular anthropometric characteristics.

3.4 GROWTH AND DEVELOPMENT

At birth we weigh some 3.3 [0.4] kg, and we are 500 [20] mm in length, of which our trunk represents some 70%. In the two decades that follow, our body length increases between three- and four-fold, our weight increases around 20-fold and our linear proportions change so that in the adult state the length of the trunk accounts for only 52% of the stature. However, the adult condition is by no means stationary. Our bodily proportions are modified by our lifestyles and the inevitable processes of ageing. The anthropometrist who wishes to chart this course (or part of it) may be tempted to do so by a *cross-sectional study* in which several samples of individuals, representative of different age bands, are measured at the same time. (A cross-sectional age-band sample is known as a 'cohort'.) However, data gathered by this means have certain limitations. In the case of children, only a very crude estimate can be obtained of the rate at which changes are taking place. Furthermore, our differences will be confounded by the effects of any secular trend that is taking place within the population. To disentangle these effects it is necessary to conduct *longitudinal studies* in which a sample of people are followed over an extended period of time.

The genetic and environmental factors that control human growth have been documented in detail by Tanner (1962, 1978), who has also published standards for the height and weight of British children which have been widely adopted in medical practice (Tanner et al., 1966, Tanner and Whitehouse 1976). The pattern of growth of a 'typical' boy and girl based on these data is shown in Figure 3.3. (The typical child is a purely fictitious individual who is average in all respects at all ages.) At ages up to 2 years measurements are made on a supine infant, and subsequently in a standing position. The rate of growth in boys is very rapid during infancy, declining steadily to reach its minimum at 11 $^1/_2$ years; it then accelerates again to reach its peak at 14 years before steadily decelerating as maturity is approached. The velocity peak around 14 years for boys (earlier for girls), known as the 'adolescent growth spurt', is associated with the events of puberty.

The peak in the chart is broader and lower than it would be for any actual child since it represents the average of a sample of children, all of whom are accelerating at different times. Hence, at 14 years some boys will have almost completed their growth spurt, whereas others will scarcely have commenced it. As a consequence the standard deviations of the bodily dimensions of samples of adolescents are very large (see the tables in Chapter 10). The typical girl is a little shorter than the typical boy from birth to puberty, but the growth spurt commences earlier in girls — at around 9 years, reaching its maximum velocity at around 12 years and growth being more or less complete by 16 years. Hence, there is a period from about 11 to 13 $^1/_2$ years when the typical girl is taller than the typical boy. The typical boy reaches half his adult stature a few months after his second birthday and the typical girl a few months before, although these figures will, of course, be subject to considerable variations in the population as a whole.

In addition to increasing in size, the human body changes considerably in shape. If the shape and composition of the body were the same throughout life, we would expect body weight to grow with the cube of stature (since weight is directly

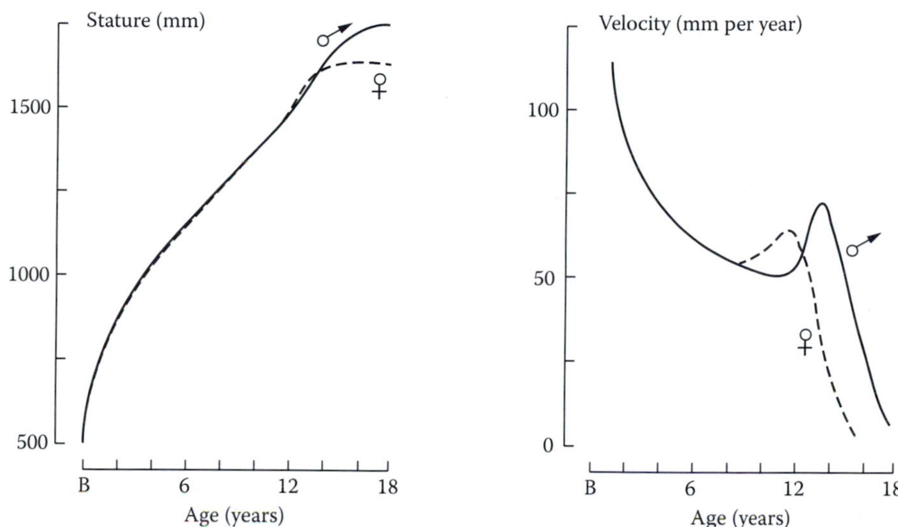

FIGURE 3.3 Growth from birth (B) to maturity of a typical boy and girl: stature (left) and velocity (i.e., rate) of growth in stature (right). (Data from Tanner, J. M., Whitehouse, R. H., and Takaishi, M. (1966). *Archives of Diseases of Childhood,* 41, 454–71; 41, 613–35.)

proportional to volume, assuming constant density). That would give an individual of average birth size, who achieved an average male adult stature of 1740 mm, a body weight of 139 kg, which is close to twice the correct figure. In reality, growth is accompanied by an attenuation of bodily proportions.

Tanner (1962) has pointed out that there are various 'maturity gradients' which are superimposed upon the growth curve of the body as a whole; hence, at any point in time the upper parts of the body (particularly the head) are closer to their adult size than the lower parts; the upper limbs are further developed than the lower, but the distal segments of the limbs (hands, feet) are ahead of the proximal segments (thighs, arms). Cameron et al. (1982) also showed differences in the timing of the adolescent growth spurt for different parts of the body. It is generally assumed that these gradients operate to give a steady unidirectional transition from the large-headed, short-legged form of the child to the typical proportions of the adult. Tanner (1962) copied an illustration of this from Medawar (1944), who in turn took his from an anatomy textbook of 1915, which in turn is based on nineteenth-century data. Medawar (1944) made the following statement: 'Just as the size of the human being increases with age, so, in an analogous but as yet unformulated way, does his shape. The property is best expressed by saying that change of shape keeps a certain definite trend, direction or "sense" in time; like size, it does not retrace its steps.' Numerous authors have fitted mathematical equations to these supposedly simple transformations, and some have attached biological significance to the constants in the equations.

Whilst compiling anthropometric estimates for British schoolchildren, Pheasant chanced on certain discrepancies which led him to believe that the assumption of a simple unidirectional change in shape was incorrect. Figure 3.4, previously published in Pheasant (1984), is based upon the cross-sectional study of the under-18-year-old

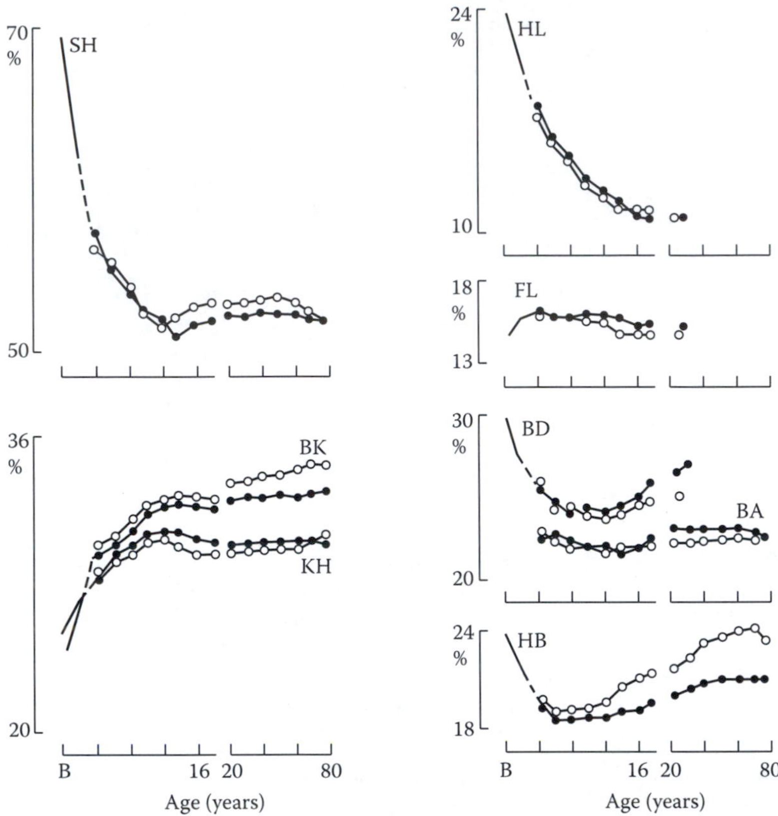

FIGURE 3.4 Effects of age on bodily proportions expressed as the relative values of various dimensions (% stature). SH = sitting height; BK = buttock–knee length; KH = knee height; HL = head length; FL = foot length; BD = bideltoid breadth; BA = biacromial breadth; HB = hip breadth; ● = boys and men; ○ = girls and women. (Original data from Snyder, R. G., Schneider, L. W., Owings, C. L., Reynolds, H. M., Golomb, D. H. and Schork, M. A. (1977). *Anthropometry of infants, children and youths to age 18 for product safety design.* Consumer Product Safety Committee, U.S. Department of Commerce, Bethesda, MD; Stoudt, H. W., Damon, A., McFarland, R. and Roberts, J. (1965). *Weight, height and selected body dimensions of adults, United States 1961-1962.* U.S. Department of Health and Human Services, National Center for Health Statistics, Hyattsville, MD; Stoudt, H. W., Damon, A. and McFarland, R. A. (1970). *Skinfolds, body girths, biacromial diameter and selected anthropometric indices of adults.* U.S. Department of Health and Human Services, National Center for Health Statistics, Hyattsville, MD; Webb Associates, (1978). Anthropometric *Source Book.* U.S. National Aeronautics and Space Administration, Lyndon B. Johnson Space Center, Houston.)

population of the United States published by Snyder et al. (1977). The mean value of each dimension for each age cohort has been divided by the mean value of stature (or supine crown–heel length for the under-2-year-olds). In some dimensions, such as head length, we may observe the smooth unidirectional approach towards adult proportions which we have been led to expect, but these are the exceptions rather than the rule. Most dimensions show what might be termed a 'developmental overshoot'.

Sitting height, for example, has achieved its adult percentage of stature by 9 years in girls and 11 years in boys; it then over-shoots and reaches a minimum at the time when the adolescent growth spurt is at its peak (12 years in girls, 14 years in boys), before climbing back to its adult proportions. Knee height, as one might expect, shows a pattern that is similar but inverted, as to a lesser degree do shoulder–elbow and elbow–fingertip lengths (not shown in the figure). Both shoulder and hip breadths are proportionally large in early infancy and pass through a proportional minimum — during adolescence in the former case and childhood in the latter. Foot length has a long plateau of elevated proportions in childhood before commencing a descent during adolescence. In summary, the data confirm the popular stereotypes of the 'dumpy' infant and the 'gangling' adolescent.

The data of Figure 3.4 are also interesting with respect to sex differences and the ages at which the bodily proportions of boys and girls first diverge. In the case of sitting height, knee height and foot length the divergence is associated with the events of puberty and the developmental overshoot. The bony pelvis of the female is broader than that of the male at birth (Tanner, 1978), and there is a slight sex difference in proportional hip breadth at the youngest age for which we have data. Hip breadth also shows a slight divergence at around 6 years and a pronounced one at adolescence, which continues well into adulthood. (Buttock–knee length is quite similar, so we are certainly dealing with soft-tissue upholstery to a large extent.) By contrast, shoulder breadth (bideltoid, biacromial) does not show any measurable divergence until as late as 17 years.

The muscular strengths of boys and girls are similar during childhood and diverge at around the time of puberty, as shown in Figure 3.5, which is based on the data of Montoye and Lamphier (1977).

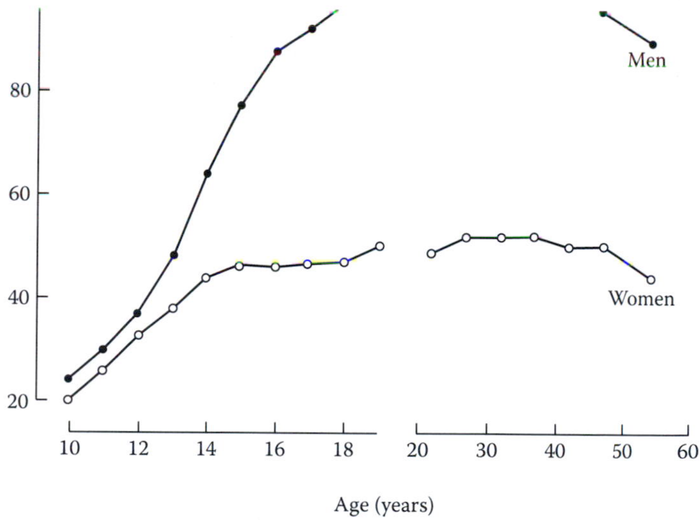

FIGURE 3.5 Effects of age and sex on grip strength. (Data from Montoye, H. J. and Lamphier, D. E. (1977). *Research Quarterly of the American Association for Health, Physical Education and Recreation,* 48, 109–20.)

The age at which we reach 'anthropometric adulthood' is by no means as clear-cut as one might suppose. Growth standards conventionally stop at 16 years for girls and 18 years for boys. The growth of a long bone occurs by cell division in plates of cartilage which separate the ends (epiphyses) from the shaft (diaphysis). When this cartilage finally turns into bone, growth ceases (eiphyseal fusion). The clavicles continue to grow well into the twenties and so, to a lesser extent, do the bones of the spine. Andersson et al. (1965) demonstrated an increase in sitting height in a majority of boys after 18 years and girls after 17 years, as well as in some boys after 20 years. A sample of Americans studied by Roche and Davila (1972) reached their adult stature at a median age of 21.2 years for boys and 17.3 years for girls, but some 10% of boys grew after 23.5 years and 10% of girls after 21.1 years. According to Roche and Davila (1972), this was partially due to late epiphyseal fusion in the lower limbs and partially to lengthening of the spine. Miall et al. (1967), in a longitudinal study of two Welsh communities, found evidence that men might grow slightly in stature well into their thirties.

3.5 THE SECULAR TREND

Human biologists use the term *secular trend* to describe alterations in the measurable characteristics of a population of human beings occurring over a period of time. Over a period of at least a century biosocial changes have been occurring in the population of much of the world which have led to:

- An increase in the rate of growth of children
- Earlier onset of puberty, as indicated by menarche (the onset of the menstrual cycle) in girls and the adolescent growth spurt in both boys and girls
- An increase in adult stature, with a possible decrease in the age at which adult stature is reached

The extensive statistical evidence concerning these changes has been reviewed by, amongst others, Tanner (1962, 1978), Meredith (1976) and Roche (1979).

Tanner (1962, 1978) summarised the available evidence and concluded that from around 1880 to at least 1960, in virtually all European countries (including Sweden, Finland, Norway, France, Great Britain, Italy, Germany, Czechoslovakia, Poland, Hungary, the Soviet Union, Holland, Belgium, Switzerland and Austria), together with the United States, Canada and Australia, the magnitude of the trend had been similar. The rate of change had been approximately:

- 15 mm per decade in stature and 0.5 kg per decade in weight at 5 to 7 years of age
- 25 mm and 2 kg per decade during the time of adolescence
- 10 mm per decade in adult stature

This had been accompanied by a downward trend of 0.3 years per decade in the age of menarche. Roche (1979) pointed out that secular changes in size at birth had been small or nonexistent.

Although the magnitudes of changes in Europe and North America have been fairly uniform, they are by no means universal. Japan, for example, has shown a particularly dramatic secular trend. The data of Tanner et al. (1982) show that in the decade between 1957 and 1967 Japanese boys increased in stature by:

- 31 mm at 6 years
- 62 mm at 14 years
- 33 mm at 17 years

In the 1967 to 1977 period, however, the rate of increase had declined to:

- 17 mm at 6 years
- 35 mm at 14 years
- 19 mm at 17 years

This suggests that the explosive biosocial forces driving the change may be beginning to wear themselves out. The effects of these changes will still, however, continue to be felt as these children progress through adulthood. In contrast, Roche (1979) cited evidence that in India, and elsewhere in the Third World, there had actually been a secular decrease in adult stature.

If people are increasing in size, are they also changing in shape? The remarkable Japanese secular trend seems to be associated with an increase in the relative length of the leg, as the data of Tanner et al. (1982), plotted in Figure 3.6, show. It is doubtful, however, whether such a change of proportion is general. Figure 3.7 shows the relative sitting heights of samples of young American males (average ages between 18 and 30 years) plotted against the year in which the measurement was taken. There is no evidence of a secular trend in adult proportions. (This conclusion has been confirmed by Borkan et al., 1983.)

It is interesting to speculate as to whether our distant forebears were as short as we might imagine from recent secular trends. Anecdotal evidence concerning a range of artefacts from doorways to suits of armour abounds. Although it is not possible to calculate stature accurately from poorly preserved skeletal remains, the long bones of ancient burials allow us to make a reasonable estimate. The archaeological evidence summarised by Wells (1963) suggests that the statures of British males from neolithic to medieval times have always fallen within the taller part of the present-day human race. Indeed, figures quoted include average heights of 1732 mm for Anglo-Saxons and 1764 mm for Round Barrow burials, the latter actually exceeding the average height of present-day young men. The secular trend then seems to be a recovery from a setback which occurred somewhere in post-medieval times. Tanner (1978) cites various evidence that in the earlier part of the nineteenth century trends were small or absent and plausibly associates them with the Industrial Revolution.

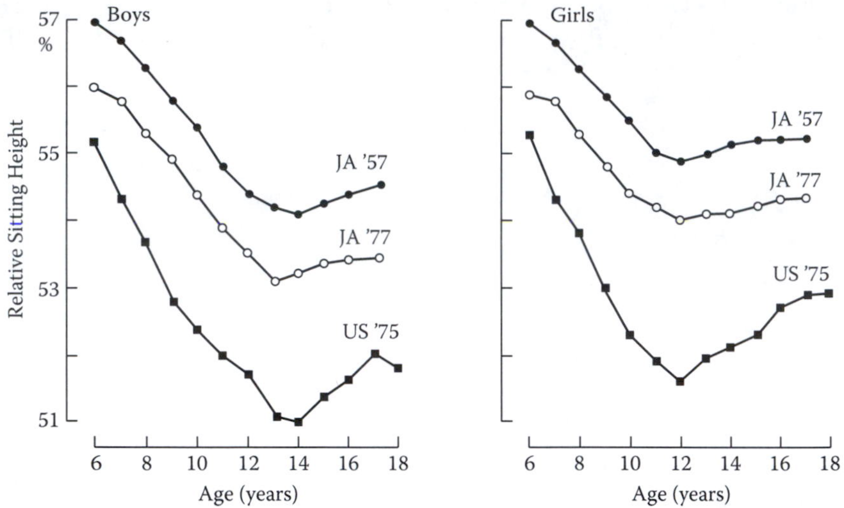

FIGURE 3.6 Secular trend in the bodily proportions of Japanese children (JA) compared with those of U.S. children. (Original data from Tanner, J. M., Hayashi, T., Preece, M. A. and Cameron, N. (1982). *Annals of Human Biology,* 9, 411–23; Snyder, R. G., Schneider, L. W., Owings, C. L., Reynolds, H. M., Golomb, D. H. and Schork, M. A. (1977). *Anthropometry of infants, children and youths to age 18 for product safety design.* Consumer Product Safety Committee, U.S. Department of Commerce, Bethesda, MD.)

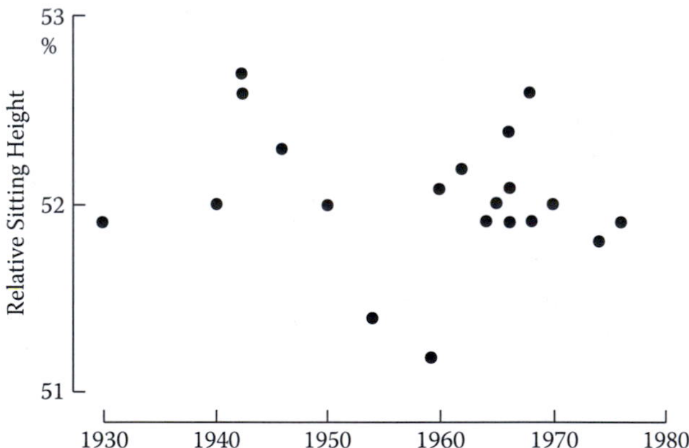

FIGURE 3.7 Relative sitting heights of samples of young adult U.S. men measured between the years 1930 and 1980. Note the absence of any detectable secular trend in bodily proportions.

What then are the determining factors that have led to the phenomenon of the secular change? Speculation has been intense on this subject, most writers maintaining a cautious tone in their conclusions. Social and environmental influences such as the improved nutritional quality of diet and the reduction of infectious disease by improved hygiene and health care are the factors that most readily spring to mind, to which we might add the effects of urbanization and reduced family size, but we cannot ignore the possible influences of genetic factors such as heterosis, the beneficial effects that are said to derive from outbreeding and the breaking up of genetic isolates. A century ago most people married and raised their children within the confines of isolated communities; today we are approaching the condition of the 'global village'. As Tanner (1962) perspicaciously observed, 'it has been shown in several West European countries that outbreeding has in fact increased at a fairly steady rate since the introduction of the bicycle'.

The consensus view amongst human biologists tends to favour the environmental rather than the genetic causes. It seems most likely that genetic endowment sets a ceiling level to an individual's potential for growth and that environmental circumstances determine whether this ceiling is actually reached. If this is indeed the case, the end of the secular trend is in sight, at least in the economically developed countries of Europe, North America and elsewhere, since we could reasonably argue that the further amelioration of environmental conditions, beyond those adequate for the achievement of full genetic potential, cannot lead to any further changes.

Considerable evidence published in the 1960s, 1970s and 1980s suggests that this limit may have indeed been reached, at least in some communities. Backwin and McLaughlin (1964) showed that Harvard freshmen from relatively modest social backgrounds increased in stature by around 40 mm from 1930 to 1958, whereas those from wealthy backgrounds showed no change. Cameron (1979) published data showing, very convincingly, that the secular trend for stature had levelled off for children attending schools in the London area by about 1960 (Figure 3.8). Tanner

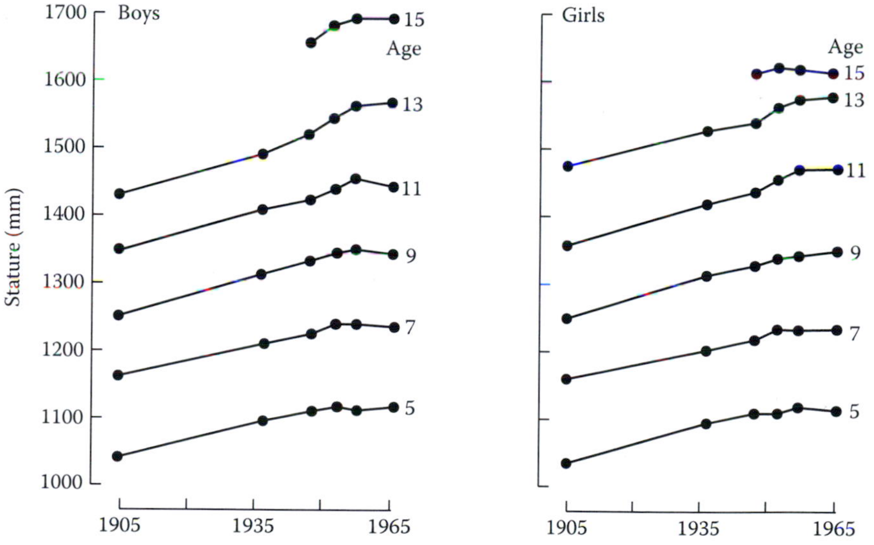

FIGURE 3.8 Secular trend in the average stature of children in the London area between the years 1904 and 1966. Note that changes were minimal after 1960. (Data from Cameron, N. (1979). *Annals of Human Biology,* 6, 505–25.)

(1978) also showed that the secular decrease in the age of menarche had come to a halt by about this time both in London and in Oslo. The subsequent national survey of Rona and Altman (1977) confirms the impression that in Great Britain the secular trend had reached a steady state. Rona (1981) was prepared to conclude that 'there is no evidence that the secular trend in growth has continued after 1959 in the UK'. Similarly, Roche (1979) reported that national surveys of U.S. children and youths in 1962 and 1974 showed constancy of stature (except at the 5th and 10th %ile levels, where small increases had occurred.) More recently, Chinn et al. (1989) analysed the evidence for a continued secular trend in the heights of English and Scottish children over the years 1972 to 1986. The conclusion that they reached was that in the 5- to 11-year-old age group the secular trend had ceased and that the probability was that the upward trend in adult stature had likewise come to a halt.

Overall then, the consensus view of anthropometrists and human biologists was that in the industrialised societies of Europe and North America, the upward secular trend in human stature had now come to a halt. Many people seem to find this surprising, however. The experiences of many schoolteachers, for example, tell them otherwise, and they are quite convinced in their own minds that the children they teach are still getting taller. It is difficult to account for this disparity between popular opinion and the available statistical evidence. In reality it is difficult to be absolutely certain either way. The differences that are known to exist between different parts of the United Kingdom and different social classes indicate that conclusions based upon small-scale or regional studies involving what may be nonrepresentative samples may be confusing. If the secular trend has indeed come to a halt, then its absence admits of two possible interpretations. The optimistic is that conditions for growth have been optimised and that all children are now reaching their genetic ceilings. The pessimistic is that the percentage of children raised under optimal environmental conditions has ceased to increase. The continued existence of significant social class differences in growth (see the next section) tends towards the pessimistic interpretation.

More recent studies indicate that there may still be a slightly increasing secular trend in economically developed countries, although less than the 10 mm per decade of the earlier part of the 20th century. Chinn and Rona (1994) brought up to date the series of surveys of English and Scottish children already mentioned and had to reverse the earlier conclusions that the secular trend in height had ceased. By 1990, statures of both boys and girls had increased, the greatest increase having occurred between 1986 and 1990. Smith and Norris (2004) reviewed the data from two studies of U.K. children's anthropometry 25 years apart, measurements made in 1970 to 1971 for the Department of Education and Science (1972) and in 1995 to 1997 for the Department of Health (1999). They found that stature had increased at all ages, on average by about 1% for boys and 1.5% for girls, but that the increase was small by age 18 years. They suggest that the increases can be accounted for by an earlier onset of menarche and puberty and an earlier growth spurt; children were reaching a given height approximately one year sooner than they were 25 years earlier, but at 18 years the boys' mean stature was only 4 mm greater and the girls' 14 mm. Peebles and Norris (1998) found that the mean stature of U.K. adults had increased by 17 mm for men and 12 mm for women between 1981 and 1995. It is possible that the U.K. population trend reached a temporary plateau around 1960, but that an upward trend has begun again since 1980. Molenbroek (1994) found that stature in the Netherlands increased between 1965 and 1980 more rapidly than at least since the mid-nineteenth century but that the growth rate decreased between 1980 and 1992 to the level of before the second World War.

In many other countries, of course, and particularly in some rapidly developing countries, a secular increase begun much later than in North America and Europe is continuing (Hauspie et al., 1996). One conclusion that can be reached from the somewhat confusing research evidence is that secular anthropometric changes can happen relatively quickly under certain circumstances and that population anthropometric and allometric characteristics are dynamic. Hauspie et al. refer to Tanner's belief that growth is "a mirror of conditions in society"; the trend tends to slow down or reverse during times of economic decline and wars.

One secular trend that has become very apparent in economically developed countries is the increase in weight and prevalence of obesity among both adults and children. High rates of obesity are also emerging in children of some developing countries, and approximately 30% of obese children become obese adults (WHO, 1998b).

The secular changes are rapid. Large-scale cross-sectional surveys carried out regularly every year between 1991 and 1998 show an increase in obesity (body mass index > 30) from 12 to 17.9% among U.S. adults (Mokdad et al., 1999). In a comparison of surveys of 16 year olds in Sweden carried out in 1974 and 1995, Westerstahl et al. (2003) found increases in mean weight of 1.9 kg (3.4%) for girls and 4.1 kg (6.6%) for boys. More of these adolescents were classed as overweight (BMI > 25) in 1995 than in 1974 (9 versus 3% for girls, 7 versus 3% for boys). Physical fitness (measured by a running-walking test of aerobic fitness, sit-ups testing dynamic abdominal and hip flexor muscle endurance, and bench-press tests of dynamic arm muscle endurance) had considerably deteriorated. Muscle strength, by contrast, had improved (as measured by a two-handed lift). Similar secular trends in weight and body mass index were found by the NHES and NHANES surveys between 1960 and 1991 in the United States (Troiano et al., 1995). Smith and Norris (2004) found that U.K. children's mean weight had increased on average by 7.9% for girls and 6.6% for boys between the 1970-1971 and 1995-1996 surveys, while the 95th %ile weight had increased considerably more (by 15.6 and 13.3%, respectively).

The results for this rapid trend are debated but, interestingly, Mokdad et al.'s (1999) analysis suggests that it cannot be attributed simply to a reduction in physical exercise. Their data showed that the level of physical inactivity had not changed substantially between 1991 and 1998. Antipatis and Gill (2001) discuss the various possible causative factors of the obesity 'epidemic' which, given the timescale, seem to be environmental rather than genetic. While improved standards of living are likely to be an important factor, societal changes (such as urban crowding, family and community breakdown, technological development leading to a sedentary lifestyle and global markets for foodstuffs) may influence people's dietary and exercise habits.

Is the continuance or otherwise of these secular trends important in terms of practical ergonomics? The increasing prevalence of obesity is certainly of significance for design of apparel and clearance dimensions in a wide range of situations, as well as for health, and may have a detrimental influence on working postures. If the upward trend in the stature of children is still increasing, it could be great enough to invalidate the anthropometric assumptions upon which design standards for school furniture and children's clothing and equipment are currently based. As Smith and Norris (2001) have shown, it is the larger children who will experience problems, the 95th %ile dimensions (for example) being significantly underestimated. Even if the upward trend in stature of young adults had now ceased, the upward trend in the stature of the adult population as a whole might well continue to increase for one or two decades. It is difficult to predict the likely magnitude of changes, since the effects of the secular trend are confounded with those of the ageing process per se and with demographic changes in the ethnic background and age structure of the population. The dynamic nature of secular trends and the current rapidity of the changes in many countries emphasise the desirability of undertaking anthropometric surveys at shorter intervals than has been common in the past. Design standards too will need to be updated at relatively frequent intervals.

3.6 SOCIAL CLASS AND OCCUPATION

Social class and occupation are inextricably linked — so much so that the latter is generally used as an operational measure of the former. The widely used system of the Office of Population Censuses and Surveys in the United Kingdom, known as the Registrar-General's Classification, divides occupations into six categories: (I) professional, (II) intermediate, (IIIA) skilled non-manual, (IIIB) skilled manual, (IV) semi-skilled manual and (V) unskilled manual. Every so often it is necessary to reclassify an occupation as its perceived status changes.

Both social class (through social and environmental effects) and occupation (through training and health effects as well as social effects) can influence anthropometry and secular trends. In a fascinating study of primiparae (women pregnant for the first time) in Aberdeen in which stature was stratified by the occupation of the woman's father, by her own occupation and by her husband's occupation, Thomson (1959) found, remarkably, that tall women had a stronger tendency to marry upwards (with respect to their father's and their own occupations) than did short ones.

Social class differences in stature remain marked. Knight (1984), in a nationwide study of the adult population of Great Britain, found an average stature of 1755 mm for men and 1625 mm for women in social classes I and II as against 1723 and 1596 mm in social classes IV and V. The differences were of similar magnitude for all age cohorts. The pattern was less clear for body weight.

The same survey also showed regional differences, ranging from an average stature of 1751 mm for men and 1619 mm for women in southwest England to 1719 and 1594 mm in Wales. Regional differences can, of course, arise from complex ethnic, social, occupational and environmental influences. Without attempting to separate these interacting effects, it is clear that the resulting differences can be considerable.

Extensive British data show social class differences in the growth of schoolchildren. Rona (1981) reviewed the evidence of British surveys over the previous 30 years or so. A difference of between 10 and 20 mm in average stature between classes I and V in the Registrar-General's Classification table already exists by the age of 2 years. By 7 years this has widened to 30–40 mm, a gap that remained constant over 30 years. In the most recent of these surveys (Goldstein, 1971; Rona et al., 1978) the differences between classes I to IV were relatively modest, which suggested that the differences in primary-school children were then mainly due to those in social class V. Rona et al. (1978) showed that the children of unemployed fathers were shorter in stature, but that the parents of these children were also shorter within each social class. Lindgren (1976) reported an extensive survey of urban children in Sweden between 10 and 18 years of age. There was no difference in height, at any age, between the social classes as defined either by the father's occupation or family income. Sweden is the only country in the world where this is known to be the case — a fact that Tanner (1978) takes to be an operational and biological measure of the existence of a 'classless society'.

Despite the close links between social class and occupation, occupation can have direct and separate influences on the anthropometric characteristics of a user population in a particular occupation or industry. In some circumstances '*self-selection*' may occur, with individuals gravitating towards occupations to which their physiques are well suited. The physical content of the occupation itself may also exert a *training effect* (sometimes referred to as *task fitness*) — or, perhaps more detrimentally, a de-training effect in the case of sedentary lifestyles. The most extreme examples of training effects are provided by athletes, as discussed by Wilmore (1976). The consequences of sedentary occupations are discussed elsewhere in this book, but the more general consequences of a sedentary lifestyle are a serious current concern in relation to the health of our populations (especially in terms of the increase in obesity). Finally, the physical aspects of work may have detrimental effects on health and functional capacity. Era et al. (1992), for example, found that elderly people who had had a higher occupational status at working age had better physical, sensory, psychomotor and cognitive functions. Savinainen et al. (2004) found similar results when comparing older people who had had a high physical workload during their working life with those who had had a low workload. The only physical capacities that they found to be better in the high workload group were flexibility of the spine and isometric trunk muscle strength.

A classic study of selection and de-training is that of Morris et al. (1956), who investigated the waist and chest girths of the uniforms of London busmen — both drivers and conductors — ranging in age from 25 to 64 years. In addition to seeing a steady increase with age in both groups, the girths of the drivers were greater than those of the conductors — even in the youngest age group. The authors postulated therefore that 'the men have brought these differences into the jobs with them'. Passing from the ridiculous (if we may so term the busmen's trousers) to the almost indisputably sublime, two studies of ballerinas are worth mention. Grahame and Jenkins (1972) measured the joint flexibility of a group of female ballet students and found it to be greater than for controls, even for joints such as those of the little finger which were not trained to be flexible. The authors therefore concluded that only girls gifted with generalised joint flexibility would undertake the rigours of ballet training. Vincent (1979) has documented the appalling, and sometimes disastrous, lengths to which these girls will sometimes go in 'competing with the sylph' — that is in pursuit of the unnaturally slender body form which is sufficiently otherworldly for their art.

In more mundane occupations, it is reasonable to expect that workers will become 'task fit' through (gradual and monitored) practice or through organised training, but it is not reasonable to accept bodily distortion or musculoskeletal disorders through requiring extreme exertions. Where extreme exertion is necessary (and only then), selection procedures may be required to identify individuals who are suitable for the job.

It is clear that anthropometric characteristics may differ between occupational groups (whether through self-selection, formal selection or training). Annis and McConville (1990), for example, note the findings of Martin et al. (1975) that law enforcement officers were larger in most dimensions than almost any other group they measured and of Reynolds and Allgood (1975), that air stewardesses were taller and lighter than average American women. Hsaio et al. (2002) reported on the differences between major occupational groups within the United States, analysing the results of the NHANES III survey. Military personnel tend to be larger, stronger and fitter than civilians. Care must therefore be taken to choose relevant survey data for design purposes.

3.7 AGEING

Figure 3.9 shows the average heights and weights of the adult civilian populations of Great Britain and the United States plotted against age. A steady decline in stature is apparent, whereas weight climbs steadily before subsequently declining at around

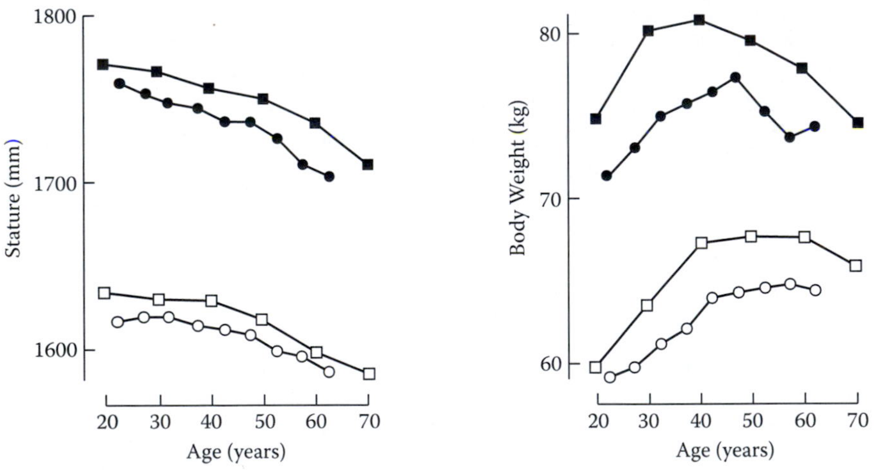

FIGURE 3.9 Average stature and weight in samples of adults of various ages. ■ = men, United States; ● = men, Britain; □ = women, United States; ○ = women, Britain.

50 years in men or 60 years in women. In analysing such a pattern, we must consider the combined effects of the ageing process and the secular trend, together with the possibilities of differential mortality (i.e., that people with certain kinds of physique may tend to die younger.) Damon (1973) showed that men of average height and weight had greater longevity than those who deviated strongly in either respect. These interactions require multicohort longitudinal studies for their elucidation. Investigations of this kind include the Welsh study of Miall et al. (1967) and the extensive Boston programme of the Veterans Administration (Damon et al., 1972; Friedlander et al., 1977; Borkan et al., 1983).

Longitudinal studies show that at around 40 years of age we begin to shrink in stature, that the shrinkage accelerates with age, and that women shrink more than men. The shrinkage is generally believed to occur in the intervertebral discs of the spine — resulting in the characteristic round back of the elderly (e.g., Trotter and Gleser, 1951) — although Borkan et al. (1983) note some decrease also in the lower limbs. Annis (1996) suggests that both this and a change in biacromial breadth may be due to changes in skeletal architecture which affect posture. The data show a longitudinal increase in weight for height until 55 years, followed by a decline. Friedlander et al. (1977) showed a steady longitudinal increase not only in hip breadth but also in the bi-iliac breadth of the bony pelvis. The mechanism of the latter is obscure but it suggests that 'middle-age spread' may not be totally due to the accumulation of fat but may also involve changes in the bony pelvis.

If we assume that no secular change has occurred in the proportions of the body, then proportional ratios calculated from cross-sectional studies should be comparable with the longitudinal results. Figure 3.4, plotted from the data of Stoudt et al. (1965, 1970), shows that this is indeed the case. The proportional decrease in sitting height is compatible with the spinal shrinkage explanation of stature decline, and the greater change in women matches the longitudinal findings of Miall et al. (1967). Dimensions with a substantial soft-tissue component such as hip breadth and buttock–knee length show a proportional increase (until 75 years) which is more pronounced in women. The proportional decline in the biacromial breadth of men over 60 years old presumably reflects the characteristic rounding of the shoulders of the elderly. It is of interest that for both sitting height and biacromial breadth, the ageing process finally abolishes the sex difference altogether.

Cross-sectional studies such as that of Stoudt et al.'s (1970) have shown an increase in skinfold thickness followed by a decline at around 40 years in men and 60 years in women. There is evidence, however, that this represents a redistribution rather than a loss of body fat. Durnin and Womersley (1974) showed that the relationship between skinfold thickness and whole body fat, as measured by densitometry, changes with age. It seems there is a transfer of fat from subcutaneous positions to deep ones (e.g., around the abdominal organs). The net quantity as a percentage of body weight continues to increase, and the longitudinal decline in weight that we see late in life is probably due, therefore, to the loss of lean tissue. Borkan and Norris (1977) found that the weight of fat was constant with age in a cross-sectional sample of middle-aged men, but that lean tissue declined markedly. Subcutaneous fat decreased on the trunk but increased on the hips, but this was accompanied by an increase of abdominal (waist) circumference indicative of a sagging of the abdominal contents (due presumably to increased internal fat and decreased muscular resistance.) A similar redistribution presumably occurs in women, but there is little numerical evidence.

The above studies are all based on the populations of the United States and Great Britain, where obesity, consequent upon an abundant food supply and a sedentary lifestyle, is prevalent. The situation in other communities will, of course, be different, and in societies where food is scarce, adult increase in body weight does not occur.

The loss of lean body weight is due principally to a wasting away of muscles (although the bones also become less dense in later life), and this leads to a decrease in muscular strength, as shown, for example, in Figure 3.5. According, for example, to Asmussen and Heebøll-Nielsen (1962), the decline is more rapid in women than in men and more rapid in lower limb muscle groups than in upper limb muscle groups. In other words, women age more rapidly than men, and the legs give out first. Both of these conclusions have been challenged, however: the first by Montoye and Lamphier (1977) and Voorbij and Steenbekkers (2001) and the second by Viitasalo et al. (1985). There is also a preferential loss of faster motor units within muscles, which reduces the ability to make powerful or rapid movements (Jones and McConnell, 1997). Jones and McConnell also note that the elasticity of tendons and ligaments changes, which makes absorption of energy less efficient (for instance when stepping down stairs).

We live in a 'greying' society. Figure 3.10 shows some demographic predictions. In 1971 about one person in six in the United Kingdom was of retirement age (i.e., 65 for men, 60 for women); by 2031, it is estimated, the figure will be closer to one person in four. The rate of increase is greatest in the oldest age groups, specifically the over-75s, who will increase dramatically in numbers. By 2025 it is anticipated that approximately one in ten of the world's population will be over 65 years old, and worldwide life expectancy will have reached 73 years (WHO, 1998a). The World Health Organization rightfully recognises the increase of life expectancy of populations in the twentieth century as 'one of the greatest achievements of all time' (WHO, 1993). Nevertheless it poses serious ergonomics questions for designers and researchers.

Beyond the middle years of life, most of us will tend to suffer from a steady diminution in our functional capacities, due partly to the ageing process as such and partly to the effects of previous disease or injury from which recovery has been incomplete. As a consequence, we experience a steady increase in the number of *critical mismatches* that we encounter in the performance of everyday tasks. The net effect of these changes is illustrated in Figure 3.11, which shows the percentage of people in different age groups having one or more specific disabilities, that is, one or more functional impairments that lead to significant difficulties in performing everyday tasks. The figure takes a dramatic upswing beyond the age of 60. As things stand at present, therefore, we typically seem to continue working until we reach the age when our bodily framework starts to pack up on us. There is something of an irony in this.

The age changes beyond 50 years of age which seem to be most important for the design of products are those in anthropometric dimensions, walking velocity and

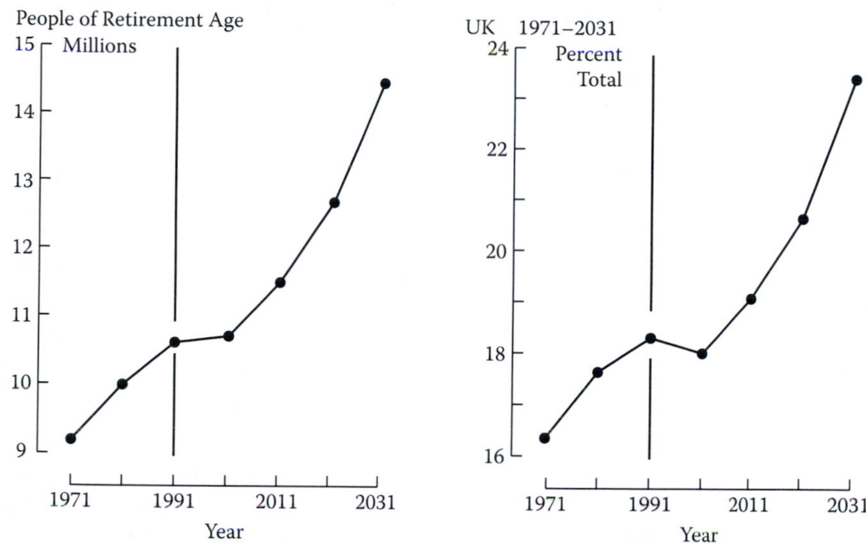

FIGURE 3.10 The ageing population: people of retirement age in the United Kingdom; (left) millions; (right) percentage of the total population. (Data from the Government Statistical Service/Central Statistical Office publication Social Trends 20, HMSO, 1990.) (From Pheasant, S. (1991). *Ergonomics, Work and Health.* London: Macmillan, Figure 16.1, p. 324. Reproduced with kind permission.)

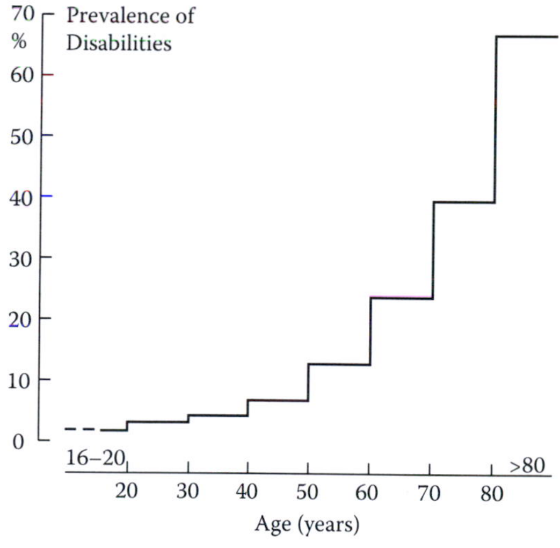

FIGURE 3.11 Prevalence of disabilities in different age groups. (Data from Martin et al., 1988.) (From Pheasant, S. (1991). *Ergonomics, Work and Health.* London: Macmillan, Figure 16.2, p. 327. Reproduced with kind permission.)

step length, forces, fine and gross psychomotor control, hearing and vision, according to Steenbekkers and Dirken's (1998) review of the findings of a series of studies at Delft University of Technology. They concluded that, for these aspects, 'ageing means becoming equal to the weaker young users; therefore more careful "trans-generational design" is advisable'. Among the changes in anthropometric dimensions, a reduced stature and greater difficulty in reaching can have particular implications for workplace design; a difference in mean stature of 39 mm was found between men at ages 30 years and 60 years in a 1978 U.K. survey (Haslegrave, 1980). Reduced joint mobility is important when considering handles and handheld products.

The rate of onset of the decrepitude which comes with old age is highly variable (Haigh and Haslegrave, 1992; Haigh, 1993). Part of this is just luck — a matter of the genes we are born with and the misfortunes we encounter along the way. Lifestyle is a major factor, however. Fisk (1993) argues that diet, working environment, social attitudes and availability of transport all contribute to a decline with age. The results of an extensive longitudinal study carried out in Finland, following 6257 ageing workers, indicates that at least four factors — biological ageing process, health, work and lifestyle — modify our capacities from the age of 45 onwards (Tuomi et al., 1991; Ilmarinen et al., 1997; Ilmarinen, 1997). The factors, however, interact strongly, and Ilmarinen (1997) concluded that 'lifestyle, work, and aging influence the severity of disease. In the absence of disease, life-style and work affect the rate of aging. The presence of disease modulates life-style and work and may influence aging as well, and both disease and aging modify lifestyle'.

It seems fairly certain that (within limits) regular physical activity can fend off the ageing process or its effects on the activities in our everyday life. Unfortunately, however, we not uncommonly get trapped in a downward spiral in which diminished functional capacity leads to a reduction of activity, which leads to a further reduction in functional capacity and so on. We encounter difficulties in doing things, so we stop doing them, and in due course are able to do less and less. The problems of the ageing society present a major challenge for ergonomics.

Part II

Application of Anthropometry in Design

4 Workspace Design

4.1 INTRODUCTION

In this chapter we shall consider the design and layout of the spaces in which people live and work, with particular reference to anthropometric considerations of:

- Clearance
- Reach
- Posture
- The influence on posture of vision and strength requirements when performing a task

The general principles explained can be applied equally to the design of products, whether equipment, tools or objects being handled. Some further examples are given in Chapter 6. ISO standard 14738, Safety of Machinery (anthropometric requirements for the design of workstations at machinery), is a useful source of more detailed guidance for both standing and seated work (ISO, 2002b). As ISO 14738 notes, an initial task analysis is essential to understand the nature of the work to be performed and to identify the factors which will have an effect on the operator, including the time aspects, force demands and need for communication or teamwork.

Many of the dimensions discussed in this chapter can be regarded as *functional* (or *dynamic*) *dimensions*, in contrast to the *static dimensions* collected in fixed, standardised postures in most anthropometric surveys. Dynamic anthropometry is the measurement of people in motion or while performing their tasks at work. It is not usually possible to measure functional dimensions with the same degree of accuracy as static dimensions, not least because of the variation in the way in which any task is performed, even by the same individual. It is also more difficult to define datum points for many functional dimensions. Maximum reach, as an example, depends on the degree to which the person is willing or able to bend forward. Some judgement therefore has to be applied when determining functional dimensions for particular tasks. All dimensional data given in this chapter are for the standard reference population as described in Table 2.5.

The rationalization of workspace layout is partly a matter of anthropometrics and partly a matter of common sense (in arranging the various elements of the workplace in relation to each other). The common sense element is embodied in the four principles set out in Table 4.1, which were first stated in a formal way by the late Ernest J. McCormick (1970). These principles are applicable to a large class of design problems which involve considerations of 'what to put where': the controls and displays on a panel, the furniture and appliances in a kitchen or the machines on a shop floor, the facilities in a large building, and so on — even perhaps to more abstract problems like the arrangement of information in a database. Examples of

TABLE 4.1
Principles of Rational Workspace Layout

- Importance principle: The most important items should be in the most accessible locations.
- Frequency of use principle: The most frequently used items should be in the most accessible locations.
- Function principle: Items with similar functions should be grouped together.
- Sequence of use principle: Items that are commonly used in sequence should be laid out in the same sequence.

Source: After McCormick, E. J. (1970). *Human Factors Engineering,* New York: McGraw-Hill.

their application will be seen in several chapters of this book. *Link analysis* is a technique which is helpful in these situations. A movement between workstations, or a shift of focus of attention between displays on a control panel, is regarded as a 'link', which can be drawn on a plan of the workplace or panel. The frequency of occurrence of each link can be established by observation, thus giving a quantitative measure when applying McCormick's third principle. (A more detailed description of this technique can be found in Kirwan and Ainsworth, 1992, pp. 118–125.)

4.2 CLEARANCE

Clearance dimensions relate to access and the space needed to perform a task, but additional clearance may be important for comfort, as discussed later. Figure 4.1 and Table 4.2 present clearance data for a variety of working positions, derived from a variety of sources (Damon et al., 1966; Van Cott and Kinkade, 1972; Department

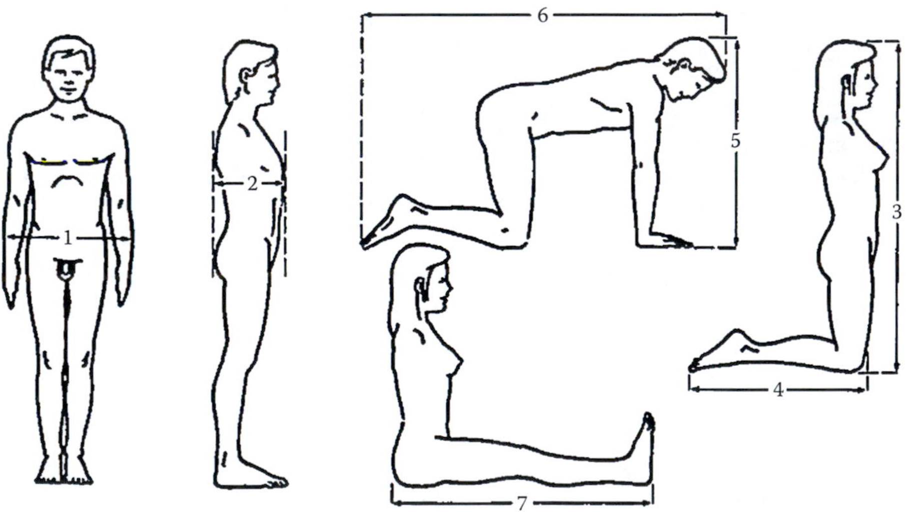

FIGURE 4.1 Clearance dimensions in various positions as given in Table 4.2.

TABLE 4.2
Clearance Dimensions in Various Positions, as Shown in Figure 4.1 (all dimensions in millimetres)

	Dimension	Men				Women			
		5th %ile	50th %ile	95th %ile	SD	5th %ile	50th %ile	95th %ile	SD
1.	Maximum body breadth	480	530	580	30	355	420	485	40
2.	Maximum body depth	255	290	330	22	225	275	325	30
3.	Kneeling height	1210	1295	1380	51	1130	1205	1285	45
4.	Kneeling leg length	620	685	750	40	575	630	685	32
5.	Crawling height	655	715	775	37	605	660	715	33
6.	Crawling length	1215	1340	1465	75	1130	1240	1350	66
7.	Buttock–heel length	985	1070	1160	53	875	965	1055	55

of Defense, 1999) and scaled, as far as possible, to match the standard population. The maximum breadth and depth of the body are overall measurements taken at the widest or deepest point wherever this occurs. The male data, based on U.S. servicemen, exceed any relevant dimensions in Table 2.5 and have been quoted directly.

Fruin (1971), in the context of an account of pedestrian movement and flow, introduced the concept of the body ellipse to define the space requirements for a standing person. In plan view the space occupied by the human body may be approximately described by an ellipse — the long and short axes of which are determined by its maximum breadth and depth. Taking the 95th %ile male data from Table 4.2 and allowing a generous 25 mm all round for clothes, the long and short axes of our ellipse become 630 and 380 mm, respectively. Figure 4.2 shows this ellipse. To give us some idea of 'elbow room', a circle has been drawn around the ellipse. The diameter of this circle is the 95th %ile male elbow span (1020 mm). Two further circles, the diameters of which are determined by the arm spans of a 5th %ile woman and a 95th %ile man, complete a first simple analysis of space requirements.

Liem and Yan (2004) showed how the body ellipse simulation approach can be combined very effectively with queueing theory when analysing the space requirements and layout of a city centre airline baggage check-in facility, thus considering the personnel and technical elements simultaneously in the design of the system. Passengers carrying luggage and pushing baggage trolleys were simulated by outline 'blocks' in a similar way to the body ellipses of those unencumbered with bags.

4.2.1 Whole Body Access

Table 4.3 and Figure 4.3 present figures for minimum dimensions for hatches or openings giving whole-body access to and egress from confined spaces, collated from various sources. The minimum dimensions given in Table 4.3 should be increased if equipment has to be worn or carried, or if the opening is intended to be used by more than one person at a time. This is particularly important for escape

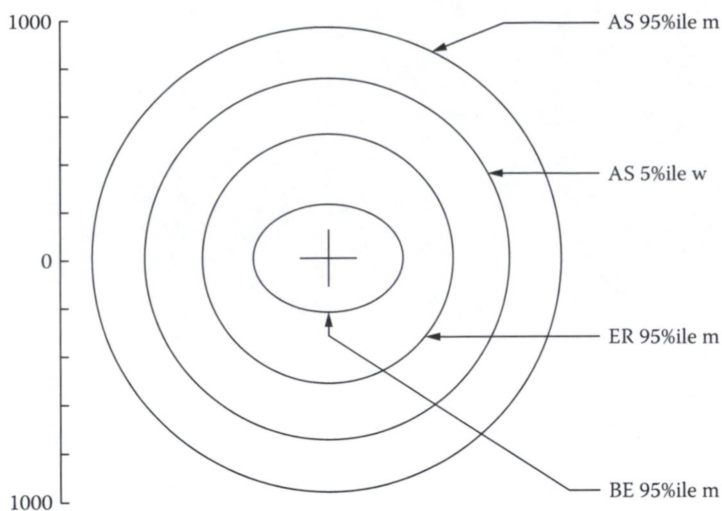

FIGURE 4.2 Simplified analysis of the space requirements for a standing person. Abbreviations: body ellipse (BE), elbow room (ER), arm span (AS). (Dimensions in millimetres.) See text for details.

hatches where a casualty might need to be rescued. MIL-STD-1472F also specifies that the 'step down' distance from a horizontal hatch should be no more than 690 mm.

EN 547-1 (CEN, 1996) is a European standard on machinery safety. The dimensions given in Table 4.3 represent the 95th %ile data for a mixed male and female European population, although the standard also includes 99th %ile values. ISO 2860 (ISO, 1992a) (an International Standard dealing with earth-moving machinery) states that the dimensions it gives are 'the smallest that will accommodate the 95th percentile operator'. Comparison of the figures in the U.S. military standard MIL-STD-1472F (Department of Defense, 1999) in Table 4.3 with the data of Table 4.2 confirms that these are also based on the 95th %ile male values of the dimensions concerned.

By inference this means that 5% of operators trying to pass through hatches of these dimensions would get stuck — which in a safety-critical application would clearly be unacceptable. Access to workspaces like pressure vessels presents a particular problem in this respect, since the aperture must be large enough to permit emergency evacuation (perhaps by two people carrying a stretcher) but as small as possible so as not to compromise structural strength. A similar problem was encountered in the design of whole-body scanners for hospitals.

If we assume that the data for body breadth and body depth in Table 4.2 are normally distributed and we apply the equation for calculating percentiles given in Section 2.2, we may calculate that:

- A 580 × 330 aperture excludes approximately 1 man in 20
- A 600 × 340 aperture excludes approximately 1 man in 100
- A 625 × 360 aperture excludes approximately 1 man in 1000
- A 640 × 370 aperture excludes approximately 1 man in 10,000
- A 660 × 385 aperture excludes approximately 1 man in 100,000

TABLE 4.3
Minimum Dimensions for Whole-Body Access (all dimensions in millimetres)

Source	Rectangular Aperture Width (W) × Depth (D)	Elliptical Aperture Width (W) × Depth (D)	Circular Aperture Diameter
For Access through an Aperture in a Horizontal Surface (i.e., floor or ceiling)			
Light/normal/working clothing			
EN 547-1 (CEN, 1996)			565
ISO 2860 (1992a)	560 × 330	580 × 330	
MIL-STD-1472F (Department of Defense, 1999)	580 × 330		
Bulky/arctic clothing			
EN 547-1 (CEN, 1996)			645
ISO 2860 (1992a)	650 × 470	690 × 470	
MIL-STD-1472F (Department of Defense, 1999)	690 × 410		760
Damon et al. (1966)	740 × 510		
For Access through an Aperture in a Vertical Surface (i.e., wall)			
Light/normal/working clothing			
ISO 2860 (1992a)		660 × 760	
MIL-STD-1472F (Department of Defense, 1999)	660 × 760		
Heavy/arctic clothing			
ISO 2860 (1992a)		740 × 870	
MIL-STD-1472F (Department of Defense, 1999)	740 × 860		
Damon et al. (1966)	780 × 500		

Note: Data from Damon et al. (1966) are also quoted in Van Cott and Kinkade (1972) and Woodson (1981).

The numbers given for the percentages of men excluded may well be underestimates, since the distributions for the dimensions concerned are likely to be positively skewed. (Note also that these figures do not allow for clothing or personal equipment.)

In the case of emergency exits and escape hatches we should expect speed of passing through to be a function of aperture size up to some critical dimension at which no further improvement is possible. Roebuck and Levendahl (1961) studied the emergency exits of aircraft and found that speed levelled off at a door width of

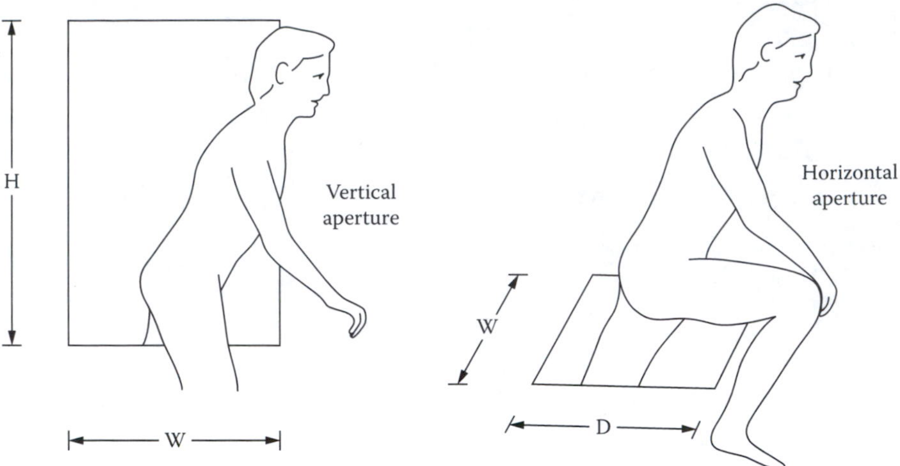

FIGURE 4.3 Whole-body access dimensions; see Table 4.3.

around 510 mm (unless steps were also involved, in which case a greater width was optimal).

Minimum dimensions for passageways in situations of limited access, such as tunnels and catwalks, are given in Table 4.4 (quoted from Damon et al., 1966, with slight modifications introduced for conformity with figures given elsewhere). Dimensions for other types of whole-body access are given in European standard EN 547 (Part 1 giving the calculations and Part 3 the anthropometric data; CEN, 1996, 1997b). EN 547-1 also gives appropriate allowances to add for apparel and body movement. EN 547-2 (in conjunction with 547-3; CEN, 1997a,b) provides minimum dimensions for access openings for upper body, head and shoulders, and arms or hands, which are particularly relevant to design for maintenance. Analysis of access for repair and maintenance is becoming increasingly important as systems become more complex, since one of the consequences of this is that equipment is packed more compactly into the available space, making access more difficult (as Majoros and Taylor [1997] recount for aircraft design).

4.2.2 Circulation Space

Minimum dimensional requirements for circulation space in buildings, passage between obstacles, and so on are summarised in Table 4.5 (from Pheasant, 1987, which in turn was based on Tutt and Adler, 1979 and Noble, 1982).

The space considered acceptable will probably depend upon the context, obviously being less in a party with a group of friends than in a public space when mixing with strangers. In the example of the baggage check-in facility mentioned earlier, Liem and Yan (2004) considered different levels of queueing density (or 'service level'), basing their analysis on space guidelines from Fruin (1971) and Wright et al. (1998). Density of pedestrians within a circulation space is characterised by interperson spacing, ranging from the obviously undesirable 'close-packed with no movement possible' to 'free circulation without disturbing other people'

TABLE 4.4
Minimum Dimensions for Passageways in Areas of Restricted Access

	Height (mm)	Width (mm)
Walking[a]		
Upright	1955[b]	
Stooped	1600	
Straight ahead		630
Crabwise (sideways)		380
Crawling on hands and knees	815	630
Crawling prone[c]	430	630

[a] For walking a trapezoidal space which is 630 mm wide at shoulder height and 145 mm wide at floor level will suffice

[b] Stature of a 99th %ile man wearing shoes and protective helmet

[c] For prone crawling, a width of 1015 mm is preferable to allow for lateral elbow movements.

Source: Data based on Damon, A., Stoudt, H. W. and McFarland, R. A. (1966). *The Human Body in Equipment Design,* Cambridge MA: Harvard University Press, with modifications.

(interperson spacing greater than 1.2 m). The degree of tolerance of the pedestrians presumably varies with the environment and duration as well as with the context. Provision of adequate circulation space and passageways is a critical safety issue when planning facilities to accommodate large crowds. Here there are likely to be cross-flows of pedestrians, in contrast to queues, which mainly move in one direction, and hazards can arise through congestion (Al-Haboubi and Selim, 1997; HSE, 1996).

4.2.3 Safety Clearances

The design of barriers to exclude people from a hazardous area has to take account of behaviour (both normal task behaviour and risk-taking behaviour) as well as the physical dimensions of the people at risk. Bottoms and Butterworth (1990) studied the design of guard rails for preventing feet and legs from contacting dangerous parts underneath agricultural machinery. They ran user trials to see the variety of ways in which a leg could be inserted under the rail (at different heights), as well as measuring the reach distance in each case. Some of the postures appear extraordinary but could easily arise when someone was working beside, cleaning or repairing the machinery. They found that, with a low barrier height, a small person might be able to reach further than a tall person, so that standard anthropometric data may be misleading when trying to estimate safety clearances. In a similar study evaluating

TABLE 4.5
Space Requirements for Circulation (all dimensions in millimeters)

Widths of Access	
One person walking normally	650 (600 restricted)
Two people passing or walking side by side	1350 (1200 restricted)
One person walking, another flattened against wall	1000 (900 restricted)
Two people passing crabwise (sideways)	900 (850 restricted)
One person carrying a suitcase	800
One person carrying a tea-tray	900
One person carrying two suitcases	1000
One person with a raised umbrella	1150
Two people with raised umbrellas	2350
One person with crutches	840
One person with a walking frame	1000
Wheelchair user — minimum	750
Wheelchair user — reasonable	800
Wheelchair user — preferred	900

Passage between Obstacles	Normal	Crabwise (sideways)
Both obstacles greater than 1000 mm in height	600	400
One obstacle greater than 1000 mm in height, the other less	600	400
Both obstacles less than 1000 mm in height	550	350
Standing in line	450 per person	

Source: After Pheasant, S. T. (1987). *Ergonomics: Standards and Guidelines for Designers,* PP 7317, London: British Standards Institution.

the safety of a consumer product, Norris and Wilson (1994) tested children's ability to crawl under a simulated swimming pool cover (heavily disguised to avoid encouraging the risk-taking behaviour that they were trying to prevent) and their strength and capability of extracting the tethering pins of the covers. Surprisingly, they discovered that it was the older children with larger heads who were more able to crawl into the smallest gaps — because they had greater strength to stretch the material taut to widen the gap, and perhaps also a greater perseverance to follow up their curiosity.

European standard EN 294 (CEN, 1992) gives safety distances to prevent adults and children over the age of 3 years from being able to contact hazards through reach by the arm, hand or fingers.

4.2.4 Personal Space

The space required directly around people will depend upon the task they are doing, as discussed in later sections of this chapter. However, even when standing or sitting still, we desire some free space for comfort. The minimum space that we accept is probably seen in aircraft seating modules, and with these space has been still further optimised by placing the seats in a 'keystone configuration' with adjacent seats facing in opposite directions, taking advantage of the fact that shoulder room plus knee room is narrower than twice shoulder room. Troy and Guerin (2004) have used digital human simulation to determine the 'human swept volume' for a seated aircraft passenger, which is the boundary surface encompassing the space used in all the usual activities performed during a flight (such as buckling a seat belt, reaching for belongings, eating, working — presumably with a laptop computer — and sleeping). This type of approach to functional space requirements will help in the design of many workplaces as well as for seating configurations.

However, the physical characteristics of the spaces we inhabit also have psychological overtones. These elusive human factors are sometimes called 'hidden dimensions' (Hall, 1969). Starting with an example, a room in a house might be 2350 mm from floor to ceiling. A man of 95th %ile stature would have a clearance of 550 mm above his head (ignoring suspended light fittings). This might lead us to suggest that, even allowing for the occasional extremely tall person, at least 20% of the room's volume is wasted. There is little doubt, however, that this space contributes in some way to psychological well-being, although there is no clear answer on how much head room is preferred.

In considering the spatial experience and behaviour of human beings, two closely related key concepts emerge: territoriality and personal space. The concept of territoriality was originally derived from observations of animal behaviour. Many species of bird and mammal will vigorously defend a home territory. By analogy, human beings (although they are essentially a gregarious species) may be described as showing signs of territoriality when they attempt to define a space as being for their own more or less exclusive use. In addition to private residences, territories in this sense might include offices and areas surrounding a chosen seat in a public place (e.g., on a train or in a library). Territory, then, can be temporary as well as permanent, which leads us to the concept of *personal space* — which has been compared with a 'psychological bubble' that surrounds us wherever we go and influences our interactions with other people. Sommer (1969) describes personal space as a portable territory — a region around the person's body, demarcated by invisible boundaries, into which the entry of other people is strictly controlled. Hall (1969) distinguished four concentric zones surrounding the individual. Each zone, defined in terms of face to face distances, was associated with a typical class of social interaction: intimate (up to 450 mm), (personal 450 to 1200 mm), social (1200 to 3500 mm) and public (over 3500 mm). Hall considered that the outer part of the personal zone, for example, is used for communication of a personal kind, while the inner part of the social zone is characteristic of the separation of people at informal gatherings or for people working together. The outer part of the social zone typifies more formal business exchanges.

We should not imagine that these zones have sharp transitions. Naturally enough they merge into one another, and experience shows us that there are significant national and cultural differences in the actual values of the distances defining the zones. Hall's figures were based upon observations of the social behaviour of Americans during the 1960s — a population which he considered to be 'middling' in their social distances. In the time that has elapsed since then, these distances appear to have shortened. None the less, they provide a model which can serve as a useful starting point for understanding spatial behaviour. The penetration of another person into a spatial zone inappropriate to the circumstances may be experienced as an unwanted and stressful intrusion. In the crowded rush-hour tube train, where total strangers are thrust into the intimate zone of personal space, people stiffen themselves to minimise bodily contact and stare at the ceiling to avoid meeting the gaze of other passengers.

A contrasting example, which illustrates the extent of cultural differences in personal space, is provided by Sen (1984). He recounts a localised custom in one Indian state for bus passengers boarding an already full bus to sit on the laps of other passengers. This is, presumably, an adaptation to local needs as well as a reflection of the more relaxed familiarity of individuals with each other in a relatively small community.

In the future, personal space may have to take account of reactions to mobile robots as well as to people. Nakashima and Sato (1999) experimented with such robots, used for transport work in hospitals, and found that the separation distance at which their subjects began to feel uncomfortable increased with the speed of the robot. By contrast, operators on manufacturing production lines who work alongside robots become familiar with them and may approach too close for safety. So, as Sommer found, context and relationships are important influences on preference for personal space.

The psychological bubble is by no means spherical; we can better tolerate the approach of strangers side by side than face to face. There is also evidence that women tolerate closer encounters than men, that opposite sex pairs approach more closely than pairs of the same sex, and that members of a peer group approach more closely than pairs of disparate age (Oborne and Heath, 1979). Little (1965) found personal space to be dependent on context; distances were greatest in an office waiting room and least outdoors. These factors are probably all important in the layout of both workspaces and public spaces.

4.3 REACH: THE WORKSPACE ENVELOPE

The area within which manual tasks can be performed easily (or at all) is defined by the workspace (or reach) envelope. Consider what happens when you reach your arm forwards. First, you raise your upper limb through 90°; this is achieved principally by a rotational movement of the arm in its socket on the shoulder blade or scapula (i.e., flexion of the gleno-humeral or true shoulder joint; see Figure 4.4). It is, however, impossible to make such a movement without a small shift in the scapula's position on the chest wall. (Anatomists call this interaction 'scapulo-humeral rhythm'.) You have now reached the position in which the static dimension of forward reach, dimension 36 in Table 2.5, would be measured. If your shoulder blades had been touching a wall at the outset they would still be doing so. As you reach farther forwards from this basic starting point, several new movements occur. Your whole shoulder girdle is thrust forwards (protracted) and you begin to incline your trunk forwards by flexion of the hip joint and spine. What determines the final limit of your forward reach? Try it and you will quickly discover that (when standing)

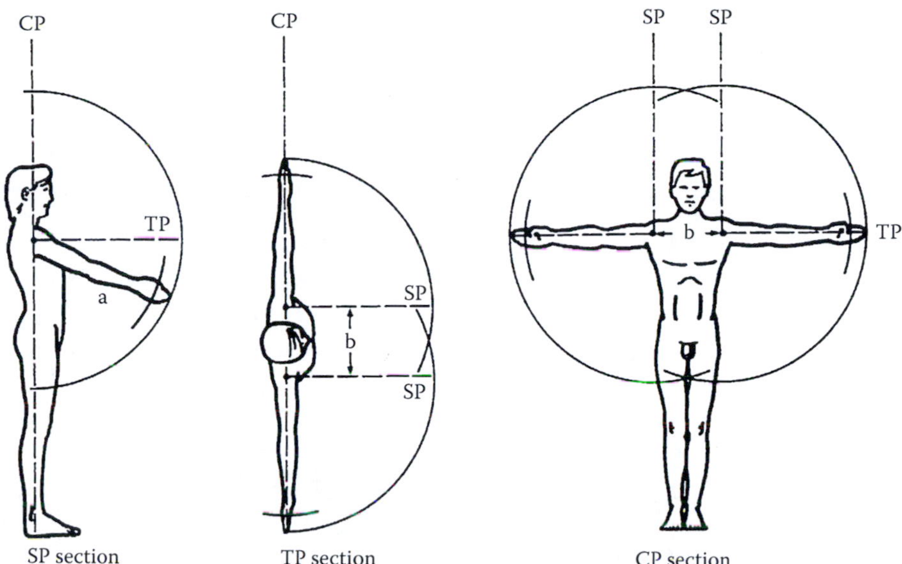

FIGURE 4.4 Zones of convenient reach (ZCR) seen in elevation and plan. Left to right: vertical section in sagittal plane (SP) passing through shoulder joint; horizontal section in transverse plane (TP) passing through shoulder joints; vertical section in coronal plane (CP) passing through shoulder joints. Each plane of a section is marked on the other two diagrams. (a, upper limb length to fingertip or grip centre; b, biacromial breadth)

it is the tendency to topple over as the horizontal position of your centre of gravity reaches the limit of the base of support of your feet. This in turn can be modified by pushing the pelvis backwards as a counter-balance.

Maximum reach therefore depends on the extent to which stretching and leaning forward is acceptable for performance of a given task (which in turn is influenced by the duration and precision of the task). It is also, of course, dependent on whether the task involves gripping by the hand, finger pinch grip or fingertip operation. Suitable increments to the basic dimension of forward grip reach for these various conditions are shown in Table 4.6.

Dynamic reach may best be characterized by the three-dimensional coordinates of a volume of space. Such a volume is referred to as a '*workspace envelope*' (or more grandly as a 'kinetosphere'). Since standing reach is essentially a matter of body equilibrium, the envelope will be modified by any factor that affects this. A weight in the hands will diminish reach. Grieve and Pheasant (1982) reported experiments showing how reach was increased by positioning the feet to increase the foot base and diminished by placing an obstacle behind the subject to limit changes in posture which might provide counterbalancing.

Similar considerations occur in seated postures although, in this case, stability is influenced by both foot and seat interfaces. Sengupta and Das (2000) showed that the maximum reach envelope is smaller when seated than when standing. Several studies of the workspace envelope of the sitting person have been published. That

TABLE 4.6
Increments to Forward Grip Reach (all dimensions in millimetres)

Dimension	Men 5th %ile	Men 50th %ile	Men 95th %ile	Women 5th %ile	Women 50th %ile	Women 95th %ile
Basic Dimension						
Forward grip reach[a]	720	780	835	650	705	755
Increments						
For a pinch grip (to the thumbtip)	35	40	40	30	35	40
For fingertip operation	105	115	125	95	105	115
For a forward thrust of the shoulder[b]	115	130	150	95	115	140
For 10° of trunk inclination	80	85	95	75	85	95
For 20° of trunk inclination	155	170	185	150	170	185
For 30° of trunk inclination	230	250	270	225	245	270

[a] Quoted directly from Table 2.5
[b] Calculated from data in MIL-STD-1472C (Department of Defense, 1981)

of Kennedy (1964) has been particularly widely quoted (Damon et al., 1966; Van Cott and Kinkade, 1972; Webb Associates, 1978). However, the reader should note that all reach envelopes are highly specific to the situation in which they were measured. The data of Kennedy (1964) were measured in an aircraft seat with the subjects securely strapped in; had the seat or the restraints been otherwise, the reach envelopes would have been numerically different.

4.3.1 Zones of Convenient Reach

At this point it is appropriate to develop the concept of a zone or space in which an object may be reached conveniently, that is, without undue exertion. Consider what it means for a control to be 'within arm's length'. The upper limb, measured from the shoulder to the fingertip (or to the centre of grip), sweeps out a series of arcs centred upon the joint (as illustrated in Figure 4.4). These define the *zone of convenient reach* (ZCR) for one hand, which extends sideways to the coronal plane of the body. The zones for the two hands intersect in the *midline* (*median* or *mid-sagittal) plane* of the body. The volume which is thus defined comprises two intersecting hemispheres. The radius of each hemisphere is the upper limb length (a), and their centres are a distance (b), equal to biacromial breadth, apart.

Many design problems are concerned with the intersection of work surfaces — vertical, horizontal or (very occasionally) oblique planes — with either the volume of the workspace envelope or that of the zone of convenient reach. This defines the boundary of the area that can be easily reached on the work surface. Typically this information is needed when laying out equipment on the work surface or controls on a control panel. Suppose we wish to locate a set of items upon the vertical wall of a control room so that they might be conveniently operated by a standing person. The items would therefore need to be within the boundary of the intersection between the plane of the wall and the zone of convenient reach. The intersection of a plane with a sphere produces a circle. The radius of this circle may be calculated by Pythagoras' theorem as

$$r = \sqrt{a^2 - d^2} \tag{4.1}$$

where r is the radius of the circle on the wall, a is the upper limb length (shoulder–grip length or shoulder–fingertip length) and d is the horizontal distance between the shoulder and the wall. Figure 4.5 shows the construction of such a zone for fingertip controls operated by a 95th %ile male or a 5th %ile female operator, assuming that the wall is at a distance of d = 500 mm. It must of course be remembered that the full design should also take account of visual concerns; optimal zones for visual displays (OVZ) have been added in Figure 4.5 according to the criteria of Section 4.6 below.

The zone of convenient reach may be similarly described for any other vertical or horizontal plane parallel to a line joining the shoulders. Requisite data for construction of the zones defining full grip reach are given in Table 4.7. If the zone for fingertip reach is needed, an appropriate increment should be added to a, and r should be recalculated using equation 4.1.

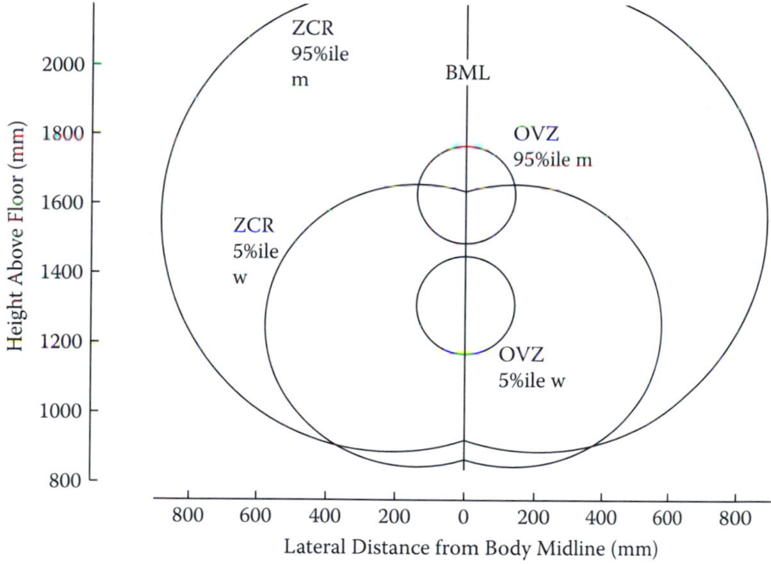

FIGURE 4.5 Zones of convenient fingertip reach (ZCR) and optimal visual zones (OVZ) on a vertical surface 500 mm in front of the shoulders. The zones are drawn for a 95th %ile man (m) and a 5th %ile woman (w). (BML, body midline)

TABLE 4.7
Zones of Convenient Reach for a Full Grip (all dimensions in millimetres)

To construct a zone of convenient reach in a vertical plane, a distance *d* in front of the shoulders, draw two circles of radius *r*; the centres of the circles are defined by standing or sitting shoulder height and biacromial breadth. To construct a zone of convenient reach in a horizontal plane, a distance *d* above or below the shoulders, draw two semicircles of radius *r*, centred upon the positions of the shoulders.

	Radius[a] (*r*) Men			Radius[a] (*r*) Women		
d	**5th %ile**	**50th %ile**	**95th %ile**	**5th %ile**	**50th %ile**	**95th %ile**
0	610	665	715	555	600	650
100	600	655	710	545	590	645
200	575	635	685	520	565	620
300	530	595	650	465	520	575
400	460	530	595	385	445	510
500	350	440	510	240	380	415
600	110	285	390			250

	Men			Women		
Body Dimensions	**5th %ile**	**50th %ile**	**95th %ile**	**5th %ile**	**50th %ile**	**95th %ile**
Biacromial breadth	365	400	430	325	355	385
Shoulder height (standing, shod)	1340	1425	1560	1260	1335	1450
Shoulder height (sitting)	540	595	645	505	555	605

[a] Figures calculated from Equation 4.1 assuming a full grip.

4.3.2 The Normal Working Area

The intersection of a horizontal plane, such as a table or bench, with the zone of convenient reach defines what work study writers call the *maximum working area* (Barnes, 1958). Within this is a much smaller '*normal working area*', described by a comfortable sweeping movement of the upper limb, about the shoulder, with the elbow flexed to 90° or a little less. Das and Grady (1983) have discussed this thoroughly. The presentation that ensues (Figures 4.6 and 4.7) is based on the original concept of Squires (1956).

A person sits at a bench or table as shown in Figure 4.6. The boundaries on the work surface define the intersections with the zones of convenient reach (maximum working area) and the easy reach normal working area. The boundaries are constructed separately for each arm and overlap in the middle where both hands can reach. This is the area where two-handed work can be done.

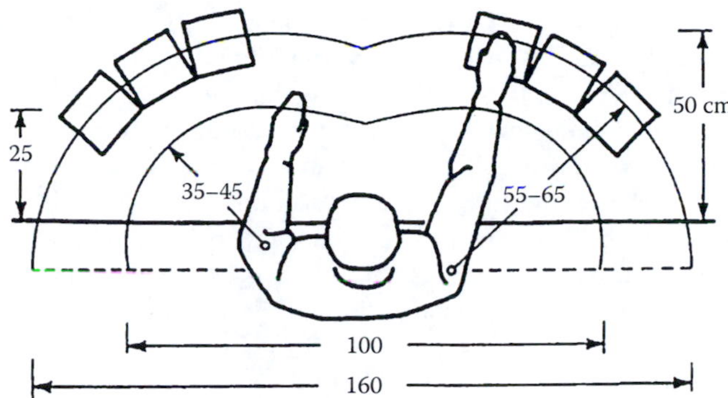

FIGURE 4.6 Horizontal arc of grasp and normal working area at tabletop height. The grasping distance takes account of the distance from shoulder to hand, the working distance only elbow to hand. The values (in centimetres) include the 5th %ile and thus apply to men and women of less than average height. (From Grandjean, E. (1988). *Fitting the Task to the Man,* 4th ed., London: Taylor & Francis, Fig. 42, p. 51. Reproduced with kind permission.)

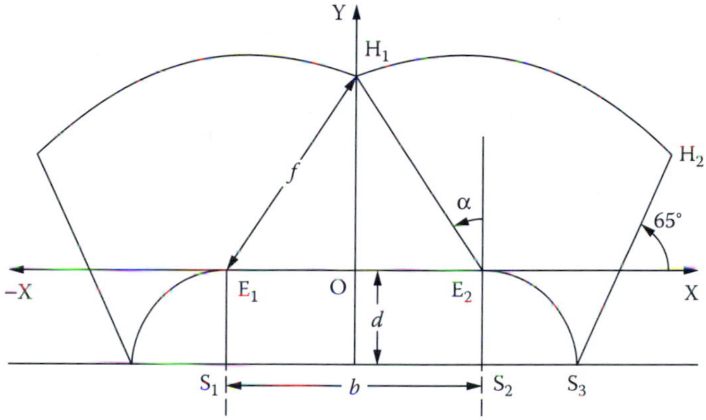

FIGURE 4.7 Construction of the normal working area. See text for details.

The normal working area is constructed in Figure 4.7, where the shoulder joints are located at S_1 and S_2, which are a distance b apart (b = biacromial breadth). The elbows are located at E_1 and E_2 at the table's edge, a distance d in front of the shoulders, such that d = abdominal depth/2. Both hands can meet at H_1, the location of which is calculated from the dimension f = elbow–grip length. Hence the angle α (between the forearm at this location and the perpendicular to the table's edge) is given by

$$\sin \alpha = \frac{b}{2f} \tag{4.2}$$

Moving from H_1, the right hand can sweep an arc out to the side as the humerus is rotated at the shoulder about its own axis. When the elbow is flexed at 90°, the comfortable limit of this outward rotation is about 25°. However, the elbow also swings out naturally, and this extends the working area further.Therefore, the path of the hand defining the outer limit of the normal working area is a prolate epicycloid, H_1H_2, formed by two simultaneous rotations: the forearm (*f*) rotates through α + 25°, whilst the elbow itself moves outwards and backwards through a circular 90° arc from E_2 to E_3. Hence, the forearm comes to lie at an angle of 90° – 25° = 65° to the table edge, with the elbow in line with the shoulders. Thus, taking a point within this movement when the radius S_2E_2 (or S_1E_1) has rotated through an angle β°, the forearm elbow height has rotated through γ° such that

$$\gamma = \frac{\beta(\alpha + 25°)}{90°} \tag{4.3}$$

because the two rotations occur simultaneously, and it can be assumed that at any instant in time they will both have moved through the same proportion of their total range.

The coordinates of the elbow with respect to the shoulder are then given by

$$X_1 = d \sin\beta \tag{4.4}$$

$$Y_1 = d \cos\beta \tag{4.5}$$

(The X-axis, with origin at E_2, runs along the table edge, and the Y-axis is perpendicular to the table edge.)

The coordinates of the hand with respect to the elbow are given by

$$X_2 = f \sin(\gamma - \alpha) \tag{4.6}$$

$$Y_2 = f \cos(\gamma - \alpha) \tag{4.7}$$

Therefore, the coordinates of the hand with respect to a point O on the table's edge in the midline of the body are given by

$$X = d \sin\beta + f \sin(\gamma - \alpha) + \frac{b}{2} \tag{4.8}$$

$$Y = d \cos\beta + f \cos(\gamma - \alpha) - d \tag{4.9}$$

Figure 4.8 and Table 4.8 give the dimensions of the normal working area (NWA). These are based on the above equations together with the anthropometric data of Table 2.5. (The figure for *d* is based on a 50th %ile, because abdominal depth is poorly correlated with limb lengths.)

The normal working area and the zone of convenient reach define the boundaries within which frequent and occasional task actions, respectively, should be performed.

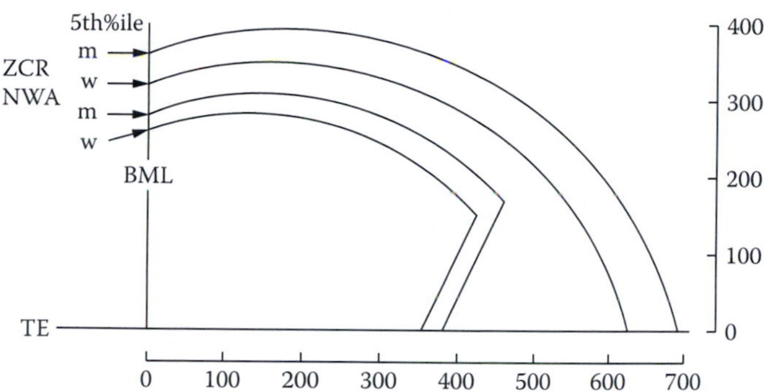

FIGURE 4.8 Zones of convenient reach (ZCR) and normal working area (NWA) on a table surface, for a 5th %ile man (m) and woman (w) (BML, body midline; TE, table edge; dimensions in millimetres).

TABLE 4.8
Coordinates of the Normal Working Area[a]

Position[b]	Degrees	5th %ile Man X	5th %ile Man Y	5th %ile Woman X	5th %ile Woman Y
H_1	0	0	281	0	257
	10	56	298	53	271
	20	114	307	105	278
	30	172	307	160	279
	40	227	300	211	272
	50	281	287	260	258
	60	333	266	307	237
	70	370	239	350	211
	80	423	206	388	181
H_2	90	460	169	421	146
I	—	380	0	354	0

[a] Origin (O) is at the table's edge in the midline of the body. The X axis runs along the table's edge; the Y axis is perpendicular. H is the grip centre of the hand at any point in the arc through which it sweeps. I is the point where the normal working area intersects the table's edge. The shape of the area is shown in Figure 4.7.

[b] H_1 and H_2 as in Figure 4.7.

An operator should not be expected to reach outside the zone of convenient reach for any regular work. The advantages for efficiency and comfort are obvious, but Sengupta and Das (2004) have further confirmed that ignoring this will lead to physiological cost and muscle fatigue. In their experiment, subjects were asked to perform a light repetitive manipulative task within each of the two zones and at more extreme reach (beyond the zone of convenient reach). Measurements of arm and back muscle activity, oxygen uptake and heart rate all showed the demands increasing from one zone to the next.

It should also be recognised that a very close reach, close to the table edge, is not comfortable or efficient because it requires the upper arm to be extended backwards and the elbow to be raised. There should therefore be an inner boundary to the normal working zone. A method for defining this has been developed by Wang et al. (1999). As a very simple rule of thumb, Tichauer (1975) recommended that no tasks should be performed within 76 mm of the edge of the table.

4.4 JOINT RANGES OF MOVEMENT

The flexibility of the human body is measured in terms of the angular ranges of motion of the joints. Joint movements are the subject of a terminology which is almost standardised (see Figure 4.9). Consider a vertical plane cutting the body down the midline into equal right and left halves. This is called the median (or midsagittal) plane. Any vertical plane parallel to it is called a sagittal plane, and any vertical plane perpendicular to it is called a coronal plane. A horizontal plane through the body is known as a transverse plane.

In general, sagittal plane movements of the trunk or limbs are called flexion and extension. (Flexion movements are those that fold the body into the curled-up foetal position.) Coronal plane movements are called abduction and adduction. (Abduction movements take a limb segment away from the midline adduction towards or across

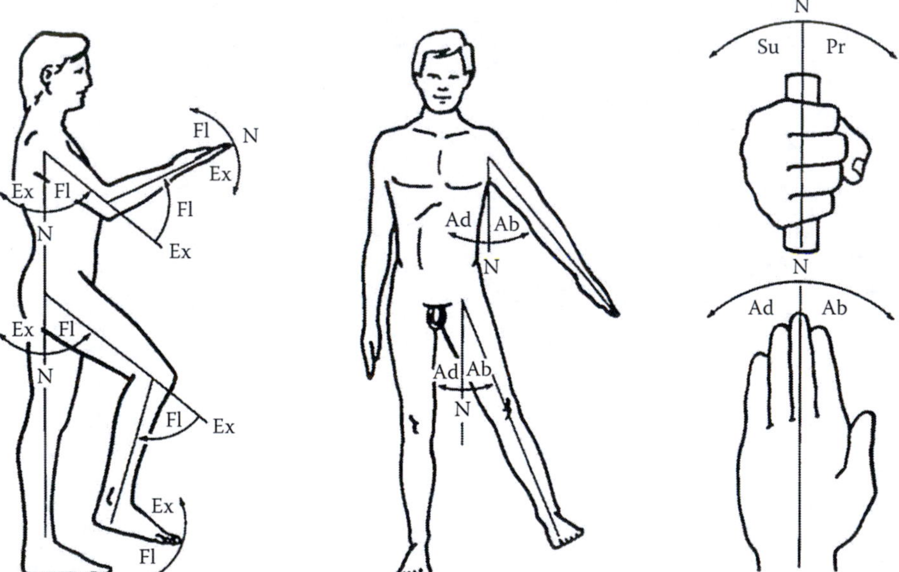

FIGURE 4.9 Terms used in the description of movements and joint ranges as given in Table 4.9. (Fl, flexion; Ex, extension; Ab, abduction; Ad, adduction; Su, supination; Pr, pronation; N, neutral position)

TABLE 4.9
Joint Ranges of Movement (degrees)

	Joint	5th %ile	50th %ile	95th %ile	SD
1.	Shoulder flexion	168	188	208	12
2.	Shoulder extension	38	61	84	14
3.	Shoulder abduction[a]	106	134	162	17
4.	Shoulder adduction	33	48	63	9
5.	Shoulder medial rotation	61	97	133	22
6.	Shoulder lateral rotation	13	34	55	13
7.	Elbow flexion	126	142	159	10
8.	Pronation	37	77	117	24
9.	Supination	77	113	149	22
10.	Wrist flexion	70	90	110	12
11.	Wrist extension	78	99	120	13
12.	Wrist abduction (radial deviation)	12	27	42	9
13.	Wrist adduction (ulnar deviation)	35	47	59	7
14.	Hip flexion[b]	92	113	134	13
15.	Hip abduction	33	53	73	12
16.	Hip adduction	11	31	51	12
17.	Knee flexion	109	125	142	10
18.	Ankle flexion (plantar flexion)	18	38	58	12
19.	Ankle extension (dorsiflexion)	23	35	47	7

[a]Accessory movements of spine increase this to 180°.

[b]Measured with the knee fully flexed. If the knee is extended, the range will be much less (approximately 60°).

Source: Data from Barter T., Emmanuel, I. and Truett, B. (1957). *A Statistical Evaluation of Joint Range Data,* WADC Technical Note 53-311, Wright Patterson Air Force Base, OH.

the midline.) Limb segments may also rotate about their own axes — inward (medially) or outward (laterally). Inward rotation of the forearm (turning the palm downwards) is called pronation; outward rotation (turning the palm upwards) is called supination.

Guidance on measuring joint ranges of movement has been produced by the American Academy of Orthopaedic Surgeons (1965), and this defines the neutral position for each joint. There are three main types of joint motion. Hinge joints, such as the elbow and knee, move in a single plane (i.e., with only one degree of freedom). Two-degree-of-freedom joints have motions in two planes, typified by the wrist, which can move in flexion, extension, radial deviation (inwards when the wrist is in pronation) and ulnar deviation (outwards when the wrist is in pronation). The major joints at the hip and shoulder have complex motions with three degrees of freedom.

There are surprisingly few joint range data available. Table 4.9 is based on a survey of male U.S. servicemen conducted by Dempster (1955), reanalysed by Barter et al. (1957) and quoted extensively (Damon et al., 1966 and elsewhere). Note that the measurements were not necessarily made in the postures shown in Figure 4.9 (refer to Damon et al., 1966 for details).

In general, women have somewhat greater flexibility than men (by about 5 to 15% on average). Decrements with age are probably small in the absence of joint disease (osteoarthritis, etc.), but since this is extremely common (universal in some populations), it is reasonable to assume greatly reduced flexibility in the elderly. Unfortunately, few statistical data are available, but Steenbekkers (1998) confirmed that there are age decrements in ranges of movement of the wrist and of the neck (except for flexion/extension).

The flexibility of one joint may be influenced by the posture of adjacent joints. The most important example of this is flexion of the hip, which is much greater when the knee is flexed than when it is extended. (Prove this by touching your toes.) The range of flexion of the knee is also strongly influenced by hip joint posture, being reduced when the hip is flexed.

4.5 POSTURE

The posture that a person adopts when performing a particular task is determined by the relationship between the dimensions of the person's body and the dimensions of the various items in his or her workspace (for example a tall person using a standard kitchen will stoop more than a short one). The extent to which posture is constrained in this way is dependent upon the number and nature of the connections (or *interfaces*) between the person and the workspace. These connections may be either physical (seat, worktop, floor) or visual (gaze direction, location of displays). If the dimensional match is inappropriate, both short- and long-term consequences for the well-being of the person may be severe.

4.5.1 Postural Loading

Posture may be defined as the relative orientation of the parts of the body in space. To maintain such an orientation over a period of time, muscles must be used to counteract any external forces acting upon the body (or in some minority of cases internal tensions within the body). The most ubiquitous of these external forces is gravity. Consider a standing person who leans forwards from the waist, as illustrated in Figure 4.10. The postural loadings on the hip extensor or the back extensor muscles are proportional to the horizontal distance between the hip or lumbo-sacral joints, respectively, and the centre of gravity of the upper part of the body (the point through which the total weight of the head, arms and trunk can be considered to act). The further the trunk is inclined, the greater this distance becomes and the greater the muscle force needed to hold the posture. Figure 4.10 shows a biomechanical analysis of the loading on the back muscles (muscle force or tension t) and lumbo-sacral joint (compressive force c), but a similar analysis could be performed for the loading on the hip extensor. Physiologists call the muscular activity that results from such loading 'static work'.

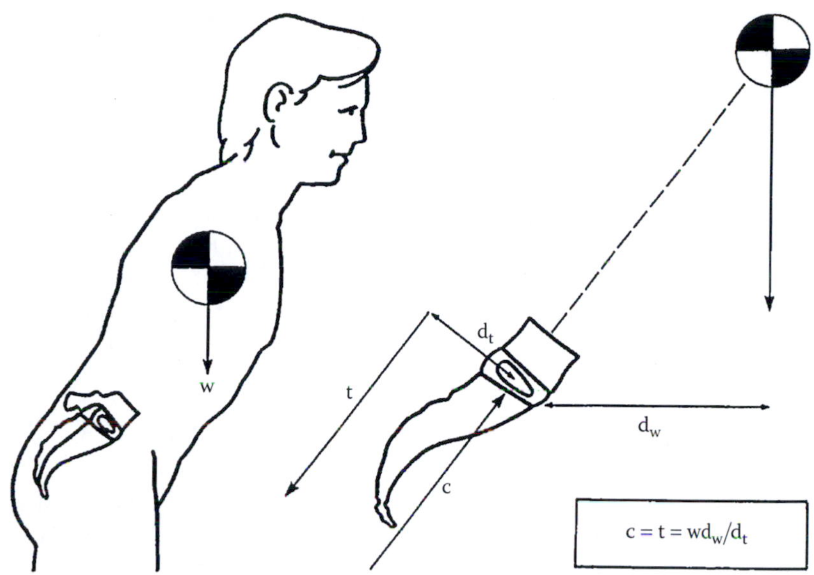

FIGURE 4.10 Biomechanical analysis of postural stress at the lumbo-sacral joint in the spine, in a forward leaning position. Note that this analysis ignores the direct effect of the weight of the trunk which the spine must support even when $d_w = 0$, standing upright with the body weight balanced over the lumbo-sacral joint. w is the weight of that part of the body above the lumbo-sacral joint; d_w is the horizontal distance between the joint and the centre of gravity of the upper body; c is the compressive force acting along the axis of the spine; t is the tension in the back muscles (erector spinae).

Muscle, as a tissue, responds badly to prolonged static mechanical loading. (The same is probably true of other soft tissues, and even perhaps of bone, but the physiology of these cases is much less well understood.) Static effort restricts the flow of blood to the muscle. The chemical balance within the muscle is disturbed, metabolic waste products accumulate, and the condition of muscular fatigue supervenes. The person experiences a discomfort which is at first vague but which subsequently develops into a nagging pain until it becomes a matter of some urgency that relief is sought by a change of position. Should you require evidence of this course of events, you should raise one of your arms and hold it out in front of you as you continue to read (or attempt to do so). Provided our workspace and working schedule allows us to make the frequent shifts of posture which are subjectively desirable, all will be well, since the physiological processes of muscular fatigue are relatively rapidly reversible by rest or change of activity (particularly if the activity involves stretching the fatigued muscle).

In general, we may think of fidgeting as the body's defence against postural stress. This mechanism characteristically operates at a subconscious level — usually we fidget before we become consciously aware of discomfort. The rate of fidgeting can be used as an index of the comfort of chairs — the less comfortable we are, the more we fidget. It is a matter of common experience that other factors are involved. Some people fidget more than others, and we all fidget more when we are bored, presumably because when we are concentrating on something else, mental activity can shut out the sensory stimuli that cause the fidgeting (or raise our threshold of discomfort). Such a hypothesis is in line with theories of the nature of pain (Melzack and Wall, 1982).

We are not usually comfortable in postures where joints are towards the outer ends of their ranges of movement. Outside the range that we do find comfortable, the level of discomfort will probably increase gradually as the joint posture becomes more extreme but will obviously depend on the length of time for which the posture is held and on the general body posture (because of the loading placed on the joint from body weight). Kee and Karwowski (2001) conducted extensive trials to obtain subjective judgements of comfort through the ranges of motion for each of the major joints in the body (when the postures were held for 60 s). They presented the resulting data in the form of isocomfort boundaries to joint deviations from neutral posture for both seated and standing postures for men and for standing postures for women. These data do not, however, take account of interactions between the postures of adjacent joints, which do occur when muscles span more than one joint. Moreover, the trials were of very short duration so that the absolute comfort levels would not apply directly to postures which have to be held for more than a minute; in this case, the isocomfort boundaries should be treated as relative zones. The results showed that people are least tolerant of joint deviations at the hip, followed by the lower back and shoulder. This is in agreement with studies of Hsaio and Keyserling (1991), who found that when changing a seated posture to perform a particular task, distal segments of the body will be moved outside their neutral (comfortable) ranges before proximal segments. Their paper defines the limits of the neutral ranges for the major upper body joints when seated.

Rebiffé (1966) gave the ranges which he considered comfortable for a seated vehicle driver in Table 4.10, basing this on analysis of data from studies by Dempster, Gleser, Trotter and Geoffrey. Since Rebiffé was applying these data to vehicle seating and control layouts, the posture suggested is more reclined than would normally be encountered in most work situations but represents a comfortable posture when supported by the backrest of a seat. The leg posture is to some extent a compromise since the driver has to be ready to apply force to the foot pedals as well as remain comfortable for long periods of time.

Suppose that the working circumstances closely constrain an operator to a particular posture and prevent postural change. The consequences may be divided into those occurring over the short term and those occurring over the long term. In the short term, mounting discomfort may distract the operator from his or her task, leading to an increased error rate, reduced output and the risk of accidents. If the postural loading is high, after a time severe muscle fatigue will ensue, and this will limit the operator's ability to continue with the task.

From the physiological standpoint, however, we are still talking about a reversible state, since the symptoms are relieved by rest or by a change of activity. At some point, nevertheless (and this point is not well defined since the transition is probably gradual rather than sharp), pathological changes in the muscle or soft tissue begin. Typically, pain comes on after increasingly short periods of postural loading

TABLE 4.10
Joint Comfort Ranges for a Seated Driver (degrees)

Joint	Comfortable Angles
Back inclination (from vertical)	20 to 30
Hip angle	95 to 120
Knee angle	95 to 135
Ankle angle[a]	90 to 110
Shoulder angle	10 to 20 extension[b], 45 flexion[c]
Elbow angle	80 to 120
Wrist angle	10 extension, 10 flexion

[a] Angle between lower leg and sole of foot, foot supported
[b] Depending on the seat back
[c] With arm support

Source: Rebiffé, R. (1966). Proceedings of the Institution of Mechanical Engineers, 181(Part 3D 1966-67), 43–50.

and rest is less certain to bring relief. At this point we are dealing not with discomfort but with physical injury and a disease process.

Back pain, neck pain and the class of conditions affecting the hand, wrist and arm which we refer to as *work-related upper limb disorders* (WRULDs) or *repetitive strain injuries* (RSIs) are all conditions that characteristically result from overuse of the muscles and other soft tissues in question. This overuse may be due to prolonged static loading, repetitive motions, acute overexertion or some combination of these. Psychological factors are also believed to be involved (possibly because psychological stress leads to hormonal and metabolic changes, increased muscle tension and changes in task behaviours). We shall return to these matters in Chapter 9.

4.5.2 Guidelines for Work Postures

From the discussion above, it is clear that, in general, variation in working posture is desirable and a fixed working posture should be avoided. If circumstances demand that work is performed in a fixed position (as in practice will very often be the case), then the deleterious effects that ensue will increase with the degree of static muscle work required to maintain the posture concerned. The following simple guidelines for reducing this are based in part upon the work of Corlett (1983); for a more detailed discussion see Pheasant, 1991a.

4.5.2.1 Encourage Frequent Changes of Posture

Sedentary workers, therefore, should be able to sit in a variety of positions. Some office chairs are now being designed with this in mind. For many industrial tasks a sit-stand workstation is to be advocated. The task is typically set at a height that is suitable for a standing person (see Section 4.7) and a high seat or perch is provided as an alternative. Moreover, there seems little doubt that most sedentary workers would be better off if their jobs required them to get up and move around once in a while. This can often be addressed by considering the layout of the workplace or enlarging the job specification to include a greater variety of tasks.

4.5.2.2 Avoid Forward Inclination of the Head and Trunk

Forward inclination of the head and trunk commonly results from visual tasks, machine controls or working surfaces that are too low or that are difficult to see sufficiently clearly. Figure 4.11 illustrates the work of sewing machinists which contains several of these problems (Li et al., 1995). In some circumstances visibility can be improved, and thus the head and neck posture also, by tilting the work surface, as shown too for reading and writing tasks by de Wall et al. (1991). ISO standard 14738 (ISO, 2002b) recommends an angle of tilt of about 15° for fine manipulative tasks with high visual demands.

4.5.2.3 Avoid Causing the Upper Limbs to Be Held in a Raised Position

Working with arms raised commonly results from a work surface that is too high (or a seat that is too low). Work surface and seat should be matched and set at heights that allow the operator to work with arms relaxed. If manipulative tasks must be performed in a raised position, perhaps for visual reasons, arm supports should be provided. In addition to causing a considerable stress to the shoulder muscles, tasks

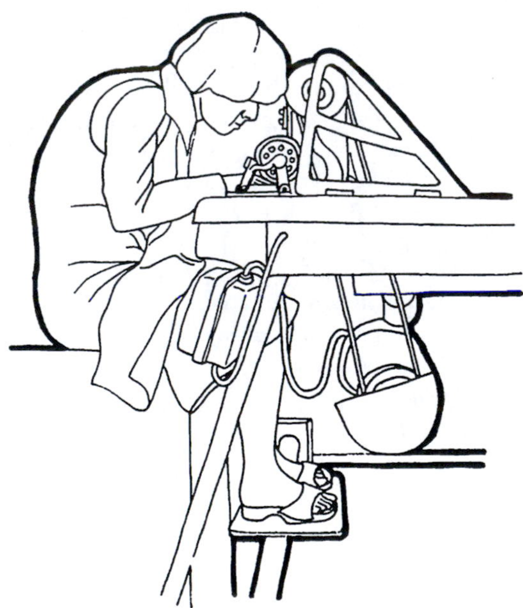

FIGURE 4.11 The sewing machinist, from an original kindly supplied by Murray Sinclair. (From Pheasant, S. (1991). *Ergonomics, Work and Health,* London: Macmillan, Fig. 1.6, p. 12. Reproduced with kind permission.)

FIGURE 4.12 Deviated wrist positions in repetitive industrial tasks, showing movements of radial and ulnar deviation with an extended wrist in a packing task, where the working level is too high. Note also the abduction of the shoulders. From an original kindly supplied by Peter Buckle. (From Pheasant, S. (1991). *Ergonomics, Work and Health,* London: Macmillan, Fig. 14.1, p. 262. Reproduced with kind permission.)

that must be performed at above the level of the heart impose an additional circulatory burden. The upper limit for manipulative tasks should be around halfway between elbow and shoulder level. Figure 4.12 shows that a poor work surface height that results in working with arms raised can have further consequences for lower arm and hand postures (as mentioned in Section 4.5.2.5).

4.5.2.4 Avoid Twisted and Asymmetrical Postures

Twisted and asymmetrical postures commonly result from expecting an operator to have eyes in the back of the head (i.e., from the mislocation of displays and controls) or from poor location of materials, controls or storage bins. They are also often the result of designers' lack of thought about the location of components and provision of access for repair and maintenance tasks. If part of the workstation has to be placed to the side, standing operators should be encouraged to move their feet rather than to lean or twist sideways; if it is a seated workstation, a swivel chair should be provided (unless the task involves exerting force).

4.5.2.5 Avoid Postures That Require a Joint to Be Used for Long Periods or Repetitively towards the Limit of Its Range of Motion

This is particularly important for the forearm, wrist and hand, which are involved in most work tasks, but equally for the shoulder, neck and other joints in the body. There can be many causes of such extreme postures. Figure 4.12 gives just one example.

4.5.2.6 Provide Adequate Back Support in All Seats

Neck, shoulder and back muscles may all be under tension if a seat does not provide adequate back support. It may be that, for operational reasons, the backrest cannot be used during the performance of the work task itself — but it will still be important in the rest pauses. What constitutes an adequate backrest is discussed further in Chapter 5.

4.5.2.7 Where Muscular Force Must Be Exerted, the Limbs Should Be in Their Position of Greatest Strength

When limbs are in their position of greatest strength (as discussed in Section 4.8), force can be exerted with the least muscular effort and thus with the least stress and (when the force exerted is high) the least risk of injury. In awkward postures or when joint postures are at the extremes of their ranges, muscles are likely to be working at a mechanical disadvantage and so be heavily loaded.

One fundamental decision is whether a particular task should be carried out while seated or standing (although of course it is best to permit an alternation between these postures if at all possible). A seated posture is normally preferable and should be used:

- When the task is of long duration
- For fine manipulative work or close visual work, when the whole body has to be kept very still
- When there are foot-operated controls

A standing posture is preferable:

- When heavy weights have to be lifted
- When handling bulky objects
- When the task involves moving around frequently

4.6 VISION AND THE POSTURE OF THE HEAD AND NECK

The visual demands of a task and the location of visual displays are important not only in themselves but also because they largely determine the posture of the head and neck. Look carefully at the printed text on this page. Fix your eyes on one particular word near the centre of the page. You will find that other words become less distinct with increasing distance from the central point of fixation and that the margins of the page are no more than an indistinct blur. Only the central part of the visual field is sufficiently sensitive for demanding visual tasks such as reading text or recognizing a face. The area of *foveal vision*, as this central region is called, is limited to a solid angle 5° about the line of central fixation. Visual work demands that the foveal regions of both eyes be directed convergently upon the task. Furthermore, the lenses of the eyes must accommodate (focus) to the appropriate distance. The processes of direction and convergence of gaze are integrated with accommodation by a set of reflexes so finely tuned that we are unaware of their existence until such times as they break down by reason of age or inebriation.

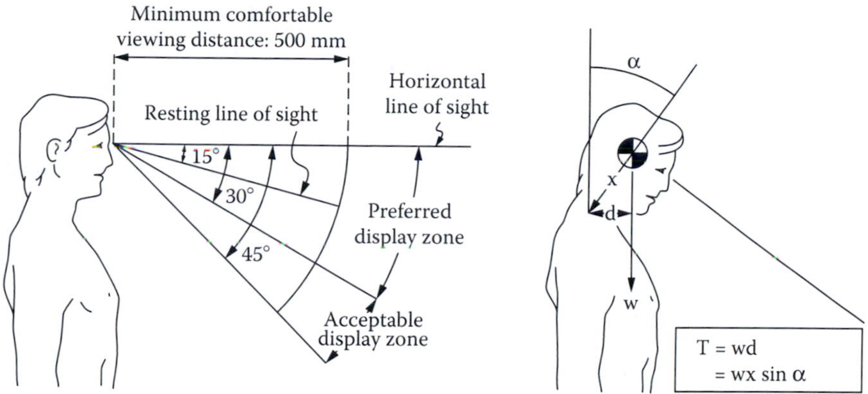

FIGURE 4.13 Vision zones and the posture of the head and neck. Left: preferred viewing conditions as described in the text. Right: postural stress to neck muscles resulting from a downward line of sight. T is the torque about the neck (at the cervical C7 intervertebral joint); w is the weight of the head and neck; x is the distance between C7 and the centre of gravity of the head and neck; d is the horizontal distance between C7 and a vertical line through the centre of gravity of the head and neck.

If we sit or stand with our head up, and look ahead, our eyes will naturally assume a slight downward gaze of 10 to 15° from the horizontal — the relaxed line of sight. The direction of gaze is altered, first, by movements of the eyeballs in their sockets (orbits) by means of the orbital muscles and, second, by movements of the head and neck. Taylor (1973) states that the eyes may be raised by 48° and lowered by 66° without head movements. In practice, only a part of this range of movement is used. Weston (1953), in his classic study of visual fatigue, suggests that, in practice, downward eye movements are limited to 24 to 27°; beyond that point the head and neck are inclined forwards and the neck muscles come under tension to support the weight of the head (see Figure 4.13).

Burgess-Limerick et al. (1998) measured the head and neck postures of seated subjects looking at targets at vertical locations ranging from 30 to 60° below horizontal eye level while performing a 1-minute visual task. They found that both head inclination and gaze angle (relative to the head) were adjusted together. While there were individual differences, the average ratio of head inclination to gaze angle change was 0.70 (with range among subjects 0.45 to 1.12). The changes in head orientation came primarily from the atlanto-occipital joint, with some contribution from a change in neck posture and a small change in trunk inclination. Delleman (1999) found that this general relation held over a number of different tasks (sewing machine operation, touch-typing, pneumatic wrenching and grinding), measuring head inclination to account for about 60% of a change in gaze inclination.

Kroemer and Hill (1986) tested the preferred angle of gaze when head movement was constrained and only eye movement permitted. For a more distant visual target (at 100 cm) they found that the preferred gaze angle was 24° below the ear-eye line (Frankfurt plane), while at a distance of 50 cm the preferred gaze angle was 33° when the person was sitting upright. Thus the preferred gaze angle becomes lower as the visual target comes closer, which Burgess-Limerick et al. (1998) suggest is due to minimising strain on the eye muscles which control vergence. Kroemer and Hill recommend that the angle of view should be 40° ± 20° below the horizontal when the viewer is sitting upright and 30° ± 20° below the horizontal when sitting more reclined.

Grandjean et al. (1983, 1984) described an experiment in which a group of computer operators were given an adjustable workstation and encouraged to set it to their own satisfaction over a period of 1 week. The preferred visual angle was 9° [4.5°] downwards from the horizontal (with a preferred visual distance of 76 [7.5] cm). Jaschinski et al. (1998) found similar results, with computer operators choosing a mean gaze angle of 8° below the horizontal and viewing distance of 80 cm (ranges 0 to 20° and 60 to 90 cm, respectively). Brown and Schaum (1980) also conducted fitting trials on computer workstations. Their results are reported in coordinate form, but it is possible to calculate that the average preferred visual angle was 18° downwards.

On the basis of the above findings we may conclude that the preferred zone for the location of visual displays extends from the horizontal line of sight downwards to an angle of 30° and that the optimal line of sight is somewhere in the middle of this zone. Given that some modest degree of neck flexion is acceptable, the zone could be extended a further 15°. The preferred and acceptable zones for locating displays are therefore as shown in Figure 4.13. Greater neck flexion is undesirable, and over a period of time neck inclination beyond 30° leads to severe muscle fatigue (Chaffin, 1973).

Some jobs require concentrated visual attention for prolonged periods of time, typified by the work of air traffic controllers, radiographers and those using microscopes or magnifying lenses. These all impose more or less fixed postures, the most extreme probably being required by the use of microscopes (Haines and McAtamney, 1993; Gray and MacMillan, 2003). In every case the workplace should be arranged to minimise the postural stress and variation in posture provided through rest breaks or performance of ancillary tasks.

Visual comfort and satisfactory posture are also dependent upon displays being located a suitable distance from the eyes. When focused on infinity, or any object more than around 6 m distant, the lens of the eye is completely relaxed. To look at closer objects than this requires effort, both of the orbital muscles for convergence and of muscles within the eye itself for accommodation. In young people the processes of convergence and accommodation reach their limits and 'near points' at around 80 and 120 mm, respectively. (The latter distance increases dramatically with age as the lens of the eye stiffens; this also reduces the speed with which the eye can accommodate to different distances.) Visual work performed excessively close to the eyes is fatiguing and leads to 'eyestrain', a poorly defined condition involving blurring of vision, headache and burning or 'gravelly' sensations around the eyes. As is the case with most criteria, there is no sharp cut-off point for minimum acceptable viewing distance, and authorities differ in the figures they recommend. Figures as low as 350 to 400 mm are sometimes quoted and indeed may be acceptable under certain circumstances. However, for most practical purposes a minimum for viewing distance of about 500 mm is probably desirable. A viewing distance of 750 mm or more may be preferable, as shown by the computer operators studied by

Grandjean et al. (1984), who adjusted their visual display units (VDUs) to an average visual distance of 760 mm (their settings ranged from 610 to 930 mm). The data of Brown and Schuam (1980) give an average preferred figure of 624 mm. This does, however, rely on the visual display being sufficiently bold and the text font size adequate to be read at this distance. (The positioning of VDUs is discussed further in Chapter 7.)

It is interesting to note that pain and spasm in the neck muscles (trapezius, sternomastoid, splenius, etc.) can lead to 'mechanical headache', experienced in various parts of the head and face and not uncommonly around or 'behind' the eyes (Travell, 1967; Dalassio, 1980; Travell and Simons, 1983). (Anatomists reading this will note the proximity of the proprioceptive supply of these muscles to the spinal nucleus of the trigeminal nerve.) The symptoms of mechanical headache and eye-strain are exceedingly similar.

The visual demands of work therefore have a considerable effect on head and neck posture and on the potential for fatigue. Moreover, task difficulty and demand for high precision both lead to poorer postures and more static postures (Li and Haslegrave, 1999; Wartenberg et al., 2004). Many tasks combine visual and manual demands, and the evidence suggests that the manual demands take precedence over the visual demands in determining reach distance and overall posture, probably because trunk inclination is avoided as far as possible (Delleman, 1999; Hsaio and Keyserling, 1991). This results in poor head and neck and shoulder postures in many visually demanding jobs such as sewing machine operation (Li et al., 1995; Li and Haslegrave, 1999). One possible solution is to introduce a sloped work surface, which can provide a clearer view of the workpiece as well as reducing reach distance slightly (Wick and Drury, 1986; Li et al., 1995).

4.7 WORKING HEIGHT

It is important to distinguish between working height and work surface height. The former may be substantially higher than the latter if hand tools or other equipment are being used in the task. In some cases the working level may actually be below the work surface. Consider the task of washing up, which, in the conventional kitchen, is performed in a recess set into the working surface (i.e., the sink). It is the working height for the task (or tasks) to be performed that should be considered when designing a workplace, rather than the height of the desk, table or bench per se.

The height above the ground at which manual activities are performed by the standing person is a major determinant of that person's posture. If the working level is too high, the shoulders and upper limbs will be raised, leading to fatigue and strain in the muscles of the shoulder region (trapezius, deltoid, levator scapulae, etc.). If any downward force is required in the task, the upper limbs will be in a position of poor mechanical advantage for providing it. This problem may be avoided if the working level is lower. One commonly hears people talk of 'using their weight' or 'getting their weight on top of the action'. This is probably a misconception: what we really mean is that a vertical force may be exerted with minimal loading to the elbow and shoulder extensor muscles. A downward force, however exerted, can never exceed body weight (unless your feet are bolted to the floor), but in some positions the muscles of your arm may lack the strength to lift your feet off the ground.

If, however, the working level is too low the trunk, the neck and head will be inclined forwards with consequent postural stress for the spine and its muscles. It may be presumed that somewhere between a working level that is too high and one that is too low there may be found a suitable compromise at which neither the shoulders nor the back are subjected to excessive postural stress.

The following recommendations concerning working height are widely quoted (see, for example, Grandjean, 1988; Pheasant, 1987, 1991a,b):

- For manipulative tasks involving a moderate degree of both force and precision: 50 to 100 mm below elbow height
- For delicate manipulative tasks (including writing): 50 to 100 mm above elbow height (but wrist support will generally be necessary)
- For heavy manipulative tasks (particularly if they involve downward pressure on the workpiece): 100 to 250 mm below elbow height
- For lifting and handling tasks: between mid-thigh and mid-chest level, preferably close to waist level (see also Chapter 9)
- For hand-operated controls (e.g., switches, levers, etc.): between elbow height and shoulder height (see also Section 4.3.1)

Few workstations are designed for one individual, however, so working height must be chosen to be adequate for the whole user population. Section 2.6.2 provides a discussion of how criteria of this kind may be applied to the dimensioning of workspaces and equipment and adequacy assessed for user populations.

Some groups need particular consideration. Paul et al. (1995) analysed the needs of pregnant women undertaking standing manual work. In the later stages of pregnancy, the changes in body shape result in the most protruding abdominal point being very close to the height of the work surface (8 [2.6] cm below elbow height). In addition, pregnant women find it more difficult to reach forward and have a smaller normal working area. As a result, they prefer a lower working surface and tend to choose the task position closer to the edge of the table. Lowering the table, however, could have adverse consequences if it means that they lean forward while working, so that the rearrangement of the workstation layout for a pregnant woman should be considered carefully in relation to the postures and task(s) being performed.

Some recommendations are available for working heights for disabled people, particularly for wheelchair users. The height of the seat of a wheelchair is typically 470 mm above the floor, but many wheelchair users sit on a cushion, so it is probably more realistic to assume a seat height of 490 mm (Goldsmith, 2000). Goldsmith provides guidance for work surface heights for living and office environments, and O'Herlihy and Gaughran (2003) recommend working heights for workshop environments (for placement of controls and for tasks including using common equipment such as a vice or drill), although their subjects were students aged 12 to 18 years and some extrapolation is needed to apply their results to industrial worker populations.

4.8 POSTURE AND STRENGTH

Studies in which strength is measured in different postures commonly show that the differences in strength between conditions are greater than between individuals. Strength is dependent on posture, first for reasons of physiology, and second for reasons of simple mechanics.

The function of a muscle is to exert tension between its points of attachment to bones, and by doing so to exert a torque or moment about the joint (or joints) that the muscle crosses. The capacity of a muscle to exert tension is dependent upon its length, which in turn is dependent upon the posture (i.e., angle) of the joint across which it acts. This defines the angle–torque relationship of the joint. In general, muscles are able to exert their greatest tension in their outer range, that is at or near their position of maximum length. Thus we should expect actions of flexion to be strongest starting from positions of extension and vice versa — at least if we measure strength in terms of torque about a joint. (For a further discussion of these matters and of muscle function in general, see Pheasant, 1991a.)

In analysing real-world problems involving the exertion of forces on external objects, however, we shall generally also have to take account of the mechanical advantage at which the muscles act through the bony levers of the limbs and trunk. Consider the action of the muscles that flex (i.e., bend) the elbow in the pulling actions shown in Figure 4.14. The torque about the elbow (Te, required to exert a force F), is given by the equation

$$Te = F \times d \qquad (4.10)$$

where d is the perpendicular distance from the line of action of the force to the fulcrum of the elbow joint. Thus the amount of effort required from a person's muscles, to exert a given force, will be very much less in the position shown on the right. In practice, leverage effects of this sort will in most cases be more important

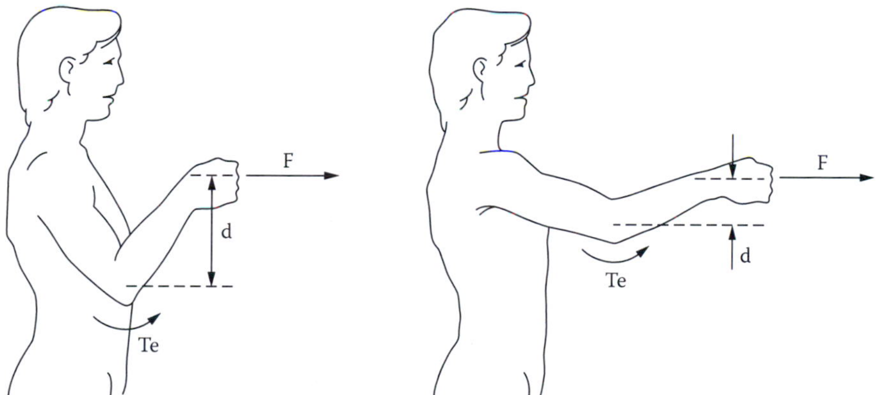

FIGURE 4.14 Torque about elbow (Te) required to exert the same pulling force (F) in two different postures.

as determinants of the force you can exert in a particular situation than the physiologically determined angle–torque relationships at the joints involved. However, both mean that strength is greatly influenced by the working posture (which is in turn determined by the layout of the workplace). The presence of any obstructions that restrict freedom to adopt the best posture will reduce the maximum strength that can be exerted or increase the load demanded of the muscle (in terms of percent of its maximum capacity) when exerting a given force (Haslegrave et al., 1997a,b).

As a general rule, for the reasons given above, pushing or pulling actions are likely to be strongest (and thus to require least effort) when exerted along the line of an almost straight limb. Bicyclists know this and adjust the saddle so that the leg is almost straight when the pedal is in its bottom position. Thus they minimise the perpendicular distance (averaged over the motion cycle) of the line of thrust from the knee joint and minimise the amount of effort required from the knee extensor muscles to do a given amount of work on the pedals. For the same reason, lifting actions will be strongest (and also safest) when the load is held close to the body, that is close to the fulcra of the articulations concerned, particularly those of the low back (see Chapter 9).

There are circumstances in which the force that a person can exert is limited by factors other than the capacity of his or her muscles, such as bodily support and stability, the deployment of body weight or the frictional resistance between the feet and the floor. These matters are discussed at length in Grieve and Pheasant (1982) and Haslegrave (2004), and more briefly in Pheasant (1991a).

Pushing and pulling actions are generally performed most easily at between shoulder height and elbow height or a little below, depending on the circumstances. According to biomechanical studies by Ayoub and McDaniel (1973), the optimum level is 70 to 80% of shoulder height, which works out at a little below elbow height or about 1000 mm for men and 900 mm for women. Fixed horizontal handles on trolleys, carts, etc., should be at this level, but vertical handles will often be a better solution in that they allow the user to find his or her own level.

Pushing actions are strongest when the feet are placed as far back as possible; pulling actions when the feet are as far forward as possible. High-friction shoes and flooring materials are important. An unobstructed floor space of 1000 mm is required; 1800 mm is preferable for pulling actions.

Tasks involving the storage of items on shelving and racks constitute an important class of handling problems. In general, the heaviest or most commonly used items should be stored in the most accessible positions. Table 4.11 provides some guidance in these matters based on the above anthropometric considerations and also on user trials reported by Thompson and Booth (1982).

In summary therefore, the main recommendations for tasks in which force has to be exerted are:

- Provide adequate space to allow the operator to adopt a good posture
- Position loads, equipment and work surfaces so that twisting and lateral bending are avoided
- Locate the working height 100 to 250 mm below elbow height for downward pushing forces and for use of many types of tools

TABLE 4.11
Recommendations for the Design of Storage, Shelving and Racks

Height (mm)	Application and Comments
<600	Reserve storage for rarely required items
	Fair accessibility for light objects; poor for heavy ones
600 to 800	Fair for heavy items; good for light items
800 to 1100	OPTIMAL ZONE FOR STORAGE
1100 to 1400	Fair — good for light items: visibility unimpeded; access fair; poor for heavy items
1400 to 1700	Limited visibility and accessibility. Most men and women will be able to stow and retrieve light items (at least on the edge of the shelves)
1700 to 2200	Very limited access; beyond useful reach for some people
>2000	Out of reach for everybody

Notes: Loads greater than 10 kg are deemed 'heavy'. Shelf depth should not exceed 600 mm at heights of 800 to1400 mm, 450 mm at heights less than 800 mm or 300 mm at heights greater than 1400 mm. Minimum acceptable unobstructed space in front of the shelves: 680 mm for small items at heights greater than 600 mm, 1000 mm at heights less than 600 mm or for bulky items at any height.

- Position loads to be lifted (or more specifically the handholds or points at which the load is grasped) at approximately waist height and as close to the body as possible
- Ensure that the floor surface and footwear give sufficient friction to be able to generate the force and to avoid the risk of slipping

4.9 ISSUES FOR BARRIER-FREE WORKSPACE DESIGN

Designing a workspace for the user population involves knowledge of the characteristics of that population and understanding of the impact that the task demands may have on different members of that population. Oliver et al. (2001) conducted a survey of older and disabled people to collect their views on the problems they experienced in everyday activities and on what they would like to be able to do more easily. Examples of some of the difficulties encountered were sitting to do tasks that other people might carry out standing, sliding heavy items along surfaces to avoid lifting them, lack of space to manoeuvre, shelves or apertures in appliances that were too low to reach, and finding surfaces at the wrong height to perform tasks. Most of these relate to the traditional ergonomic concerns of reach, working level, clearance and access, and as such they are soluble. Ambulant people will find fixtures such as light switches, electrical outputs and window catches operable at any height from floor level to somewhere between 1800 and 2000 mm. A wheelchair user (whose upper limbs are unimpaired) can reach a zone from around 600 to 1500 mm in a sideways approach but considerably less head on. It may be that the location of fittings within this limited zone will prove entirely acceptable for ambulant users of a building. However, in the case of working heights such a compromise is not so easy. An extensive collection of recommendations related to these matters is to be found in Goldsmith (2000) and Lacey (2004a).

Some user populations are very diverse, particularly when public facilities or consumer products are concerned. The ultimate goal is barrier-free design in which all potential users are accommodated, but, in reality, there will often be conflicts between different people's needs so that compromises have to be made in the design to satisfy these as well as possible for all users (not just for the majority and not just for the 5th to 95th %ile range). This is a considerable challenge to designers and demands innovation to find design solutions.

A study to improve the accessibility of post office counters by Ellis and Parsons (2000) illustrates some of the issues involved. In the U.K., post offices offer various services to the general population as well as handling the mail. These services include the payment of benefits and pensions, so that post office counters are used by many elderly and disabled customers. The user population also includes teenagers and parents with children. Ellis and Parsons identified the disabilities which will affect the design of a 'point of sale' such as the post office counter as mobility difficulties, visual impairment, hearing impairment, dexterity limitations and learning difficulties. The fact that this will create potentially conflicting constraints on the design is illustrated by the contrast between the counter height that would suit a wheelchair user and one that would suit a tall person who has difficulty bending. Another question that might arise in facilitating wheelchair access is whether this would extend counter depth, which could pose difficulties for those with visual or hearing impairments. Solutions can be found for these conflicting needs. The new counter design included, among many other features (including of course those needed by the staff working at the other side of the counter):

- A deep overhang to facilitate wheelchair access
- An edge profile with recesses which provided shelves at different levels for writing
- A raised profile (or 'lip') at the edge of the shelf to help those with manual dexterity problems pick up stamps and coins
- Radiused edges and contoured undersides of shelves to minimise the risk of children being hurt if they hit their heads on them
- Strong contrast at the edges of the counter to help visually impaired users

This case study shows that designers can provide good solutions which reduce the barriers faced by disabled users, but the design problems are difficult and it has to be recognised that it will often not be possible to meet all needs. However, as Ellis and Parsons' (2000) study shows, considerable improvements can be made that satisfy many disabled users (without disadvantaging nondisabled users).

The following is a brief checklist of some of the people we might need to consider if we are to design for a broad spectrum of users:

- Wheelchair users
- Children: Is the workspace or product usable by children? If so, should it be?

- Left-handed as well as right-handed people
- Anthropometric extremes
  - A woman in the advanced stages of pregnancy
  - An obese person
  - An adult who is outside the arbitrary 5th to 95th %ile range of stature
- People with disabilities
  - Hands that are stiff and painful, weak or with tremor
  - Inability to bend or twist without discomfort or pain
  - Inability to reach far
  - Weakness of leg muscles, making it difficult to raise the body against gravity
  - Impaired vision
  - Impaired hearing
  - Learning difficulties, loss of memory, confusion

Barrier-free workspace design is a challenge to the skills of designers, but there are publications that offer useful guidance on accessibility, such as British standards BS 8300 Access and Facilities for Disabled People (BSI, 2001) and BS 4467 Dimensions in Designing for Elderly People (BSI, 1991), and manuals for architects, "Universal Design" (Goldsmith, 2000) and "Designing for Accessibility" (Lacey 2004a). The World Health Organization's International Classification of Functioning, Disability and Health (WHO, 2001) may also be helpful when considering information related to health conditions and impairments and their consequences for functional abilities. Culver and Viano (1990) provide anthropometric data for pregnant women (within the U.S. population) and estimate the effects on clearance dimensions for seated women at various stages in pregnancy. Yamana et al. (1984) provide similar information for Japanese women. Sources of anthropometric data relevant to elderly or disabled populations include Imrhan (1994), Smith et al. (2000) and Steenbekkers and van Beijsterveldt (1998), although there are still many gaps in the data, particularly in functional measurements. Nowak (1989) has shown how the zone of convenient reach can be calculated for a wheelchair user and applied this to analyse the needs of school and college students between the ages of 15 to 18 years.

5 Sitting and Seating

5.1 FUNDAMENTALS OF SEATING

The purpose of a seat is to provide stable bodily support in a posture that is:

- Comfortable over a period of time
- Physiologically satisfactory
- Appropriate to the task or activity in question

All seats are uncomfortable in the long run, but some seats become uncomfortable more rapidly than others, and in any particular seat some people will be more uncomfortable than others. Comfort may also be influenced by the task or activity that the user is engaged while sitting. In other words, comfort (or more strictly the rate of onset of discomfort) will depend upon the interaction of *seat characteristics*, *user characteristics* and *task characteristics* (the main factors which influence this are shown in Table 5.1).

TABLE 5.1
Determinants of Sitting Comfort and Discomfort

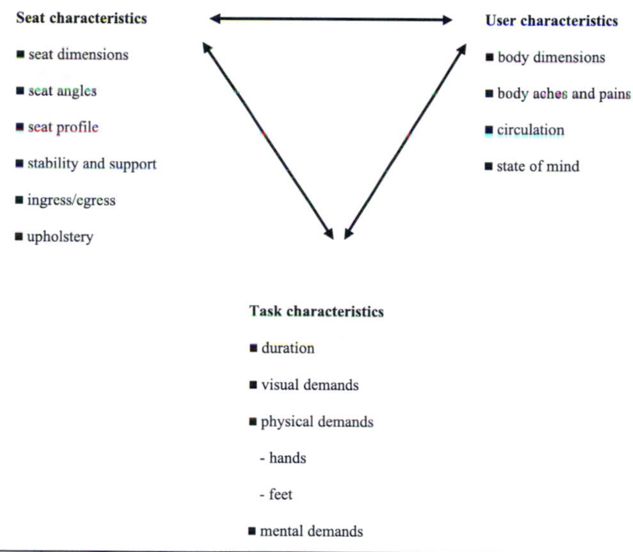

The seat characteristics listed in Table 5.1 influence the posture that will or can be adopted and the areas which provide support for the trunk, shoulders, head and lower body. If the body is not adequately supported, the sitter's posture will only be maintained through muscle effort and, if the seat profile does not match the sitter's body shape and size, additional pressure may be imposed on soft tissues not designed for load bearing. A frequent cause of discomfort is a raised or hard edge at the front of a seat, causing pressure on the underside of the thigh, and this is exacerbated when the seat is even slightly too high for the user. Muscle effort and localised pressure lead in a relatively short time to physiological responses of muscle fatigue, impeded (or even occluded) blood circulation and venous blood pooling resulting in oedema (swelling) at the ankles and feet.

Task characteristics are equally important (Eklund, 1986); the visual and physical demands of a task have a strong influence on the posture which has to be adopted (as shown for example in sewing machine operation, discussed in Section 4.5.2.2), so the task demands influence the seat characteristics which are appropriate to provide support while performing the task. For example, if the sitter needs to perform an intricate assembly task, a reclined backrest will be ineffective and the posture tiring. If force must be exerted during the task, the seat must be stable and provide a surface against which the force can be reacted.

Task duration is a very important factor. The physiological responses already mentioned will increase with time, as will the discomfort perceived. In order to properly assess the comfort (or discomfort) of a seat, it is necessary to sit in it for a period. First impressions of a seat, or showroom appeal, can be very misleading. There are varying opinions about the length of time needed to make a good judgement of a seat's comfort but it probably lies between 5 and 30 minutes. Fernandez and Poonawala (1998) showed that a stable rating of the comfort of a work seat could be obtained after three hours. A comparison between seats (to rank them in order of comfort/discomfort) can probably be made very quickly, within a few minutes (Helander and Zhang, 1997).

However, sitting still for long periods is never healthy. If circulation is impeded and blood flow slowed by sitting in cramped conditions over a long period of time, oedema occurs in the lower legs and there may even be a danger of a venous thrombosis (blood clot) occurring. This has become a worry for long distance air travellers (Hirsch and O'Donnell, 2001), but long bouts of work at a computer or other seated task could have similar effects. Beasley et al. (2003) and Lee (2004) have reported cases of sudden collapse and death, respectively, without previous evidence of disease, which they attribute to extended periods of sitting at a computer (over a period of 3 to 4 days). Such cases appear to be leisure-related rather than work-related disorders but do draw attention to the risks of sitting in immobile postures for prolonged periods. The dangers of sitting in inadequate seats for extended periods of time were, in fact, recognised as long ago as the 1940s when Londoners sat in deck chairs in air-raid shelters during the Second World War (Simpson, 1940). The more recent potential hazard of the computer age, which is becoming called 'eThrombosis', is still being investigated, but, in any case, for sound physiological reasons seat design should avoid constraining posture (as opposed to providing stable support) and wherever possible give freedom for the user to vary his or her posture. We all respond to physiological signals but may not always heed them to the extent of varying posture sufficiently.

Branton (1969) noticed unconscious variation in posture when he observed train passengers on long journeys; they moved through recurring sequences of postures with a period of 10 to 20 minutes, with more systematic behaviour than fidgeting. This sequence started by sliding forward in the seat with a backward slumped posture, resisting this by using the arms as props, then crossing the knees and finally stretching the legs forward to end up in a nearly horizontal posture. As Branton put it, 'The seat slowly and repeatedly ejected the sitter'. In this case, while such posture cycling may have reduced the severity of discomfort and physiological harm, one feels that it is unlikely to have led to comfort.

Branton attributed the observed behaviour to the excessive depth of the seat and the lack of lumbar support, exacerbated by the slipperiness of smooth upholstery (as well of course as by the vibration and sway of the train). It can be seen therefore that, in matching the seat to the user, anthropometric factors are of major importance — but by no means uniquely so. An appropriate match between the dimensions of the seat and those of its users is necessary for comfort, but not sufficient. We shall return to the anthropometric aspects of seating in due course.

In general, a seat that is comfortable in the (relatively) long term will also be physiologically satisfactory. In one sense it is difficult to see how this could not be the case, given that the neural events that tell us that we are 'uncomfortable' may in physiological terms be regarded as warning signs of impending tissue damage. We might suppose therefore that, in the absence of such warnings, no damage is imminent. It may not be as simple as this, however. There are those who believe that extensive covert damage due to 'poor sitting posture' may occur in the absence of subjective discomfort. This is actually a very difficult argument to settle either way. To gain some further insight into these matters, we turn to a consideration of the physiology and biomechanics of the sitting posture, with particular reference to the structure and function of the lumbar spine.

5.2 THE SPINE IN STANDING AND SITTING

The human vertebral column (spine or backbone) consists of 24 movable bony vertebrae separated by deformable hydraulic pads of fibrocartilage known as intervertebral discs. (Up to 10% of people possess a greater or lesser number of vertebrae, but these 'anomalies' seem to have little functional consequence.) The column is surmounted by the skull and rests upon the sacrum, which is firmly bound to the hip bones at the sacro-iliac joints. The vertebrae can be naturally grouped into seven cervical vertebrae (in the neck), twelve thoracic vertebrae (to which the ribs are attached) and five lumbar vertebrae (in the small of the back, between the ribs and the pelvis) (Figure 5.1). The spine is a flexible structure, the configuration of which is controlled by many muscles and ligaments.

In the upright standing position, the well-formed human spine presents a sinuous curve when viewed in profile. The cervical region is concave (to the rear), the thoracic region convex and the lumbar region again concave. A concavity is sometimes known as a '*lordosis*' and a convexity as a '*kyphosis*'. These are enclosed by the convexities

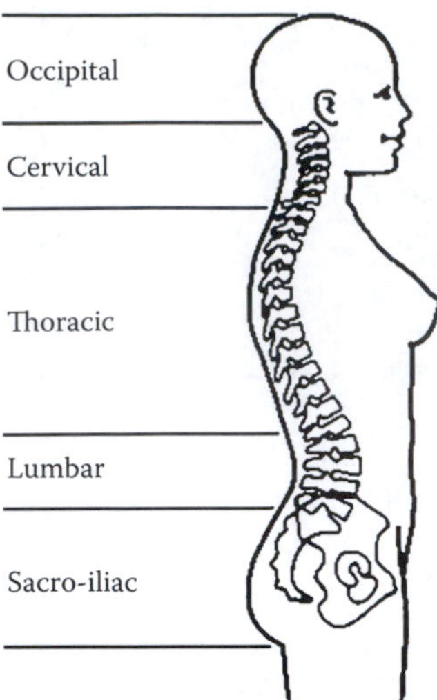

FIGURE 5.1 The well-formed human spine presents a sinuous curve when viewed in profile.

of the occiput (back of the head) above, with the sacro-iliac region and buttocks below, making five curves in all.

In the upright standing position, the pelvis is more or less vertical, and the lowest lumbar vertebra and sacrum make angles of about 30° above and below the horizontal, respectively (at what is known as the L5/S1 intervertebral joint; see Figure 5.2). Consider what happens when you sit down on a relatively high seat (such as a dining-room chair). You flex your knees through 90° and make another 90° angle between your thighs and trunk. Most of your weight is taken by the ischial tuberosities (IT in Figure 5.2), two bony prominences of the pelvis which you can feel within the soft tissue of your buttocks if you sit on your hands. Only part of the right angle between the thighs and trunk is achieved by flexion at the hip joint. After an angle of 60° is reached this movement is opposed, unless we are very flexible, by tension in the hamstring muscles (located in the backs of the thighs). Hence we tend to complete the movement by a backward rotation of the pelvis of 30° or more, as shown on the left-hand side of Figure 5.3.

This backward rotation must be compensated by an equivalent degree of flexion in the lumbar spine if the overall line of the trunk is to remain vertical. Hence in sitting down we tend to flatten out the concavity (lordosis) of the lumbar region. The greatest amount of curvature occurs at the lower intervertebral joints L5/S1 and L4/L5 (Bridger and Bendix, 2004).

In relaxed unsupported sitting, the lumbar spine may well be flexed close to the limit of its range of motion, giving the slumped posture shown on the left of Figure 5.3. In this position, the muscles will be relaxed, because the weight of the trunk will be supported by tension in passive structures such as ligaments. This is achieved, however, at the expense of a considerable degree of deformation of the intervertebral discs, the pads of fibrocartilage, or 'gristle', which separate the bony vertebrae (see Figure 5.4). When the lumbar spine is flexed, the intervertebral disc is compressed

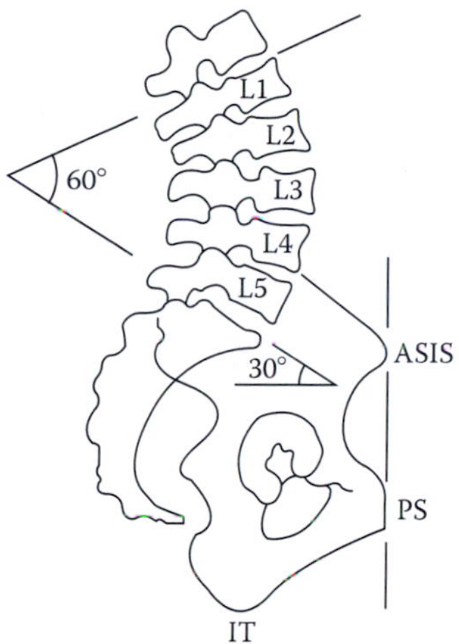

FIGURE 5.2 Typical orientation of the lumbar spine and pelvis in the standing position. (ASIS, anterior superior iliac spine; PS, pubic symphysis; IT, ischial tuberosity) (From Pheasant, S. (1991). *Ergonomics, Work and Health*, London: Macmillan, Figure 5.2, p. 102. Reproduced with kind permission.)

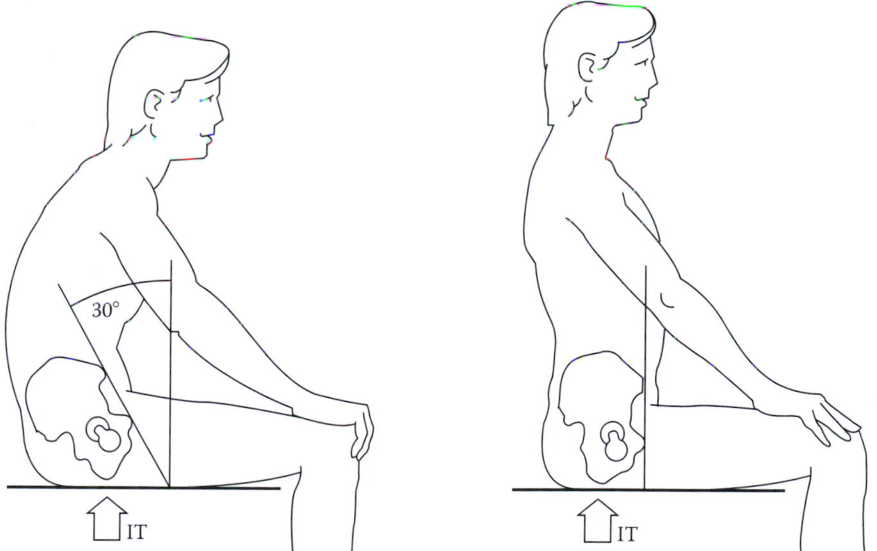

FIGURE 5.3 In relaxed sitting (left) the pelvis rotates backwards and the spine is flexed. To sit up straight (right) requires muscular exertion to pull the pelvis forward. The ischial tuberosities (IT) act as a fulchrum.

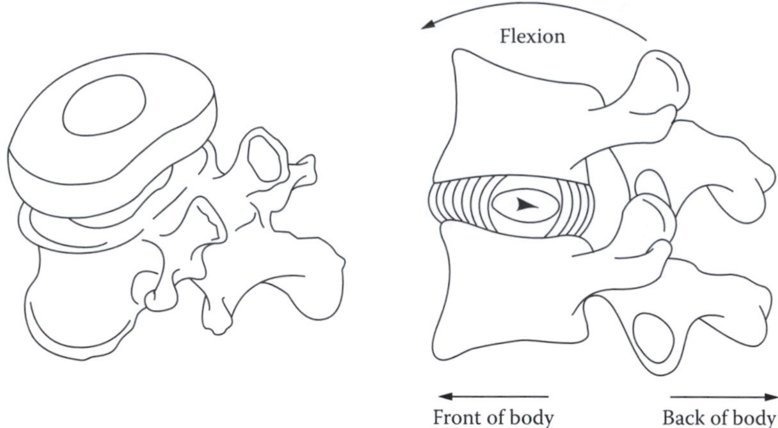

FIGURE 5.4 Structure of the intervertebral joint. Left: lumbar vertebra surmounted by intervertebral disc, showing outer annulus and inner nucleus pulposus. Right: deformation of the disc during flexion of the spine. (Redrawn from Kapandji, I. A. (1974). *The Physiology of the Joints,* Edinburgh: Churchill Livingstone.)

at the front and extended at the back, as shown on the right-hand side of Figure 5.4. This is widely thought to be a bad thing. (The full reasons that this should be so are beyond the scope of the present discussion, although some will become apparent later. Suffice it to say that in the authors' view the reasons are good ones.)

In order to 'sit up straight' and regain our lost lordosis, we must make a muscular effort to overcome the tension in the hamstrings. (The effort probably comes from a muscle deep within the pelvis called the iliopsoas.) We cannot merely relax the hamstrings, since their tension is a passive one, caused by the stretching of tissue (just like an elastic band) rather than by actual muscular contraction. We shall probably also need to activate our back muscles to support the weight of our trunk. If sitting up straight is prolonged, this static muscle loading is tiring and may become a major source of postural discomfort, particularly in someone who has a preexisting tendency to suffer from back trouble.

In designing a seat, therefore, the objective is to support the lumbar spine in its neutral position (i.e., with a modest degree of lordosis) without the need for muscular effort, thus allowing the user to adopt a position that is *both* physiologically satisfactory *and* comfortably relaxed. In general this will be achieved by:

- A semi-reclined sitting position (to the extent that this is permitted by the demands of the working task), so that the hip angle is greater than 90° and some of the upper body weight can be supported by the backrest
- A seat that is neither lower nor deeper than necessary (see below)
- A backrest that makes an obtuse angle to the seat surface (thus minimizing the need for hip flexion) and is contoured to the form of the user's lumbar spine (thus providing support in the lumbar region)

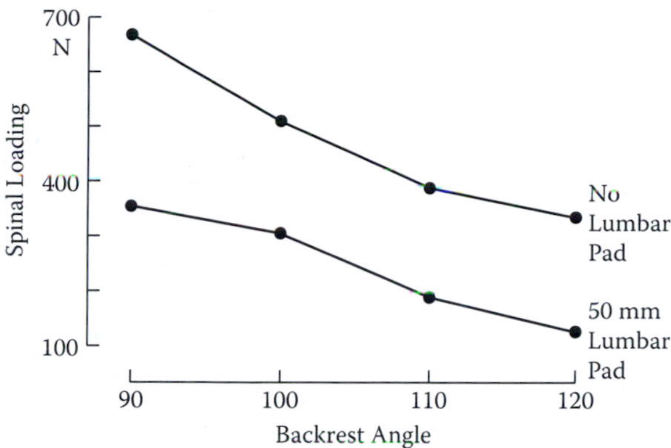

FIGURE 5.5 Spinal compression (calculated from intra-discal pressure which was measured directly by a needle-mounted transducer) at seat back angles from vertical (90°) to reclined (120°), with and without a pad in the lumbar region. (Data from Andersson, G. B. J., Örtengren, R., Nachemson, A. and Elfström, G. (1974). *Scandinavian Journal of Rehabilitation Medicine*, 3, 104–14. From Pheasant, S. (1991). *Ergonomics, Work and Health*, London: Macmillan, Figure 11.3, p. 218. Reproduced with kind permission.)

The extent to which the backrest of the seat supports the weight of the trunk (and thus reduces the mechanical loading on the lumbar spine) is a direct function of its angle of inclination to the vertical. This may be predicted theoretically (as a simple matter of cosines), and it has been confirmed by Andersson et al. (1974) in a series of experimental studies in which the hydrostatic pressure within the nucleus pulposus of the intervertebral disc was measured directly using a needle-mounted pressure transducer. The results show that intervertebral disc pressure (reflecting compressive loading on the spine) is very much higher when sitting upright than when supported in a reclined posture against a backrest (Figure 5.5). Although not shown here, the intra-discal pressure was found to be still higher, by some 35%, when the sitter adopted the slumped posture on the left of Figure 5.3 and did not contact the backrest at all. This shows the importance of maintaining a lordotic lumbar curve. Andersson et al. (1974) also found that, for any given angle of backrest inclination, the intra-discal pressure was measurably less if the backrest was contoured with a pad in the form of the lumbar spine.

Grandjean (1988) reported the results of a series of fitting trials using what he called a 'sitting machine' to establish the best contours for a seat. This was an adjustable test rig, by means of which it was possible to determine the preferred seat profiles of experimental subjects (or more specifically, the profiles that minimised reported aches and pains during sitting). The reported preferences of subjects who suffered from back trouble were much the same as those of people who did not. Figure 5.6 shows the overall preferred profiles (for both groups of subjects) for a 'multi-purpose' chair and an 'easy chair'.

Andersson's pressure measurements and Grandjean's fitting trials confirm that a seat that enables the user to adopt a semi-reclined position and has a backrest that is contoured to the shape of the lumbar spine will *both* minimise the mechanical

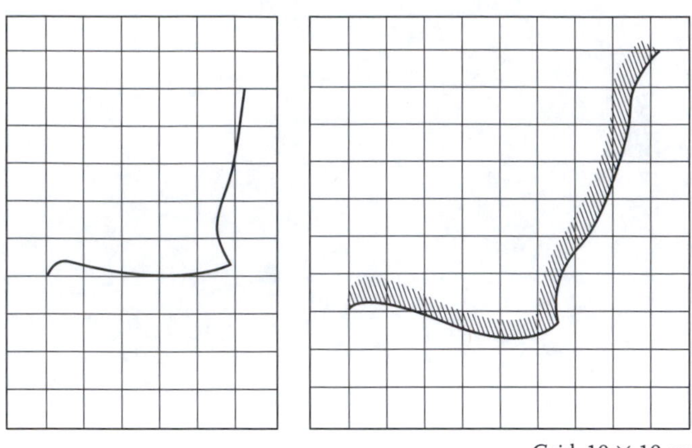

FIGURE 5.6 Seat profiles of a multipurpose chair (left) and an easy chair (right), both of which caused a minimum of subjective complaints. (From Grandjean, E. (1988) *Fitting the Task to the Man*, 4th ed., London: Taylor & Francis, 1988, Figure 52, p. 60. Reproduced with kind permission.)

loading on the lumbar spine *and* maximise the overall levels of reported comfort (both for users who suffer from back trouble and for those who do not).

A problem arises, however, in tasks such as writing, which entail forward leaning and in which the support of the backrest will tend to be lost. The backrest remains important in these activities, however, during rest pauses. Grandjean (1988) describes a study of office workers, using time-lapse photography, which showed them to be in contact with the backrest for 42% of the time.

5.3 FORWARD TILTING SEATS AND 'SIT-STAND' SEATS

A radical new approach to seat design has recently been proposed. Mandal (1976, 1981, 1991) argued that the seat surface should slope forwards, hence diminishing the need for hip flexion (particularly in tasks such as typing and writing) and encouraging lumbar lordosis. With such an 'open' hip angle (at about 135° according to Keegan [1953]), the spine can assume a profile closer to that when standing, as shown in Figure 5.1, and without the need for excessive back muscle tension. The intra-discal pressure is therefore reduced. Nachemson and Elfstrom (1970) found intra-discal pressure in a standing posture to be about 30% lower than when sitting upright with hips flexed at 90°.

A number of seat designs now incorporate a tilt mechanism (see Figure 5.7). The potential disadvantage of such a design is that if you sit on the chair without thinking, you will tend to exert a backward thrust with the feet in order to stay in the seat. This is a particular problem if the chair is on castors. High-friction upholstery is not really an answer since female attire (in particular) generally provides a low-friction interface between the outer and inner garments. Women, therefore, tend to slide out of their skirts. Experience suggests that balancing correctly on the

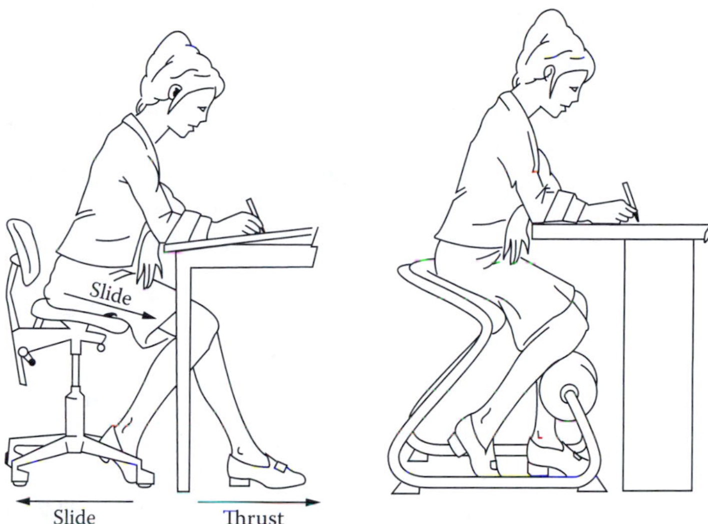

FIGURE 5.7 Two radical approaches to seat design: the forward tilting seat (left) and the kneeling chair (right).

forward slope is a skill that needs to be learned. According to Mandal (1981), users make take 1 to 2 weeks to become used to such chairs.

These difficulties should be overcome with the 'kneeling chair', which provides a seat sloping forward at some 30° to the horizontal, combined with a padded support for the knees. Brunswic (1981) evaluated these concepts. She found that:

- When seat angle and knee angle were independently varied, forward tilted seat positions did not result in a lumbar posture that was significantly different from that obtained with a horizontal seat and the knees at right angles
- The lumbar posture of subjects using a kneeling chair in writing and typing tasks was not significantly better than when they used a conventional office chair

Bendix et al. (1988) also found that, although the lumbar spine was more lordotic than when sitting on a conventional office chair, the lordosis was still less than when standing, and the change could largely be explained by the lack of a backrest.

Drury and Francher (1985) evaluated a kneeling chair by means of a user trial of considerable sophistication. Subjects typed or operated a computer terminal. They concluded that the comfort was 'no better than conventional chairs and could be worse than well-designed office chairs'. The principal complaints were difficulties of access and egress, pressure on the shins, and discomfort in the knee region (due presumably to impairment of circulation resulting from the acute angle). There was 'little or no decrease in back discomfort', and, in spite of training, subjects 'often slumped forward to give a kyphotic spine'. Presumably, they did this to rest their back muscles by 'hanging on their ligaments'.

The subjects in this trial all had 'normal' backs. Atherton et al. (1982) tested a kneeling chair, along with a number of conventional office chairs, on a group of subjects, all of whom had musculoskeletal problems of one sort or another and half of whom had bad backs. The kneeling chair came about half-way down the list in their rank order of preferences.

The kneeling chair has two obvious disadvantages: standing up and sitting down is necessarily difficult, and it fixes the lower limbs with the knees in a position of flexion well past the mid-range. This reduces the possibilities for fidgeting and changes of posture (except by a backwards and forwards rocking of the pelvis). Since the chair (in its basic form) has no back, the loading on the lumbar spine cannot be any lower than it is in upright standing, whereas when a person leans back fully on the backrest of a conventional seat, the loading on the lumbar spine may be very much less than in standing (Andersson et al., 1974; Pheasant, 1991a).

Overall, such scientific studies as have been done on this subject do not suggest that the kneeling chair offers any particular material advantages as compared with a well-designed conventional chair, either with regard to sitting in general or with regard to office use in particular. Having said this, however, one must also add that on the basis of clinical experience it is quite clear that *some* people who suffer from back trouble find the kneeling chair helpful. One's impression is that these people are only a minority of back-pain patients (and possibly a fairly small minority), but experience suggests that those back-pain patients who like the kneeling chair often like it very much indeed. It would be interesting to know why; perhaps these are patients who have an unusually poor tolerance of spinal flexion.

The sloped chair shown on the left of Figure 5.7 was developed further as the 'Nottingham' chair by Eklund, Corlett and Gregg (Eklund, 1986; Corlett and Gregg, 1994; Corlett, 1999). They showed that the tendency to slide forward on a sloped seat can be reduced by contouring of the seat surface and counteracted without discomfort — provided that the load on the feet does not exceed one third of average body weight. (Even on a conventional seat, part of the body weight is carried by the feet.) The function of the contour on the seat surface is to have a flat surface under the ischial tuberosities (to support the body weight) and a sloped surface under the thighs so that the hip angle is not flexed. The sitter is still able to recline against a backrest. The angle of slope of the seat can be varied (up to the critical point at which foot loading becomes uncomfortable). The seat can therefore be used in its highest position as a '*sit-stand seat*'.

Sit-stand seats are useful for work at high work surfaces and to facilitate seated work with a reasonable reach distance in a situation where knee room is restricted. Conversely, a high sit-stand seat will permit longer reach distances than an ordinary seat because the legs can be moved to give a wider foot base, and thus greater stability, whilst the sitter is still fully supported by the seat. This can be useful for tasks such as those in a mail sorting office or at a supermarket checkout. However, these are only possible when the sit-stand seat has been designed to fully support the sitter in the semi-standing posture. This means that it must have proper foot support, maintain a good spinal posture, avoid muscular effort in maintaining the posture, and not exert under-thigh pressure. Stools or 'perch' seats, while useful for a momentary rest during standing work, do not provide adequate support for performing work tasks.

5.4 ANTHROPOMETRIC ASPECTS OF SEAT DESIGN

The design factors influencing the major dimensions of a seat (shown in Figure 5.8) are discussed in the following sections.

5.4.1 Seat Height (H)

As the height of the seat increases beyond the popliteal height of the user, pressure will be felt on the underside of the thighs. The resulting reduction of circulation to the lower extremities may lead to 'pins and needles', swollen feet and considerable discomfort. As the height decreases the user will (a) tend to flex the spine more (due to the need to achieve an acute angle between thigh and trunk); (b) experience greater problems in standing up and sitting down, due to the distance through which his or her centre of gravity must move; and (c) require greater leg room. In general, therefore, the optimum seat height for many purposes is close to the popliteal height, and where this cannot be achieved a seat that is too low is preferable to one that is too high. For many purposes, therefore, the 5th %ile female popliteal height (380 to 400 mm depending on style of shoes worn) represents the best compromise. If it is necessary to make a seat higher than this (e.g., to match a desk height or because of limited leg room), the ill effects may be mitigated by shortening the seat and rounding off its front edge in order to minimise the under-thigh pressure. It is of overriding importance that the height of a seat should be appropriate to that of its associated desk or table.

5.4.2 Seat Depth (D)

If the depth is increased beyond the buttock–popliteal length (which for a 5th %ile woman is 435 mm), the user will not be able to engage the backrest effectively without unacceptable pressure on the backs of the knees or leaning back without proper lower back support. Furthermore, the deeper the seat, the greater the problems

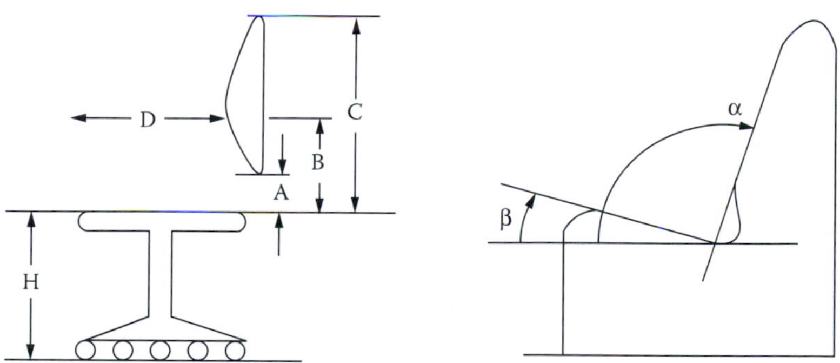

FIGURE 5.8 Seat dimensions.

of standing up and sitting down. The lower limit of the seat depth is less easy to define. As little as 300 mm will still support the ischial tuberosities and may well be satisfactory in some circumstances. Tall people sometimes complain that the seats of easy chairs are too short; an inadequate backrest may well be to blame (see below).

5.4.3 Seat Width

For the purposes of support, a width that is some 25 mm less on either side than the maximum breadth of the hips is all that is required — hence 385 mm will be adequate. However, if there are armrests or sides to the seat, the clearance between these must be adequate for the largest user. The hip breadth of the 95th %ile woman unclothed is 435 mm. In practice, allowing for clothing and leeway, a minimum clearance of 500 mm is required.

5.4.4 Backrest Dimensions

The higher the backrest, the more effective it will be in supporting the weight of the trunk. This is always desirable, but in some circumstances other requirements such as the mobility of the shoulders may be more important. We may distinguish three varieties of backrest, each of which may be appropriate under certain circumstances: the low-level backrest, the medium-level backrest and the high-level backrest.

The *low-level* backrest provides support for the lumbar and lower thoracic region only and finishes below the level of the shoulder blades, thus allowing freedom of movement for the shoulders and arms. Old-fashioned typists' chairs generally had low-level backrests, as do many general purpose stacking chairs. To support the lower back and leave the shoulder regions free, an overall backrest height (C) of about 400 mm is required (remembering that this should be measured from the compressed seat surface).

The *medium-level* backrest also supports the upper back and shoulder regions. Most modern office chairs fall into this category, as do many 'occasional' chairs, auditorium seats, etc. For support to mid-thoracic level, an overall backrest height (C) of about 500 mm is required, and for full shoulder support, about 650 mm (95th %ile male values rounded up). A figure of 500 mm is often quoted for office chairs (see Chapter 7).

The *high-level* backrest gives full head and neck support. For the 95th %ile man, an overall backrest height (C) of 900 mm is required.

Whatever its height, it will generably be preferable and sometimes essential for the backrest to be contoured to the shape of the spine, and in particular to give 'positive' support to the lumbar region in the form of a convexity or pad (see above). To achieve this end, the backrest should support you in the same place as you would support yourself with your hands to ease an aching back. Andersson et al. (1974) found that a lumbar pad that protrudes 40 mm from the main plane of the backrest at its maximum point will support the back in a position that approximates that of normal standing. However, the variability between individuals is considerable — in both depth and height of the spinal curve (Lueder et al., 1994). Coleman et al. (1998) studied preferences for the position of lumbar support on office chairs and found that the mean preferred height was 190 mm above the compressed seat surface, and the mean preferred depth from the front of the seat was 387 mm. However, individuals varied greatly in their preferences. There was a relationship between preferred height and body mass index (BMI), with higher lumbar supports being chosen by users with greater BMIs. They also found that older people were more sensitive to lumbar support position and more likely to adjust it to suit themselves. The positioning of the backrest is therefore very important, so adjustment should be provided for the backrest to suit all individuals. Coleman et al. (1998) recommend a range of adjustment between 150 and 250 mm from the compressed seat surface.

For work chairs, an adjustable backrest is not only desirable but in some contexts essential. Some typical recommendations are summarised in Table 5.2.

To use the lumbar support to its full advantage, it is necessary to provide clearance for the buttocks, so in some kinds of chair (including work chairs) it may be appropriate to leave a gap between the seat surface and the bottom edge of the backrest. Similarly, in high backrests, it is important to leave free shoulder space. If the contour of the backrest is too far forward in the scapular region, it is no longer possible to gain advantage from support in the lumbar region (Goossens et al., 2003). Experience suggests that this also applies to backrests that are as high as the neck or head of the sitter. Goossens et al. recommend that scapular support should be a minimum of 6 cm to the rear of the lumbar support.

A medium- or high-level backrest should be flat or slightly concave above the level of the lumbar pad, but the contouring of the backrest should in no cases be excessive. A curve that is too pronounced is probably worse than no curve at all.

TABLE 5.2
Typical Recommendations Concerning Backrest Dimensions of Work Chairs

	EN 1335-1 (CEN, 2000) Office Work Chair	HSE (1997)	Coleman et al. (1998) Office Chair
Seat surface (compressed) to bottom of backrest (A)		100–200	
Seat surface (compressed) to foremost point of backrest (B)			
Fixed height backrest:			
Within the height range	170–220		
Adjustable height backrest:			
Range provided	170–220 (minimum)	170–300	150–250 (recommended)
Range of adjustment	50		100
Vertical height of backrest (C-A)		200–550	
Fixed height backrest	260		
Adjustable height backrest	220		

Note: All dimensions in mm.

5.4.5 Backrest Angle or 'Rake' (α)

As the backrest angle increases, a greater proportion of the weight of the trunk is supported. Hence the compressive force between the trunk and pelvis is diminished (and with it the discal pressure). Furthermore, increasing the angle between trunk and thighs improves lordosis. However, the horizontal component of the compressive force increases with the backrest angle. This will tend to drive the buttocks forward out of the seat unless counteracted by (a) an adequate seat tilt; (b) high-friction upholstery; or (c) muscular effort from the sitter. Increased rake also leads to increased difficulty in standing up and sitting down, particularly for the elderly.

The interaction of these factors, together with a consideration of task demands, will determine the optimal rake; this will commonly be between 100° and 110°. A pronounced rake (e.g., greater than 110°) is not compatible with a low- or medium-level backrest, since the upper parts of the body become highly unstable.

5.4.6 Seat Angle or 'Tilt' (β)

A positive seat angle (backwards tilt) helps the user maintain good contact with the backrest and helps counteract any tendency to slide out of the seat. Excessive tilt reduces hip–trunk angle and ease of standing up and sitting down. For most purposes 5 to 10° is a suitable compromise (see also Chapter 7). The alternative forward sloping seat design has been discussed in Section 5.3.

5.4.7 Armrests

Armrests may give additional postural support and be an aid to standing up and sitting down. The latter is particularly important for elderly people and for women in the later stages of pregnancy. An observational study showed that, by using armrests, pregnant women can maintain a much more stable and less forward leaning posture when sitting down or rising from a chair (Hirao and Kajiyama, 1994). Interviews confirmed that they worried about stability and felt safer when the chair had armrests and also that they had particular difficulty in leaning forward (and thus rising from chairs) in the third trimester of pregnancy.

An elbow rest 200 to 250 mm above the seat surface is generally considered suitable. European standard EN 1335-1 recommends this as the minimum range when armrests are height adjustable (CEN, 2000). An elbow rest that is somewhat lower than sitting elbow height is probably preferable to one that is higher, if a relaxed posture is to be achieved.

Armrests should support the fleshy part of the forearm, but unless very well padded, they should not engage the bony parts of the elbow where the highly sensitive ulnar nerve is near the surface; a gap of perhaps 100 mm between the armrest and the seat back may, therefore, be desirable. European standard EN 1335-1 recommends a minimum 'useful' length of 200 mm for armrests (CEN, 2000). If the chair is to be used with a table, the armrest should not restrict access under the table (which would lead to an extended reach distance to work on the table). The armrest should not, in these circumstances, extend more than 350 mm in front of the seat back.

5.4.8 Leg Room

In a variety of sitting workstations the provision of adequate lateral, vertical and forward leg room is essential if the user is to adopt a satisfactory posture.

Lateral leg room (e.g., the kneehole of a desk) must give clearance for the thighs and knees. In a relaxed position the legs are somewhat separated. ISO 14738 recommends a width of 790 mm (ISO, 2002b).

Vertical leg room requirements will, in some circumstances, be determined by the knee height of a tall user (95th %ile shod man = 620 mm). Alternatively, thigh clearance above the highest seat position may be more relevant (95th %ile man = 185 mm). The vertical space required — adding the 95th %ile male popliteal height shod and thigh thickness — is at least 700 mm.

Forward leg room is rather more difficult to calculate. At knee level, clearance is determined by buttock–knee length from the back of a fixed seat (95th %ile man = 645 mm). If the seat is movable, we may suppose that the user's abdomen will be in contact with the table's edge (although, in practice, most people will choose to sit further back than this). In this case clearance is determined by buttock–knee length minus abdominal depth, which will be around 425 mm for a male who is a 95th %ile in the former and a 5th %ile in the latter. At floor level, at least an additional 150 mm clearance for the feet is needed, which gives a figure of 795 mm from the seat back or 575 mm from the table's edge. All of these figures are based on a 95th %ile male sitting on a seat that is adjusted to approximately his own popliteal height, with his lower legs vertical. If the seat height is lower than this, he will certainly wish to stretch his legs further forward. A rigorous calculation of the 95th %ile clearance requirements in these circumstances would be complex, but an approximate value may be derived as follows.

Consider a person of buttock–popliteal length b, popliteal height p and foot length f sitting on a seat of height H (as shown in Figure 5.9). He stretches out his legs so that his popliteal region is level with the seat surface (i.e., his thighs are approximately horizontal). Ignoring the effects of ankle flexion, the total horizontal distance between buttocks and toes (d) is approximated by

$$d = b + \sqrt{p^2 - \mathrm{H}^2} + f \tag{5.1}$$

Hence, in the extreme case of a man who is a 95th %ile in the above dimensions, sitting on a seat that is 400 mm in height requires a total floor level clearance of around 1190 mm from the seat back or 970 mm from the table edge (if he is also a 5th %ile in abdominal depth). Such a figure is needlessly generous for most purposes; most ergonomics sources quote a minimum clearance value of between 600 mm and 700 mm from the table edge. ISO 14738, however, recommends leg space of 547 mm at knee height and 882 mm at foot level (ISO, 2002b).

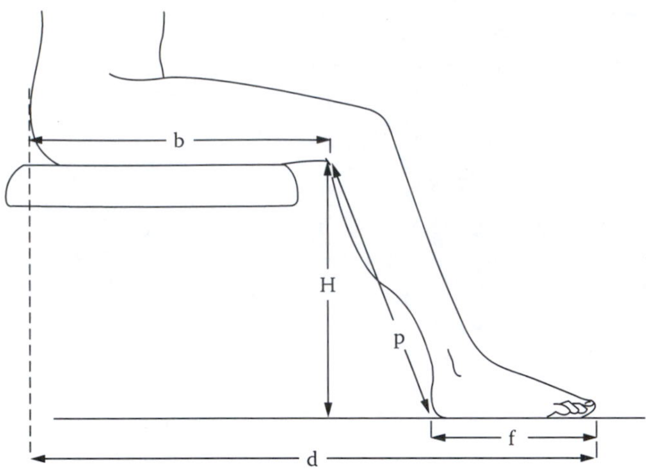

FIGURE 5.9 Calculation of forward leg room.

5.4.9 Seat Surface

The purpose of shaping or padding the seat surface is to provide an appropriate distribution of pressure beneath the buttocks. The consensus of ergonomic opinion suggests the following:

- When compressed, most of the seat surface should be more or less flat rather than shaped, although a rounded front edge is highly desirable (sometimes called a 'waterfall' contour)
- Upholstery should be firm rather than soft (it is sometimes said that a heavy user should not deform it by more than 25 mm)
- Covering materials should be porous for ventilation and rough to aid stability

The traditional Windsor chair can be surprisingly comfortable in spite of its total absence of upholstery. Its basic form was probably developed by the craftsmen of the Chiltern beechwoods sometime around the beginning of the eighteenth century. A critical feature seems to be the subtle contouring of the seat known as its 'bottoming'. This was hand carved, using first an adze, then a variety of shapers, by a man known as the 'bottomer', whose specialised trade was considered the most skilled of all the activities that contributed to the chair-making process. He worked by eye without recourse to measurements; contemporary machine-made versions are said to be less satisfactory.

5.4.10 Seats for More than One

When considering benches and other seats in which users sit in a row, it is necessary to bear in mind that the breadth of a 95th %ile couple is less than twice that of a 95th %ile individual. (The chance of two people, each 95th %ile or more, meeting

TABLE 5.3
Sitting in a Row

Number of Persons	Width Required (mm) Mean	SD	95th %ile
1	480	28	526
2	960	40	1026
3	1440	48	1519
4	1920	56	2012

at random on a bench is only 1 in 400.) In general, *n* people sitting in a row have a mean breadth of *nm* and a standard deviation of *sn*, where *m* and *s* are the parameters of the relevant body breadth, which will usually be that of the shoulders. Table 5.3 gives values based on male data and including a clothing correction of 15 mm. Rear seats in cars may have a little less than these widths if it is assumed that the outer passengers will be happy to turn slightly sideways to recline into the corner of the seat. Bench seats in public transport, on the other hand, should be a little wider to allow the travellers some personal space.

However, if the row of seats is divided by armrests, the problem is more complex. Assume each user sits in the centre of the seating unit. A little reflection will tell us that the minimum separation of seat centres will be determined by the distribution of pairs of half-shoulder breadths: mean 480 [SD 40] mm; 95th %ile = 545 mm, taking the values from Table 5.3. Since, in the presence of armrests, the minimum seat breadth is 500 mm (see earlier discussion of seat width in Section 5.4.3) and an armrest cannot reasonably be less than 100 mm wide, 600 mm between seat centres will satisfy all criteria.

5.5 EVALUATING A SEAT

From the foregoing analysis, it is clear that anthropometric considerations are important but are far from the only criteria against which a work or leisure seat should be judged. Different tasks call for different seats, and sitting has both physiological and biomechanical effects. The adequacy of seating is influenced by all of these (Figure 5.1).

The range of assessment techniques which can be used to evaluate seating is discussed thoroughly by Corlett (2005). Techniques include empirical studies (such as fitting trials and observations of actions and behaviour), biomechanical measurements and analysis (including measurements of force, pressure and spinal shrinkage), electromyography and subjective judgements using various psychophysical techniques. Discomfort in different regions of the body can be assessed by using a body map to locate the site (or sites) of the discomfort and then a pain scale to record the severity (Corlett and Bishop, 1976; Borg, et al., 1981). Comfort can similarly be measured by the use of subjective rating scales. Shackel et al. (1969) developed a General Comfort Rating specifically for the assessment of seating. This has most usefully been employed in conjunction with Drury and Coury's (1982) Chair Feature Checklist (a set of nine rating scales) which identifies the seat parameters which contribute to any discomfort experienced by the sitter.

Physiologically, comfort is the absence of discomfort. There do not appear to be any nerve endings capable of transmitting a positive sensation of comfort from a chair. Comfort is a state of mind which results from the absence of unpleasant bodily sensations, although it can also be affected by other factors. Recent studies have shown that perceptions of comfort and discomfort of seating are in fact separate entities and are influenced by different factors; they are not, as commonly assumed, opposites on the same scale (Zhang et al., 1996; Helander and Zhang, 1997). Discomfort is associated with biomechanical factors and fatigue, manifesting itself (in the context of seating and depending on its source) in reports of low back pain, excessive pressure, oedema in the lower leg, or neck and shoulder pain and in feelings of stiffness, numbness, soreness or pain. All these effects tend to increase over time spent sitting. Comfort, by contrast, appears to be related more to aesthetics, to feelings of relaxation and well-being, and to a neutral feeling or absence of discomfort. Comfort–discomfort rating scales cannot therefore any longer be treated as unidimensional. The outcome of Helander and Zhang's (1997) study is their Chair Evaluation Checklist, which can be used to measure comfort and discomfort independently. Their field study indicated that discomfort is primarily determined by sitting (and increases with the duration) but that the features of the seat have little influence (unless the seat is very poorly designed).

5.6 DYNAMIC SEATING

Some modern office chairs incorporate a rocking mechanism in the seat such that it may be tilted forwards and backwards. Bendix and Biering-Sørensen (1983) report a trial in which the subjects preferred a seat that was free to tilt between an angle of 5° forwards and 5° backwards compared with seats fixed in either position. Experience indicates, however, that many users actively dislike tilting seats; it is important that the user should be able to lock the tilt mechanism in place if he or she wishes.

One intended purpose of a tiltable or rocking seat is to help trunk movements, as when reaching forward, but it might be thought that this would require continuous muscular effort for balancing. Bendix et al. (1985, 1988), however, did not find any difference in lumbar muscle activity from that when using a more conventional seat — nor did they find that trunk flexion differed. Another aim of these chairs is to avoid postural fixity when sitting for long periods and to encourage leg movements, which will stimulate blood circulation. Stranden (2000) tested a chair with a 'free floating tilt' mechanism in which the seat and backrest moved together, raising the front of the seat when the user leaned against the backrest. He found that it did stimulate venous pumping through movements of the leg and so counteracted formation of oedema.

Other types of 'multidynamic' chairs are available, but there have been few evaluations of their comfort or effectiveness. In some, the seat back is linked to the seat pan so that it moves in a fixed ratio (typically 2°:1° or 3°:1°) to the seat tilt, but a fixed ratio does not reflect people's postural preferences, which Dainoff (1994) found to vary in a nonlinear relationship with tilt angle. The tilt can be provided in a variety of ways, with pivot points being located below the hip joint, just behind the knees, or at a virtual point near the ankles. The location of the pivot point raises some complex anthropometric questions, since some geometries are more effective than others in maintaining the seat pan and seat back contact interfaces in the right positions relative to the sitter's body size and spinal curvature throughout the 'recline path' (Grant and Goldberg, 1994). A related design parameter is the force required to tilt the seat, overcoming the tension set for the seat; this may vary across the range of seat tilt angles if the mechanical advantage changes with the tilt or with the sitter's posture.

Another innovation is 'continuous passive movement' lumbar support, in which this portion of the backrest inflates and deflates over a slow cycle (Reinecke et al., 1994). The objective is to promote spinal movement and nutrition of the intervertebral discs. Continuous passive movement has also been introduced for the whole seat, with a small ±0.6° swivelling motion at a frequency of 0.08 Hz, again in an attempt to both reduce static loading on the discs in the spine and improve venous blood circulation (van Deursen et al., 2000).

Studies carried out on dynamic seating show that many, very different postures can be adopted for seated work and, indeed, be comfortable and healthy. The traditional, nearly upright seat need no longer be chosen as the norm. However, Dainoff and Balliet (1991) make the point that new designs of chairs may take time to be accepted. Many people may be unwilling initially to explore the new postures with unfamiliar points of balance, and the active behaviour required of the sitter may be at the expense of relaxation. It also has to be remembered that dynamic chairs will be unsuitable for certain types of tasks, particularly if force has to be exerted.

5.7 THE EASY CHAIR AND ITS RELATIVES

The foregoing discussion has applied largely to work seats, and some further design considerations for office work are addressed in Chapter 7. An easy chair may have very different design criteria, since, in ergonomics terms, it should be designed for very different tasks (see Table 5.1). The function of an easy chair is to support the body during periods of rest and relaxation. If not actually dozing or engaged in peaceful contemplation, the user may be reading, watching television or in conversation. The form of the chair follows naturally from these functions and from the considerations of Sections 5.2 and 5.4.

Grandjean (1973) recommends a seat tilt (β) of 20 to 26° for an easy chair and an angle between seat and backrest of 105 to 110°. This gives a backrest rake (α) of as much as 136°, which is really only suitable for resting and requires a degree of agility for standing up and sitting down. Le Carpentier (1969) found a tilt of 10° with a rake of 120° to be suitable for both reading and watching television. The latter recommendation is probably better, with the caveat that both tilt and rake should be much less for people, such as the elderly, who have difficulty in getting in or out of chairs. For elderly users, a rake of more than 110° may cause problems.

Holden et al. (1988) recommend particular chair and wheelchair features for elderly people with different levels of mobility, including a chair with contoured support for those with poor balance and trunk strength. It is also worth remembering that the difficulties of standing up and sitting down will be reduced if the space beneath the front of the chair is unimpeded, allowing the user to place his or her feet beneath the body's centre of gravity, hence achieving a more vigorous upward thrust and a more controlled descent.

A high-level backrest is virtually essential to the proper role of an easy chair in providing support for the trunk. Its shaping is something of a challenge (as has already been alluded to in relation to work seats). It is possible to design a gentle lumbar curve that will suit most users, but an equivalent pad for the neck and occiput (back of head) is more problematic. Ideally, this should give you similar support to the natural action of clasping your hands behind your head. A sensible way of achieving this is to incline the upper part of the backrest forwards from the main rake by around 10° and to provide a movable cushion. (This solution has been adopted on certain train seats, but, unfortunately, the range of adjustment of the cushion often seems not quite adequate for the shorter person.)

The fundamental problems of designing an easy chair had essentially been solved by around 1680, as the collection of almost any English country house will testify. Ergonomics research has merely confirmed the intuitions of the designers of the past. However, the present-day furniture showroom typically presents chairs in a range of styles that, in ergonomic terms, are rarely better than just adequate and not infrequently fall short on numerous criteria. There are, of course, exceptions, but these are commonly either reworkings of traditional types (such as the ever popular 'William and Mary' shown in Figure 1.2) or else chairs that are described as 'orthopaedic' and sold more as 'aids' than as the furnishings of a stylish home.

The most common failings in the contemporary armchair are a seat that is too deep and a backrest that is too low (and frequently lacks lumbar support). One may suppose that this is due to an attempt to make the seat and back equal in length in the interests of visual symmetry (like the Mies Van der Rohe 'Barcelona' chair of 1929 (Figure 1.2) or to an even more misguided attempt to fit the entire chair into a cubic outline (like Frank Lloyd Wright's 'Cube' chair of 1985 or 'Le Grand Confort' by Le Corbusier and Charlotte Perriaud of 1928–1929). Combined with the weighty stylistic influence of these modern masters is a marketing need to incorporate the armchair into a three-piece suite (or some other combination). With the exception of a few historical types, such as the William and Mary love-seat, high-backed settees are virtually unknown. In reality, as anthropometric data quite clearly show, the backrest height needs to be around twice the seat depth if an easy chair is to perform its proper function.

Tall people sometimes complain of seats being insufficiently deep (i.e., too short from front to back). Observation suggests that, on engaging the backrest and finding that it only reaches mid-shoulder level, they move down into the seat in an attempt to gain head support. As a result their buttocks slide forwards until they are in danger of dropping off the front of the seat. (This also leads to the flexed position which is physiologically least satisfactory.) Hence, for an easy chair, the problem stems from an inadequate backrest rather than a seat that is not deep enough.

A common misconception, held by designers and consumers alike, is to equate depth and softness of upholstery with comfort. The luxurious sensation of sinking into a deep over-stuffed sofa is indicative of an absence of the support necessary for long-term comfort in the sitting position. In functional terms, we are now dealing with something more amorphous than a seat per se; it is in fact an object for sprawling or reclining on, rather than for conventionally sitting on. Structurally, however, the object retains the form of the seat. A seat supports its user in a sitting position and a bed supports him or her in a recumbent position, but there are a whole variety of intermediate sprawling postures which can be perfectly satisfactory, especially when, supported by mounds of cushions, one has the opportunity for frequent posture changes. Taken to its logical conclusion, the concept of 'amorphous furniture', which does not dictate any posture in particular, leads to items such as the 'sag bag' — a sack full of polystyrene beads, which enjoyed a brief vogue among young homemakers in the 1970s. A whole family of all but extinct furniture types, which generically we could call couches, are essentially designed for sprawling. Notable members of this family are the 'day bed' mentioned in Shakespeare (*Twelfth Night*, II.v) and the chaise-longue. A steeply raked easy chair can double as a couch when used in conjunction with a footstool — as in the ergonomically excellent Charles Eames lounge chair and ottoman of 1956 (Figure 5.10). The three-piece suite aims to serve for both sitting and sprawling. It commonly does both tolerably but excels at neither. There is considerable scope for design innovation in changing this state of affairs.

FIGURE 5.10 The Charles Eames lounge chair and ottoman (1956) give good support in a wide variety of postures.

6 Hands and Handles

6.1 INTRODUCTION

The previous chapters have looked at the design of workplaces and at how best their design can be optimised to fit the variation in anthropometry among their user populations. It is time to consider in more detail the tasks that are performed in those workplaces and the tools and equipment used. One of our distinctive features as humans, compared to most other species, is how we have evolved to use our hands. Most physical aspects of work involve the dexterity, precision, psychomotor control or strength of hand movements. We first, therefore, need to consider these anthropometric characteristics and then to understand the related parameters which influence the effectiveness, efficiency, comfort and health of work performed by the hands, particularly during the use of tools.

What do we mean by a handle? In the most general terms, we could regard a handle as any part of an object that is held by any part of the hand. The adequacy of a handhold can be judged by much the same criteria as, say, a suitcase handle, a control knob or the handle of a hand tool. Too often, such mundane design details are overlooked. Interestingly, when Helander and Furtado (1992) compared the assembly of manufactured components by robots and by human operators, they found that humans could outperform the robots on certain tasks *when* the components had been designed for robot assembly. In other words, the components' designers had neglected to consider ergonomic aspects of handling and assembling these until they discovered that robots were not tolerant of inadequate design.

6.2 ANTHROPOMETRY OF THE HAND

Table 6.1 gives anthropometric data for the adult hand, gathered from a number of sources. It may be assumed that these figures are for a population of British adults equivalent to that of the 'standard reference population' as described in Table 2.5. The dimensions are illustrated in Figure 6.1. Hand length and hand breadth (dimensions 1 and 12) are from Table 2.5; dimensions 2 to 11, 13 and 15 are from Kember et al. (1981); dimensions 16, 17 and 19 are from Gooderson et al. (1982); dimension 20 is from Davies et al. (1980) for women and estimated by scaling for men; dimension 18 is scaled down from Garrett (1971). More detailed dimensions of finger segments and various hand spans can be found for adults in ADULTDATA (Peebles and Norris, 1998), while Porter (2000) gives finger lengths and diameters of British children between the ages of 6 months and 7 years.

The fingers (or digits) of the hand are referred to by name: starting with the thumb, then the index, middle (long), ring and little fingers. The ranges of movement at the complex joints in the fingers and at the wrist vary greatly between individuals

TABLE 6.1
Anthropometric Estimates for the Hand (all dimensions in millimetres)

	Men				Women			
Dimension	5th %ile	50th %ile	95th %ile	SD	5th %ile	50th %ile	95th %ile	SD
1. Hand length	173	189	205	10	159	174	189	9
2. Palm length	98	107	116	6	89	97	105	5
3. Thumb length	44	51	58	4	40	47	53	4
4. Index finger length	64	72	79	5	60	67	74	4
5. Middle finger length	76	83	90	5	69	77	84	5
6. Ring finger length	65	72	80	4	59	66	73	4
7. Little finger length	48	55	63	4	43	50	57	4
8. Thumb breadth (IPJ)[a]	20	23	26	2	17	19	21	2
9. Thumb thickness (IPJ)	19	22	24	2	15	18	20	2
10. Index finger breadth (PIPJ)[b]	19	21	23	1	16	18	20	1
11. Index finger thickness (PIPJ)	17	19	21	1	14	16	18	1
12. Hand breadth (metacarpal)	78	87	95	5	69	76	83	4
13. Hand breadth (across thumb)	97	105	114	5	84	92	99	5
14. Hand breadth (minimum)[c]	71	81	91	6	63	71	79	5
15. Hand thickness (metacarpal)	27	33	38	3	24	28	33	3
16. Hand thickness (including thumb)	44	51	58	4	40	45	50	3
17. Maximum grip diameter[d]	45	52	59	4	43	48	53	3
18. Maximum spread	178	206	234	17	165	190	215	15
19. Maximum functional spread[e]	122	142	162	12	109	127	145	11
20. Minimum square access[f]	57	67	77	6	51	59	66	5

[a] IPJ is the interphalangeal joint, i.e., the articulation between the two segments of the thumb.
[b] PIPJ is the proximal interphalangeal joint, i.e., the finger articulation nearest to the hand.
[c] As for dimension 12, except that the palm is contracted to make it as narrow as possible.
[d] Measured by sliding the hand down a graduated cone until the thumb and middle fingers only touch.
[e] Measured by gripping a flat wooden wedge with the tip end segments of the thumb and ring fingers.
[f] The side of the smallest equal-sided aperture through which the hand will pass.

and generally reduce with age and onset of arthritic disease. The average range of movement at each joint and functional ranges of movement used in a variety of activities of daily living can be found in Norkin and White (1995). Brumfield and Champoux (1984) found that everyday activities such as eating, reading and making a telephone call can be accomplished with motions of the wrist between 5° flexion and 35° extension, while for personal care activities (washing, dressing, etc.) the range is 10° flexion to 15° extension.

With the complexity and variety of hand actions that are possible, many functional dimensions of hand and fingers may be needed for the design of tools, controls and other objects to be handled (including components being assembled in manufacturing production). Information on such functional dimensions is relatively scarce, and it may sometimes be necessary to conduct a small-scale survey to collect the data directly (see section 2.7.2). One helpful source of anthropometric data for thumb

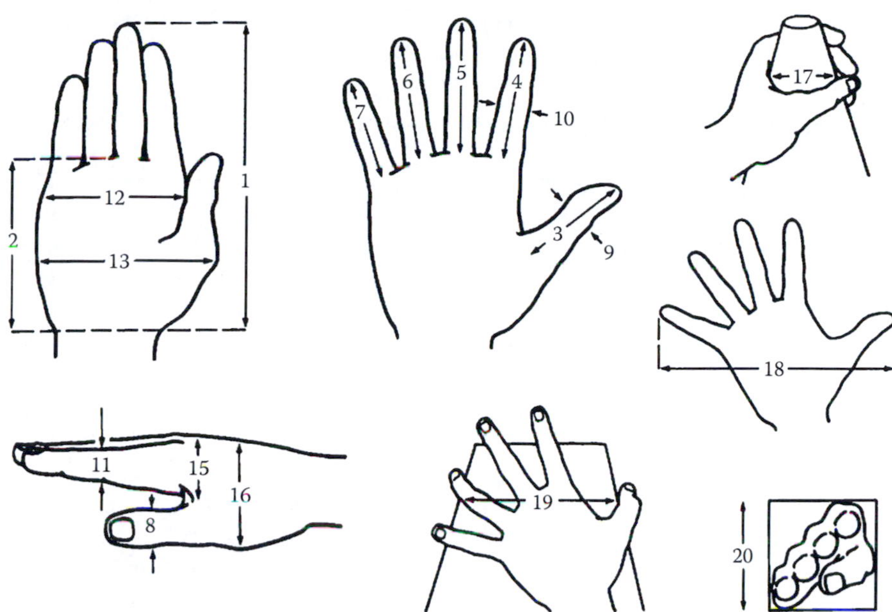

FIGURE 6.1 Anthropometry of the hand, as given in Table 6.1.

motions is Gilbert et al. (1988). However, thumb anthropometry may have to be reviewed in the near future when teenagers with years of practice in mobile phone text messaging enter the adult population.

6.3 HAND DOMINANCE (HANDEDNESS)

For the majority of people, their right hand is dominant and they carry out most tasks with this hand. A significant minority (approximately 10%) have their left hand dominant, although this number probably underestimates the true percentage of left-handed people. Historically, during certain social periods and in certain cultures (and still today), children have been strongly discouraged from using their left hands. Even individuals with strong hand dominance will perform some tasks with the nondominant hand, but the common suggestion that left-handed individuals are more flexible and better able to use their nondominant hand is not true. Garonzik (1989) has shown clearly that left-handed people are at a considerable disadvantage in using workstations and equipment designed for right-handed users.

There only seem to be very slight differences in the relative proportions of right- and left-hand dominance between men and women or between ethnic groups (other than differences due to cultural pressures) (Hardyk et al., 1975). It used to be thought that humans were unique in having hand preferences, but recent studies have shown that animals do also. Elephants in the wild display side preferences in their trunk actions (Martin and Niemitz, 2003) and rats have paw preferences, with 73% right-pawed, 20% left-pawed and 8% using either paw (Guven et al., 2003).

Ingenious detective work on scratchmarks on teeth from fossil remains of Neanderthals in Spain in the early and middle Upper Pleistocene era suggests that humans were predominantly right-handed even 350,000 years ago (Bermùdez de Castro et al., 1988). This study included what must be a unique user trial with a flint tool — cutting pieces of meat gripped in the teeth. This reproduced the characteristic scratches on (simulated) teeth which were identical to those observed on the Neanderthal teeth. In a later study, Lalueza Fox and Frayer (1997) reviewed the evidence from their own investigation of Neanderthal remains in Croatia (dated to 130,000 years ago) together with evidence from other studies. Of the 20 sets of teeth studied, 18 were judged to be those of right-handed individuals and two of left-handed individuals, which, despite the small sample, is remarkably similar to the present-day proportions.

Observations of very young children suggest that hand lateralization develops during their first year, becoming firmly established for most children by about the age of 5 years. Hinojosa et al. (2003) found that, between the ages of 7 and 11 months, infants strengthened their hand preferences (whether for right hand or for left hand) when reaching and grasping. During that time the infants who had shown no preference at 7 months showed some increase in the use of the right hand.

Use of the dominant hand gives the greatest psychomotor skill as well as feeling more 'natural'. Grip strength also seems to be slightly greater in the dominant hand (Edgren and Radwin, 2000). However, too many products and tools are designed for right-handed operation (whether deliberately or unconsciously), and left-handed people are at a significant disadvantage when they attempt to use them.

6.4 ANATOMICAL TERMINOLOGY

Standard anatomical terms that are used to describe the position and movements of the forearm, wrist and hand are illustrated in Figure 6.2. The movements of flexion, extension and radial and ulnar deviation occur at the wrist joint complex, that is, at the 'true' wrist (radiocarpal) joint and at the various articulations which are present between the eight small bones of the wrist (intercarpal joints). Ulnar deviation is sometimes also known as 'adduction' of the wrist and radial deviation as 'abduction', but the terms are confusing and are best avoided. (Extension is also sometimes called dorsiflexion, with flexion then termed palmar flexion.)

The forearm has two long bones — the radius and ulna — which run from the elbow to the wrist and articulate with each other at their top and bottom ends. When the hand is in its palm up or *supine* position, these two bones are parallel. (The radius is on the thumb side; the ulna is on the little finger side.) As the hand is turned into the palm down or *prone* position, the lower end of the radius rotates about the axis of the ulna and the shafts of the two bones cross — a movement which can be felt when lightly holding your arm just above the wrist. Note then that the movements of *pronation* and *supination* occur at the two articulations between the radius and ulna rather than at the wrist as such. In practice, however, the natural hand movements we use in everyday life often entail actions of pronation and supination in combination with movements occurring at the wrist.

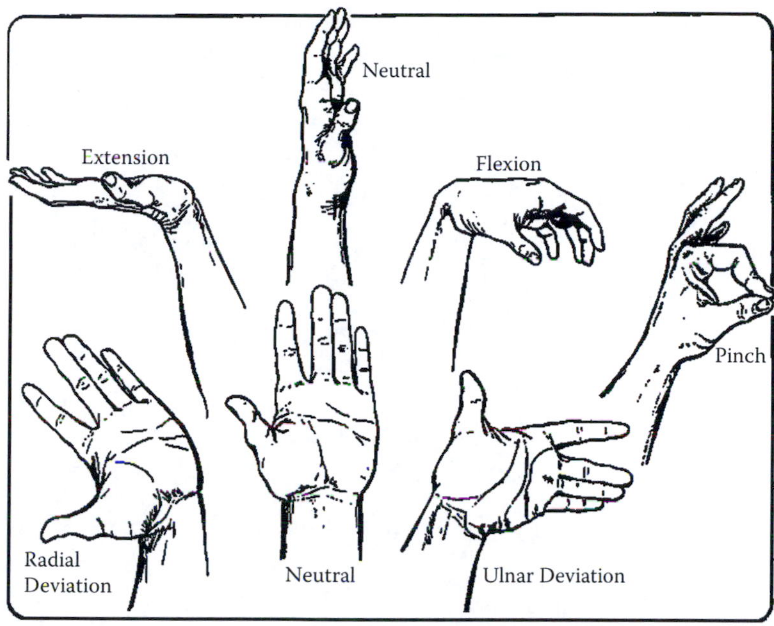

FIGURE 6.2 Hand and wrist postures. (From Putz-Anderson, V. (1988). *Cumulative Trauma Disorders,* London: Taylor & Francis, Fig. 15, p. 54. Reproduced with kind permission.)

Place your hand in your lap in a palm up (supine) position and allow it to relax completely. It will naturally adopt what anatomists call the *position of rest* (Figure 6.3), in which the fingers and thumb are slightly flexed. This is the position in which the resting tension in the muscles that respectively flex (i.e., bend) and extend (i.e., straighten) the fingers are in equilibrium.

The most distinctive feature of human hand motions is the *opposable thumb*, which permits objects to be grasped between the thumb and other fingers or between the thumb and the palm of the hand. Each finger has three joints, providing great flexibility. The joints of the thumb, starting at the base, are termed carpometacarpal, metacarpophalangeal and interphalangeal. The joints of the other four fingers are known as metacarpophalangeal, proximal interphalangeal and distal interphalangeal. The segments of the fingers differ in length and, when they are flexed to grip a small object, the pads of the fingers are arranged approximately around the circumference of a circle. When fairly relaxed they, roughly speaking, occupy a 60° arc of a circle of diameter 125 to 175 mm.

Anatomists have made a number of attempts to classify the infinite variety of actions of which the human hand is capable. The most basic distinction is between gripping (or 'prehensile') actions of various kinds, and nongripping actions (such as poking, pressing, stroking, slapping, etc.). In a gripping action the hand forms a *closed kinetic chain* which encompasses the object in question and holds it in place through the mechanical opposition of parts of the hand; in a nongripping action the hand is used in an 'open chain' configuration. A few common everyday actions fall between these two categories, in that the kinetic chain of the hand is on the point

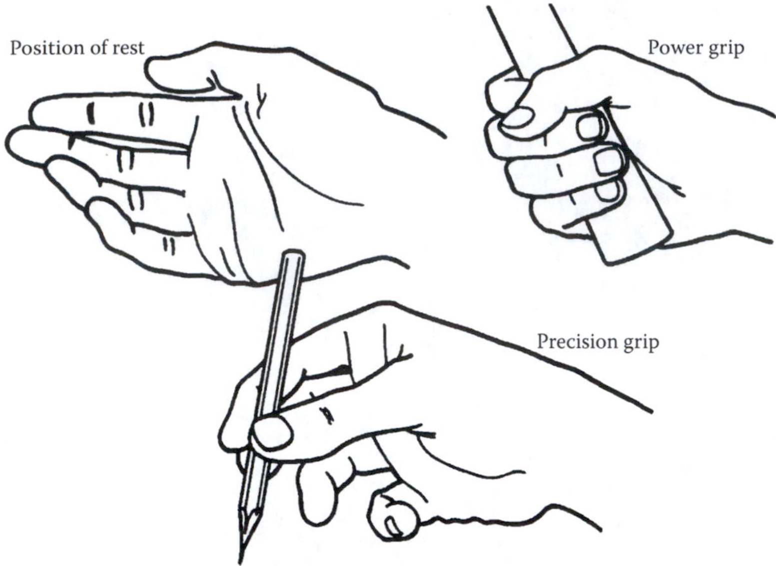

FIGURE 6.3 The position of rest, the power grip and the precision grip.

of closing — for example, the hooking action that we use to carry a heavy suitcase and the action by which we scoop up a handful of small objects.

In a classic and widely quoted paper on the subject, Napier (1956) divided gripping actions ('*prehension*') into two main categories (see Figure 6.3):

1. *Power grips*, in which the fingers (and sometimes the thumb) are used to clamp the object against the palm
2. *Precision grips*, in which the object is manipulated between the tips (pads or sides) of the fingers and thumb

Note that both entail a closed kinetic chain.

Although this classification will take us quite a long way in understanding hand function, it is something of an oversimplification. In the basic power grip shown in Figure 6.3 the thumb wraps around the back of the fingers to provide extra stability and gripping force. As the need for precision increases, however, the thumb moves along the shaft of the tool handle — providing extra control and the possibility of both power gripping and precision manipulation as the situation may demand. For a further discussion see Pheasant (1991a).

6.5 HAND STRENGTH

Hand and wrist strength vary between individuals by as much as the strength of other muscle groups — by a factor in the region of 1:3 among able-bodied adults. Figure 3.5, for example, shows the average difference in grip strength between men and women. Voorbij and Steenbekkers (1998) found that the grip force of the dominant hand is stronger by 6.5% than that of the nondominant hand. If gloves are worn, the force that can be exerted is likely to be reduced due to poorer contact and increased likelihood of slipping at the handle–handle interface. However, on occasions well-fitting gloves can help if the grip surface is too smooth or is uncomfortable to hold. The wearing of gloves can reduce tactile feedback, which in turn causes a tendency to grip with unnecessarily high force.

Hand strength for many gripping and twisting actions is strongly influenced by grip span (Pheasant and Scriven, 1983; Shivers et al., 2002; Hallbeck and Kadefors, 2004). The effect of this on handle and hand tool design is discussed later in Sections 6.7.1 and 6.7.2. Radwin et al. (1992) have shown that the strongest fingers in precision grips are the index and middle fingers but that contributions from fingers change with external force demands. Imrhan and Sundararajan (1992) measured pull strengths of fingers when gripping small objects (using a lateral, chuck or pulp pinch grip). This is relevant to many everyday activities such as opening ring-pull cans, tearing plastic or paper strips off cartons or opening plastic bags. The lateral pinch grip (with the pad of the thumb opposed to the lateral aspect of the index finger) was found to be the strongest for pulling, so that a sufficient grip area should be provided to permit this. Kinoshita et al. (1995) found similar variations in finger strength and contribution when lifting objects (like rods, coins, balls or cups) with a precision grip requiring coordination of several fingers.

Strength decreases with age. Static torque strength for the two-handed twisting action of unscrewing a lid of a jar reduces by about 35 to 40% for healthy adults between the ages of 20 to 30 years and 70 to 80 years, while grip strength reduces by about 30% (taking mean values for the age groups; Voorbij and Steenbekkers 1998). Average wrist-twisting torque strength among elderly men is in the region of 1.5 to 7.9 Nm (depending on lid diameter) for opening circular lids with a rough grip surface, while finger torques when opening small 12 mm tubes are in the range of 0.6 to 0.7 Nm, depending on sex (Imrhan, 1994).

However, people of all ages find their strength further reduced by injury or disease (arthritis is a very common condition), so many people within the general population will have much weaker hand strengths than indicated by published tables of adult strength. They experience many problems with everyday activities, from twisting the lids of jars to opening food packaging.

One difficulty with the design of bottles and screw-top containers (including medicine containers) is that they have to be child-proof as well as easily accessible by elderly people. Imrhan (1994), in a review of the hand strength data available, concludes that larger diameter lids are better for child safety than smaller ones. He also recommends short-lever arm handles rather than large ridged spheroid handles for taps on wash basins which will be used by the elderly. Pinch grip is also difficult for the elderly. Imrhan found that lateral pinch grip deteriorated less than the other types of finger pinch, so that objects to be gripped should have large finger contact surfaces to facilitate this.

Operation of controls should not, of course, demand anything in the region of maximum strength. A widely used design guideline is that when a force has to be exerted continuously for a period of time it should not exceed 10 to 15% of maximum strength (but it should be noted that this force level is not necessarily acceptable in the context of a regular occupational task or when the action is repetitive, particularly when there are risk factors present for musculoskeletal disorders). Forces that are exerted over a short period of time or at frequent intervals should not exceed 30% of maximum strength. Forces that are exerted only occasionally and for a brief moment should not exceed 60% of maximum strength. The resulting design criterion limit for a control force should of course relate to the strength of weaker users. However, Imrhan (1994) urges caution in applying these strength guidelines to controls or handles which are used by elderly people, since little is yet known about their tolerance to force exertions.

6.6 FUNDAMENTALS OF HANDLE DESIGN

The purpose of a handle is to facilitate the transmission of force from the musculoskeletal system of the *user* to the *tool* or object being used in the performance of a *task* or for a purpose (see Figure 1.1). As a general rule, we can say that to optimise force transmission is to optimise handle design. The principles of design for a wide variety of tools are discussed by Greenberg and Chaffin (1976) and Freivalds (1987). Freivalds highlighted four main anatomical concerns, in addition to effectiveness of the tool itself; these are the avoidance of static muscle loading, awkward wrist and finger postures, tissue compression and repetitive finger action.

The following guidelines for handle design stem as much from common sense as from scientific investigation. Unfortunately they are commonly violated in the handles and objects around us.

1. Force is exerted most effectively when hand and handle interact in compression rather than shear. Hence, it is better to exert a thrust perpendicular to the axis of a cylindrical handle than along the axis (Fb in Figure 6.4 rather than Fa). If the latter is necessary, a knob on the end will give extra purchase.
2. All sharp edges or other surface features, which cause pressure hot spots when gripping the handle, should be eliminated. These include:
   - 'Finger shaping,' unless designed with anthropometric factors in mind
   - The ends of tools such as pliers, which may dig into the palm if the handle is short
   - The edges of raised surfaces (e.g., for the application of labels or logos)
   - 'Pinch points' between moving parts (such as triggers)

   In general, wherever two planes meet (within the area that engages the hand), the edges should be rounded; there are no exact figures, but a minimum radius of curvature of about 25 mm seems reasonable.
3. Handles of circular cross-section (and appropriate diameter 30 to 50 mm) will be most comfortable to grip since there will be no possibility of hot spots — but they may not provide adequate purchase for exertion of high force. Rectangular or polyhedral sections will give greater purchase but will be less comfortable.
4. Where high precision is required, handles should have a smaller diameter (8 to 16 mm).

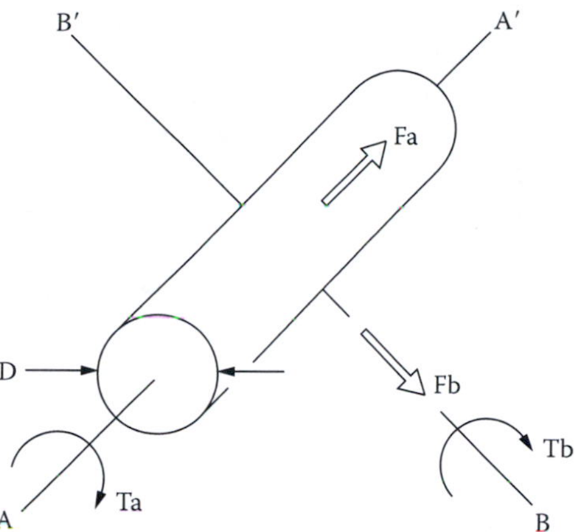

FIGURE 6.4 A cylindrical handle showing the long axis A-A' and the perpendicular axis B-B'. F is the component of force exerted along an axis and T the torque exerted around an axis. D is the diameter of the handle.

5. Surface texture and quality should neither be so smooth as to be slippery nor so rough as to be abrasive. The frictional properties of the *'hand/handle interface'* are complex since the skin is both visco-elastically deformable and lubricated. Heavily varnished wooden handles give a better purchase than metal or plastic of similar smoothness. The explanation is possibly in their resilience (elastic compliance). Rubber is similar but becomes tacky. The subject is worthy of more extensive investigation.
6. For a handle which is to be held in a power grip, the handle length should be sufficiently long to provide space for the whole palm (generally 100 mm, but 110 to 120 mm may be preferable). Diagonal hand width is a good approximation to define the minimum length, but the handle should not be so long as to contact the wrist.
7. If part of the hand is to pass through an aperture (as in a suitcase or teacup) adequate clearance must be given. It is remarkable how often this perfectly obvious design principle is violated. The following spaces will accommodate virtually all users, with a slight leeway:
   - For the palm, as far as the web of the thumb (as in gripping the handle of a suitcase), allow a rectangle 115 mm × 50 mm
   - For a finger or thumb, a circle 35 mm in diameter will allow insertion, rotation and extraction

   These dimensions must be increased if users will be wearing gloves.
8. When a handle is used in a gripping action, its shape should reflect the curves of the hand — a concave surface described by the fingers opposed to a convex surface formed by the heel of the palm and base of the thumb — but with a gentle curve which does not risk exerting excessive pressure on the palms of users with small hands.

9. Support surfaces at appropriate points, such as thumb grips, will help both with force exertion and with stability for precision tasks, but their location and contouring need careful consideration in relation to both the hand anthropometry of the user population and the demands of the specific task.
10. The tool should be able to be used equally comfortably and efficiently by left- and right-handed users.

Depending on the function of the tool and type of exertion required, it may sometimes be difficult to satisfy both guideline 10 and guideline 8 or guideline 9, in which case the tool should be designed for supply in two versions (or perhaps in modular form with a pair of interchangeable handles [Bobjer, 1989]).

6.7 BIOMECHANICS OF TOOL DESIGN

6.7.1 Gripping and Squeezing

An important group of cutting and crushing tools, from pliers and wire cutters to nutcrackers and secateurs, are operated by a forceful squeezing action across two pivoting arms. The fingers curl around one arm of the tool and the heel of the palm butts against the other. The effective cutting and crushing force is determined by the mechanical advantage of the tool and the user's grip strength. The latter is determined inter alia by the distance across the two arms (D in Figure 6.5). The optimal handle separation is 45 to 55 mm at the position in which the force is being applied for both men and women.

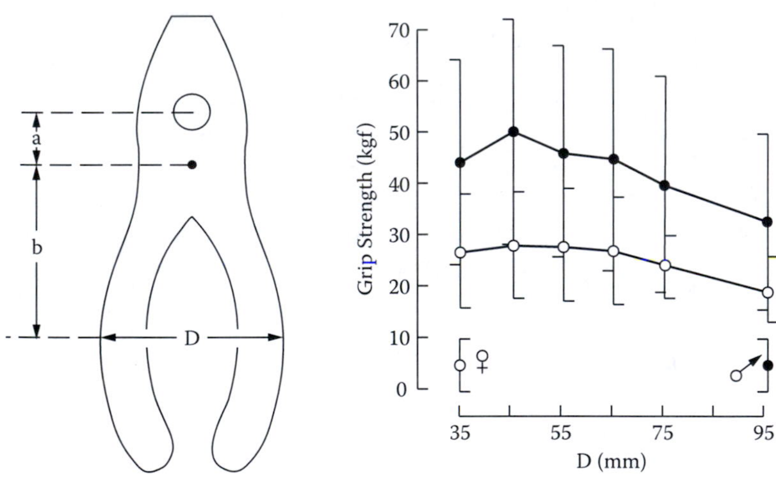

FIGURE 6.5 Grip strength (G) as a function of the handle span (D). The vertical lines in the graph indicate the 5th to 95th %ile limits of grip strength measured in samples of 22 men and 22 women. The tool is a lever of the first class, with a mechanical advantage of b/a. Hence, the effective cutting or crushing force is (G × b)/a. (Data from Pheasant, S. T. and Scriven, J. G. (1983). In K. Coombes (Ed.), *Proceedings of the Ergonomics Society's Conference 1983,* London: Taylor & Francis, pp. 9–13.)

6.7.2 Gripping and Turning

Consider a cylindrical handle as shown in Figure 6.4. It may be gripped and turned about its own axis A-A′ or about a perpendicular axis B-B′. Screwdrivers employ the former action, T-wrenches the latter.

When the handle is employed as a T-wrench, the available torque (Tb) depends upon the anthropometry and strength of the user and is, within reasonable limits, independent of the design of the handle. When the handle is employed as a screwdriver (rotating about axis A-A′), the strength of the action is no longer determined by the user's capacity to generate torque but by the ability to transmit it across the hand–handle interface. It is, therefore, strongly dependent upon handle design. Torque Ta about axis A-A′ is exerted by a shearing (frictional) action on the cylinder's surface. This is illustrated in Figure 6.6. Hence,

$$Ta = G \times \mu \times D \tag{6.1}$$

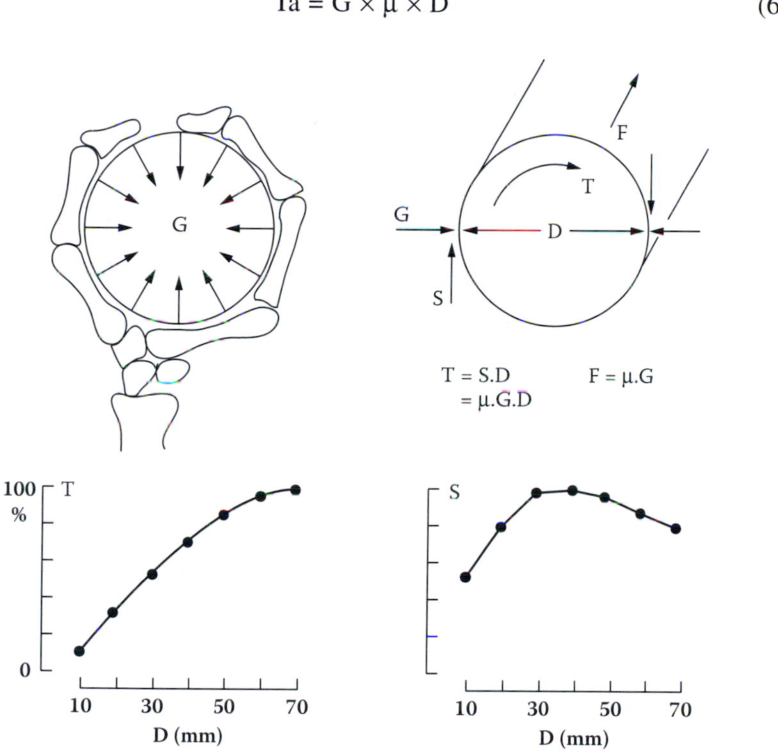

FIGURE 6.6 The mechanics of the gripping and turning action, using a cylindrical handle. Note that torque (T) is greatest on the 70-mm handle, whereas both shear ($S = T/D = \mu \times G$) and thrust (F, not plotted) are greatest on handles in the 30 to 50 mm size range. (Data from Pheasant, S. T. and O'Neill, D. (1975). *Applied Ergonomics,* 6, 205–8. From Pheasant, S.T. (1991). *Ergonomics, Work and Health,* London: Macmillan, Fig. 14.5, p. 267. Reproduced with kind permission.)

where G is the net compressive force (i.e., grip), D is the diameter of the cylinder, and μ is the coefficient of limiting friction between the hand and the handle. For any handle of circular cross-section (i.e., cylinder, sphere or disc), Ta will increase with diameter. We should also expect G to be dependent upon diameter (the optimum value of which can only be determined empirically). Figure 6.6 summarises the results of such an experiment.

Few real handles are actually circular in cross-section, but quite substantial irregularities in shape seem to make surprisingly little difference. Hence, commercially available screwdrivers (London pattern, cabinet makers', engineers, etc.) perform no better in these tests than do knurled steel cylinders of equivalent diameter (Pheasant and O'Neill, 1975; Pheasant and Scriven, 1983). Subsequent (unpublished) experiments have shown that the same is true for a variety of devices such as taps and doorknobs.

However, torques exerted about axis B-B′ (in Figure 6.4), as in using T-shaped or L-shaped devices, are much greater. The torque that may be exerted using a typical L-shaped lever-type door handle is in the order of twice that available from any cylinder, sphere or disc turning about its own axis.

The strength of a thrusting action along the axis A-A′ of a cylindrical handle is generated by a frictional force and is given by

$$Fa = G \times \mu \tag{6.2}$$

The diameter is only relevant as a determinant of G. Hence, we find that the optimal handle diameter for axial thrusts is somewhat less than that for turning actions.

Optimal handle sizes for the different types of forceful action are summarised in Table 6.2. It is also worth noting that the maximal hand–handle contact area — which will minimise the surface stress to the skin (Pheasant and O'Neill, 1975) — occurs on handles 50 to 60 mm in diameter.

However, the most appropriate size and shape of handle may depend on the task to be performed. For example, while a screwdriver produces maximum torque when held in a power grip with a handle of optimum diameter to match the user's hand size, a smaller handle which can be held and manipulated by the fingers will be more efficient and usable for making fine adjustments.

6.7.3 Pushing, Pulling, Pressing and Lifting

Depending on the location of the handle interface and resulting postures which can be adopted, forces in pushing, pulling, pressing and lifting may be augmented by use of body weight and leverage and do not depend solely on the strength of the muscles of the forearm and hand. The surface of the handle (size, shape, coefficient of friction) will influence the ability to transmit the force. The shape and size of the contact surface also affect the precision with which the force can be exerted.

Pushing actions can be performed with the flat of the hand, with a fingertip or when gripping any of a wide variety of types of handle or control knob or lever. The force that can be exerted will depend on the type of grip, on the design of the

TABLE 6.2
Handle Sizes that Allow the Greatest Force/Torque in Operation

Pivoting Tools	Distance across Arms (mm)
Turning force or squeezing grip force (G)	45–55
Handles of Circular Cross-Section	**Diameter (mm)**
Cylinders	
Axial thrust (Fa)	30–50
Axial rotation (Ta)	50–65
Spheres	
Axial rotation (Ta)	65–75
Discs	
Axial rotation (Ta)	90–130

Note: For cylindrical handles used to exert force or torque perpendicular to the handle axis (Fb, Tb), the diameter is not critical; a diameter of 30 to 50 mm is suitable.

grip surface or control and on the strength of the muscles that can be brought into play.

The ability to exert a pulling, pressing or lifting force will generally be greatest with a power or hook grip and least if the handle is so small that only a pinch grip can be used (as sometimes found with decorative cupboard handles, for example). Design limits for forces on typical handles and controls can be found in Corlett and Clark (1995).

6.8 THE NEUTRAL POSITION OF THE WRIST AND HANDLE ORIENTATION

Grip strength is greatest when the wrist is in its neutral position — reducing progressively as the wrist moves away from the neutral position in any direction (i.e., flexion, extension, radial deviation, ulnar deviation). (A more detailed discussion of the extent to which strength is reduced with different types of grip and with deviations from neutral wrist posture can be found in Hallbeck and Kadefors (2004) and Wells (2004).) Grip strength is least when the wrist is flexed. This is because when the wrist is flexed, the finger flexors (which are the prime movers in the gripping action) are shortened, and their capacity to generate tension is thus diminished (see Section 4.8).

For this reason alone, it would seem desirable that the handles of tools should be designed in such a way that, when the tool is in use, the wrist should remain as close as possible to the neutral position, since the less the strength of the gripping action in a given position, the harder the muscles will have to work to maintain a given level of gripping force.

There are also other reasons. The tendons of the various forearm muscles that act on the fingers and hand run around a variety of bony and ligamentous 'pulleys' where they cross the line of the wrist joint. When the wrist is in a nonneutral position, the mechanical loading on the tendons at these points of contact will be increased. (This is a matter of basic mechanics.) This increase in loading may lead to an increase in the wear and tear on the tendons which the working task entails and to the development of conditions like tenosynovitis, carpal tunnel syndrome and other work-related musculoskeletal disorders attributable to overuse (see Chapter 9).

When the wrist is in its neutral position, the long axis of a cylindrical handle that is held firmly in the hand makes an angle of 100 to 110° to the axis of the forearm (Figure 6.7). This is because the carpal bones in the palm of the hand are different lengths. This so-called natural angle of the wrist is seen in the traditional designs of carpenters' saws (for example). When using such a tool, the cutting edge of the blade is parallel to the axis of thrust of the forearm when the wrist is in its neutral position. Thus the neutral position of the wrist is preserved in use (see Pheasant 1991a for a further discussion).

For other types of tools, Tichauer (1978) argued that it was better to bend the handle of the tool than to bend the user's wrist and proposed an angled-handle design of pliers to replace the more conventional type shown in Figure 6.5. Subsequently, angled handles have been devised for many other types of tools, including soldering irons, hammers, butchers' knives, files and masons' trowels (Tichauer, 1966; Knowlton and Gilbert, 1983; Konz, 1986; Armstrong et al., 1982; Bobjer, 1989, Hsu and Chen, 1999; Strasser et al., 1994). Some, such as the bent-handled soldering iron, may have even greater benefits through reducing arm abduction and consequent static loading. Not all these tool modifications have been produced commercially, but increasingly innovative designs continue to appear in tool manufacturers' catalogues, as tool design parameters become better understood.

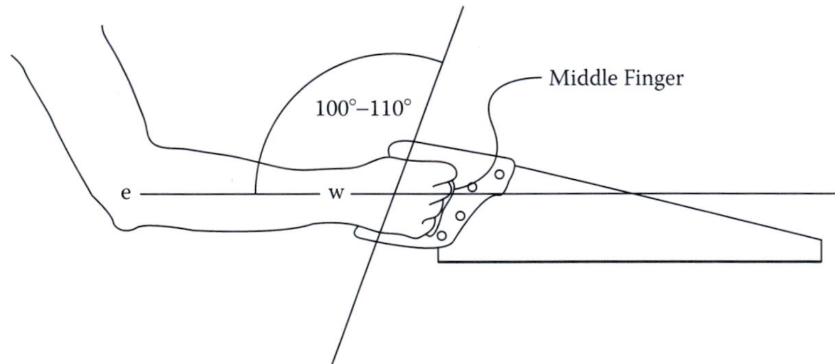

FIGURE 6.7 The neutral position of the wrist is preserved if the axis of grip makes an angle of 100 to 110° with the axis of the forearm (e-w).

Several of the studies mentioned have looked at the degree of bend that should be provided for different tools. This cannot simply be decided by aligning the tool axis with the neutral wrist posture when the tool is held statically in the hand (as in the illustrations in Figures 6.3 and 6.7). The use of the tool may involve complex motions, and some tools are used for a variety of tasks. Konz (1986), for instance, refined the initial design of the bent-handled hammer through a series of experiments. Han (2003) evaluated bent-handled pliers and found that they reduced wrist deviation in all four of the most common tasks for which they are used in electrical assembly work (gripping a small component, bending a wire, twisting wires and turning a component such as a bolt head).

A design compromise may be needed to achieve good usability and adequate speed of work (and thus productivity) at the same time as minimising wrist deviation. Some studies have also found that subjects preferred traditional designs of tools to the redesigned ones. While this may, in part, have been due simply to lack of time to become familiar and practised with them, it may also have been because the redesign failed to consider usability issues, such as the ease with which the handle can be gripped or the degree of control users have in a precision task (as found by Dempsey et al., 2002 and Duke et al., 2004). Päivinen et al. (1999/2000) and Dempsey and Leamon (1995) present helpful discussions of usability criteria, mostly in the context of plier-like hand tools. A thorough discussion of the biomechanical aspects of hand tool design can be found in Chaffin et al. (1999).

6.9 WORK TASKS USING HANDHELD TOOLS

6.9.1 Posture and Workstation Design

The effects of posture on whole-body strength have already been mentioned (Chapter 4), and similar effects have to be considered for hand and arm strength when using tools and handling other objects. Schulze et al. (1991) found distinct differences in postural loading between male and female operators using pneumatic screwdrivers in furniture assembly. The female operators were disadvantaged by both strength and stature and had to adopt postures with greater arm abduction and trunk flexion, leaning into the task to assist in exerting force.

Both grip strength and torque strength can be significantly reduced when an individual is forced to adopt a nonoptimal posture, as shown in many studies, including those of controls such as the handwheel valve (Shih et al., 1995) and large handwheel (Wolstad et al., 1995) as well as of hand tools.

Changing the location of the handle or control — or changing the orientation at which a tool must be used — leads to a change in arm posture, so that muscle strength may be reduced and the muscle may be working at a poorer mechanical advantage. (Similar considerations apply to handle or handhold positions on boxes which are lifted or carried [Drury et al., 1985]). Changes in wrist posture, in particular, have a considerable effect on strength, as discussed in Section 6.8. Mital and Channaveeraiah (1988) and Habes and Grant (1997) demonstrated the effects on the use of a screwdriver over a range workstation layouts and handle diameters and orientations. Dempsey et al. (2002, 2004) in two studies, one of the use of pliers in a wire-twisting task and the other of a repetitive screw driving task, showed that productivity can also be affected by workpiece orientation and work height. Han (2003) found similar results for both pliers and powered screwdrivers with in-line and pistol-type handles. Increased wrist deviation resulting from poor workpiece location, increased arm and shoulder elevation at higher work heights, and increased arm abduction or elevation at awkward workpiece orientations can increase the risk of developing musculoskeletal disorders (as discussed below in Section 6.9.2). Workstation layout must therefore be considered in conjunction with the selection or design of the tool for a particular task. The studies above confirm that, when this is done, a good ergonomic solution can be found to promote both productivity and health.

6.9.2 Risk of Musculoskeletal Injury

The problems of musculoskeletal injuries incurred during industrial work are discussed in more detail in Chapter 9, but many of the tasks concerned involve intensive use of the hand and wrist. In some vehicle assembly plants, for example, operators may use powered or nonpowered hand tools for 4 hours a day (Garmer et al., 2002), and the same is probably true in electrical component assembly and meat processing. The problems are not confined to the use of nonpowered hand tools, because the holding and operation of powered hand tools often involves static muscular effort even when the operating forces themselves are low.

The U.S. Bureau of Labor Statistics reported that, of the 522,528 musculoskeletal disorders reported in the United States in 2001, 8.9% involved the wrist and 1.6%, the hand or fingers, while 2.9% were due to the use of hand tools (statistics for nonfatal occupational injuries and illnesses with days away from work; BLS, 2001). The total incidence of injuries resulting from the use of hand tools is much greater than this — around 10% according to Mital and Channaveeraiah (1988) — and includes categories of injury other than musculoskeletal disorders.

It is generally recognised that the jobs that carry the highest levels of risk are ones in which forceful gripping actions are combined with turning actions or are made with a deviated wrist. In practice, turning actions will generally involve wrist deviation, but you can have a deviated wrist without making a turning action (e.g., when using certain poorly designed hand tools).

Anatomical and biomechanical considerations would lead us to predict that there will be certain relatively consistent associations between particular disorders and particular patterns of movement. For example, we should expect clothes wringing types of action (in which flexion and ulnar deviation are combined with supination of the forearm) to result in tenosynovitis or peritendinitis affecting the tendons of the extensors (particularly those that act on the thumb) and likewise that repetitive flexion and extension of the wrist would lead to carpal tunnel syndrome (either directly due to mechanical irritation of the median nerve or secondary to a flexor tenosynovitis). Carpal tunnel syndrome may also be caused by repeated impact as in using a hammer. Tenosynovitis of the finger flexors may likewise be caused by repeated blunt trauma, for example, when using badly designed hand tools that have pressure hot spots on their handles.

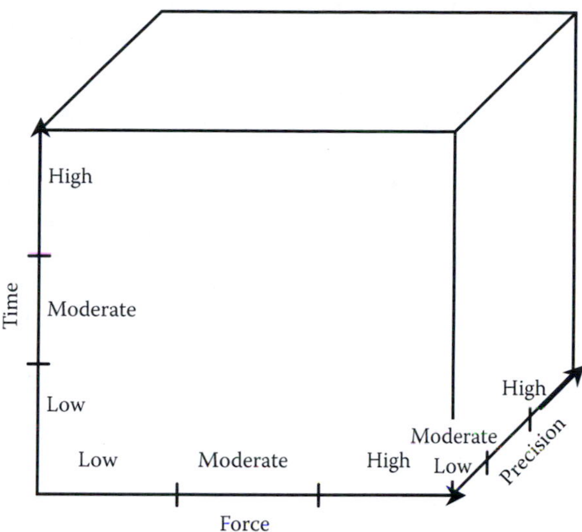

FIGURE 6.8 Cube model for classification of work with hand tools (from Sperling L., Dahlman, S., Wikström, L., Kilbom, Å. and Kadefors, R. (1993). *Applied Ergonomics,* 24(3), 212–220).

Experience suggests that these predictable associations do indeed occur reasonably consistently, and such little epidemiology as there is would tend to confirm this (Pheasant, 1991a). But they do not by any means occur infallibly and, in practice, just about any of the disorders in question may be associated with any of the motion patterns. In one way this is not particularly surprising. The functional anatomy of the hand is complex. All of the forearm muscles, whose tendons cross the wrist to insert on the bones of the hand, have multiple functions — as prime movers, synergists or muscles of stabilisation. It would seem probable, furthermore, that the various anatomical structures of the 'muscle tendon unit' (i.e., the muscle itself and its soft-tissue attachments at either end) may be subject to overuse injury both when they repeatedly contract and when they are repeatedly stretched by their antagonists. Thus the number of potentially injurious permutations and combinations is considerable.

The particular type of gripping action that the task entails is also a factor. Pinch grips and claw-like grips (i.e., 'precision grips') both entail a higher internal biomechanical loading for a given externally applied force (and thus a higher level of risk) than full grasping actions (i.e., 'power grips'). The over-spreading of the hand, when gripping an object or when squeezing the arms of a pivoting tool to exert an external force, is also a risk factor.

Apart from posture, three of the most important factors influencing musculoskeletal loading during operation of a hand tool are considered to be force, time and precision. The latter is important because the degree of control required influences the type of grip (and thus the muscles involved), the grip force and the static muscular effort for stabilisation of the hand. Sperling et al. (1993) have proposed a 'cube' model, shown in Figure 6.8, to represent these factors and to assist in assessing the severity of demands from a hand tool. Each of the three factors can be rated as high, moderate or low demand, so that the cube is subdivided into 27 small subcubes and a given tool can be categorised as falling within one or other of these. The critical levels separating the low, moderate and high demands represented by each subcube have been defined, and each subcube has been given a given rating of more (or less) acceptable, on the basis of evidence from research literature (the details of the classification scheme are explained in Sperling et al., 1993). They distinguish, for example, between typical work with assembly pliers (low force, high time, moderate precision) pop riveting tongs (high force, high time, low precision) and bolt clippers (high force, low time, moderate precision). According to the cube model, the second of these would be classed as unacceptable and the first and third as needing further investigation. By identifying the severity of demands on these three dimensions, it is possible to see whether improvement is needed and, if so, whether the tool factor (force), task factor (precision) or organisation of the work (duration, repetition) should be changed.

7 Ergonomics in the Office

7.1 INTRODUCTION

The basic office workstation typically consists of a desk, chair and computer at which the user will undertake:

- *Paper-based tasks*: reading, writing and conversation
- *Screen-based tasks*: involving use of keyboard, mouse and possibly other input devices

Andersson et al. (2000) suggest a categorisation of office tasks which will help to focus on the specific task demands in the evaluation and improvement of individual office workers' workstations and work environments:

- Telephone work
- Computer work
- Meetings and visitors
- Hard copy writing and reading work
- Other kinds of manual work sitting by the desk
- Work done away from the desk
- Miscellaneous (including breaks)

With this categorization, a 10-day diary is proposed as a useful way of collecting information on any individual office worker's tasks, which can then be analysed under the seven categories to prioritise their workstation needs.

Only a few years ago it was confidently predicted that the 'paperless office' was just around the corner and that before long all information would be handled solely by means of electronic media. Although this goal has not yet been achieved, office work has become steadily more screen-based, to the extent that the office workstation that does not have a visual display screen (VDU or VDT or DSE [display screen equipment]) is now very much a rarity.

Apart from traditional office work, many other jobs are now undertaken at very similar workstations and involve a similar range of tasks. A few diverse examples of such jobs are bank teller, telephone operator in a call centre (whether for sales or helpline), CAD (computer-aided design) engineering designer, shopfloor production controller in a manufacturing company, television production director, security CCTV (closed circuit television) monitor and air traffic controller. (Most control rooms now have moved to VDU-based interfaces, and their operation has become very similar to office work.) There are differences in the demands of the work and the ancillary equipment used, depending on the circumstances, but the basic unit of chair, desk and computer are similar in all these jobs. Many ergonomic factors in such jobs are task-specific (for example to support face-to-face communication, provide shielding from speech interference in an open plan office, facilitate frequent ingress and egress from the workstation or permit simultaneous monitoring of several display screens) and thus cannot be fully covered in this chapter. The basic design principles for screen-based work will apply, but further and more detailed analysis of the specific task requirements will be necessary.

Traditional office 'concepts' are also changing, with current moves for teleconferencing, 'hot-desking', rapid (standing) meetings and 'cocon' (communication-concentration) offices. These have all introduced new equipment and variations in the layout of offices. Teleconferencing requires at least the use of a web camera but may involve several people interacting at a single computer workstation. Hot-desking has been introduced to make maximum use of computer workstations when staff spend part of their day elsewhere, as for example in the newsroom of a newspaper or television company; here staff share the workstations and there is a predominant need for easy and rapid adjustability to fit all users. Current management 'concepts' have suggested that 'standing meetings' may be more productive (given some credence from the experiments of Vercruyssen and Simonton [1994], who found reaction time to be faster when people were standing than when they were sitting down and tentatively suggested that this may apply to rapid decision-making, although the effect is small). Staff will need to have a writing surface and, perhaps, consult a display screen during such standing meetings. A 'cocon' office is characterised by separate small workstations for carrying out individual tasks with a large communal room reserved for meetings. The specific ergonomic requirements for each of these office 'concepts' cannot all be discussed in this chapter (not least because the implications of such work have been little studied to date; for a review see de Croon et al., 2005), and task analyses should be carried out when designing the workstations.

The following widely accepted principles apply to office work in the interests of comfort and the avoidance of adverse long-term effects on health:

- For writing, the working surface (i.e., desktop) should be somewhat above the user's elbow height, as measured in the standard upright sitting (or standing) position (see Sections 2.7 and 4.7). This is because, in order to write with a relaxed and natural action, the arm must be abducted and flexed somewhat at the shoulder (i.e., raised sideways and forwards).
- For keyboard work, the shoulders should be relaxed, with the upper arms hanging freely at the sides, the forearms more or less horizontal and the wrists as far as possible in a neutral position (i.e., not bent forwards, backwards or sideways). Thus the so-called 'home row' of keys (ASDFG, etc.) on the computer keyboard should be at or close to elbow height.
- For comfortable sitting in a conventional office seat, the user's thighs should be approximately horizontal, with the feet resting flat on the floor. Thus the seat should be at or close to the user's popliteal height — but preferably slightly below it.

- For keyboard work there is an element of controversy as to whether it is preferable that the user's trunk should be upright or reclined. We shall return to this last point in Section 7.8 below).

International standard ISO 9241 Ergonomic Requirements for Office Work with Visual Display Terminals (VDTs) gives recommendations for the design of screen-based and office workstations. This consists of 18 parts, but the most relevant to workstation arrangements are ISO 9241 Part 1:1997/Amd 1:2001 General Introduction, Part 2:1992 Guidance on Task Requirements, Part 3:1992/Amd 1:2000 Visual Display Requirements, Part 4:1998 Keyboard Requirements, Part 5:1998 Workstation Layout and Postural Requirements and Part 9:2000 Requirements for Non-Keyboard Input Devices (ISO, 2001, 1992b, 2000a, 1998a, 1998b and 2000b, respectively). Several other standards are also useful. British standard BS 3044:1990 Guide to Ergonomics Principles in the Design and Selection of Office Furniture (BSI, 1990a) covers the general principles and gives guidance on how to match appropriate furniture and equipment to task requirements, as well as on the environmental and personal factors which have an impact on this. ANSI/HFS 100-1988 American National Standard for Human Factors Engineering of Visual Display Terminal Workstations (ANSI/HFS, 1988) also gives similar guidance. This has recently been updated, and the revised version is currently issued as Draft Standard BRS/HFES 100 Human Factors Engineering of Computer Workstations (BRS/HFES, 2004). In the United Kingdom, such standards are regarded as advisory but not mandatory. Screen-based office work, however, is subject to the provisions of the Health and Safety (Display Screen Equipment) Regulations 1992 (as amended in 2002) under European Directive 90/270/EEC. The reader who is concerned with compliance with these regulations is referred to the relevant Health and Safety Executive guidance document (HSE, 2003b), in which the provisions are discussed in detail and which contains a VDU workstation checklist that can be used as an aid to risk assessment.

7.2 THE OFFICE DESK

The user has three points of physical contact with his or her workstation and environment: the desk (or keyboard), the seat and the floor. If a range of users, who vary in both size and shape, are to attain the same desirable working position (as defined by the anthropometric criteria set out above), then two out of these three must be adjustable. Nowadays, virtually all office chairs are adjustable for height. In the United Kingdom, however, adjustable office desks remain something of a rarity (although there are some signs of gradual change in this respect). Elsewhere in the world adjustable desks are more common. This is the case for example in Australia, where the so-called 'RSI epidemic' of the 1980s (see Chapter 9) acted as a major stimulus to the improvement of working conditions in offices. (In the following discussion, it is assumed that work will be carried out while seated. The same principles can be applied to the height of standing desks simply by substituting standing elbow height for the corresponding seated dimension.)

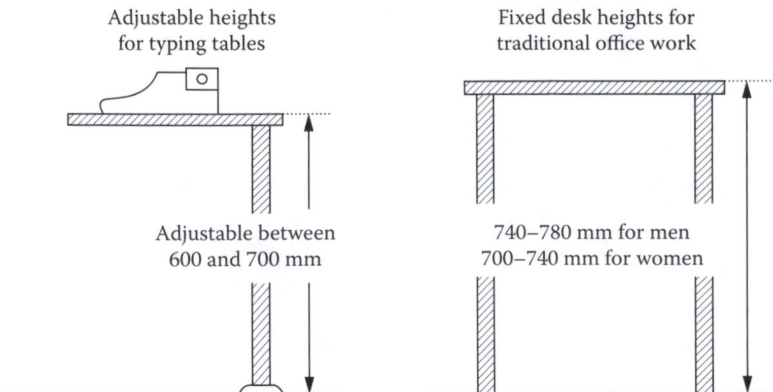

FIGURE 7.1 Recommended desktop heights for traditional office jobs. Left: range of adjustability for typing desks; right: desktop heights for reading and writing without a keyboard. (From Grandjean, E. (1988). *Fitting the Task to the Man,* 4th ed., London: Taylor & Francis, Fig. 37, p. 42. Reproduced with kind permission.)

The heights of desks for seated work at different office tasks which were recommended by Grandjean (1988) are given in Figure 7.1. Grandjean noted that these were slightly higher than recommended by most other authors because he felt that a low desk was particularly undesirable for tall users.

Ergonomically, adjustable height desks are the preferred solution for office work, particularly if this is intensively screen-based. A fixed height desk may be regarded as an acceptable second best, provided that the floor level is adjustable, which in practice may be achieved (in part) by the provision of a footrest where required.

Given that the desk is to be of fixed height, what height would represent the best possible compromise for a working population of adult men and women? Supposing we say that for *paper-based office work* the desktop should ideally be 75 mm above the user's sitting elbow height (SEH), and the seat should be 50 mm lower than the user's popliteal height (PH). Then optimal desk height = (SEH + 75) + (PH – 50) mm. Since we are concerned with finding the best possible compromise (which minimises the number of people falling outside whatever tolerance bands about this optimum we care to propose), then 50th %ile anthropometric values are required (see section 2.6.2). On this basis the best compromise general purpose desk height for men would be 735 mm and for women 705 mm, giving an overall compromise figure of 720 mm (although this does not take into account Grandjean's suggestion that the height should be slightly biased towards the taller members of the population).

For *keyboard work*, however, the anthropometric criteria given above indicate that the best possible compromise height would be a good deal lower — first, because the home row of the keyboard should be at sitting elbow height rather than above it and second, because the home row will itself be some 30 to 50 mm above the desk surface. However, the difference may not be as important as it seems, in that although we may in theory regard a seat height of 50 mm below popliteal height as ideal, in practice, a seat of popliteal height or even a little above would probably be

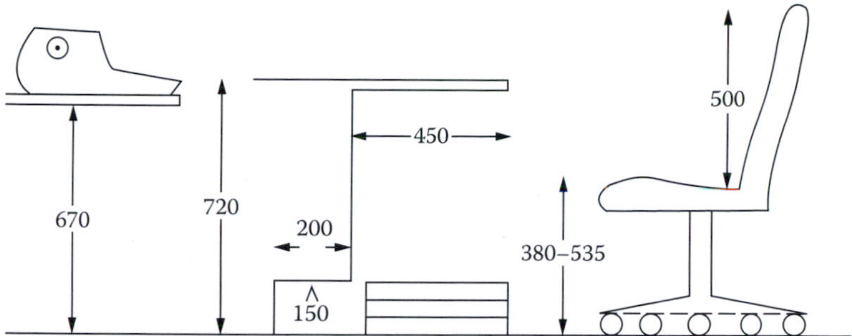

FIGURE 7.2 Compromise dimensions for office furniture (mm).

entirely acceptable to the majority of users. In reality, therefore, a standard desk height of 720 mm or thereabouts is probably just about as good a compromise as any other for screen-based office work, although we must bear in mind that, as a compromise, it is somewhat biased in favour of the taller half of the user population. This, however, as we shall see, is probably no bad thing. The compromise dimensional recommendations for fixed height office desks are summarised in Figure 7.2.

Office desks are made to a standard height; office workers are not (see Table 1.1). The standard desk is satisfactory for the average person, but for people who are markedly shorter or taller than average it can cause serious problems (particularly for intensive screen-based keyboard work, where the potential long-term consequences of a mismatch are likely to be greater).

To reach an appropriate working height in relation to the keyboard, a woman with short legs working at a standard height desk will have to adjust her chair to a level that is too high for comfort. As a consequence she will tend to perch on the front edge of the seat, thus losing the support of the backrest. This may lead to back problems and worse (see Section 7.8 below). If she lowers the seat, however, she will commonly end up working with her shoulders hunched and her arms abducted (i.e., the elbows raised out sideways). The static muscle loading which results may lead to neck and shoulder problems. Abduction of the arms at the shoulders calls for a compensatory ulnar deviation of the wrists, in order to maintain the alignment of the fingers at the keyboard (i.e., the wrists are bent sideways in the direction of the little finger). This is very unsatisfactory indeed, the ulnar deviated wrist being an important causative factor in the aetiology of upper limb disorders (see Chapter 9). Alternatively, instead of abducting her arms, the short user may incline her forearms upwards, which leads to flexion (i.e., forward bending) of the wrists, or she may work with her wrists in extension (i.e., backwards bending), both of which are undesirable.

The problems of the short user are solved relatively easily with a footrest. As a rough guide, anyone of around 1600 mm or less in stature will probably need a footrest when working at a standard height desk. A footrest should be adjustable in height (or sloped to allow the feet to rest at any height) and have a sufficiently large surface for the feet to be moved around and leg posture varied.

The problems of the unusually tall user are more difficult to solve. At a standard height desk he will tend to find himself working with his spine in flexion however he adjusts his chair. The only real solution is a nonstandard desk or one of adjustable height — or a standard desk which is raised up in some way. In the final analysis, an adjustable desk may be regarded as the ergonomically preferable solution to the postural problems of the keyboard user. The range of adjustment required may be calculated from the combined dimension of shod popliteal height plus sitting elbow height, which has a distribution of 710 [42] mm in men and 690 [40] mm in women. We may calculate therefore that, for keyboard work, the desktop should adjust from about 600 mm to 750 mm to accommodate the 5th %ile female and 95th %ile male user, assuming a keyboard thickness of 30 mm; for writing the range should extend somewhat higher. In practice, however, the bottom part of this range may be of limited use because of knee-room problems, and a range of 650 to 750 mm is probably about right.

The distance between the undersides of the elbows and the tops of the thighs in the 'standard sitting position' (see Section 2.6) is only around 80 to 85 mm on average and in many people it will be a good deal less. If the user is to adopt the recommended keying position, as described above, this space must accommodate the thickness of the keyboard (generally some 30 to 40 mm) plus that of the desktop itself. It follows therefore that the recommended keying position will be a physical impossibility for a proportion of users. It is a matter of utmost importance furthermore that the thickness of the desktop (and its supporting structures) should be kept to an absolute minimum, commensurate with the requirements of structural strength. In particular, desks that have obstructions below the working surface, such as 'knee-hole' drawers, must be regarded as wholly unsuitable for keyboard work. (See the story of Janice at the beginning of Chapter 1.)

An important but frequently neglected ergonomic aspect of desk design is the adequacy of its surface area. The desktop must be large enough to allow the screen to be placed at a suitable viewing distance (see Section 7.4) and permit the user adequate flexibility in placing the keyboard and mouse (see Section 7.7.2). The overall space that is needed will of course depend also on what other items live on the desk. Clutter expands to fill the available space (Pheasant's principle of ergonomic decay).

7.3 THE OFFICE CHAIR

Chair design has already been discussed in Chapter 5, but a few requirements specific to seating for office work will be mentioned here. Dimensional recommendations for office chairs are summarised in Figure 7.2.

7.3.1 Seat Height

To meet the requirements of a range of users, the height of the seat should be easily adjustable from the sitting position. The height range that is required will, in principle, depend upon whether the seat is to be used with an adjustable desk or a fixed height desk (presumably of standard height). If the former, then a range of 5th %ile female to 95th %ile male shod popliteal height would seem appropriate, which works out at 380 to 515 mm (assuming 25 mm heels for both sexes). Given a 720 mm desk, it is unlikely that anyone will want a seat higher than 535 mm (720 minus 5th %ile female sitting elbow height). A height range of 380 to 535 mm should thus in principle meet all eventualities. In practice it may be a little over-generous.

7.3.2 The Backrest

Typists' chairs traditionally had low-level backrests, whereas executive chairs had medium-level or even high-level backrests. The supposed justification for this was that a typist needed freedom of movement for his or her shoulders. In reality, however, it was probably more a matter of the differentiation of status, combined perhaps with a puritanical distrust of comfort in the workplace. With the old-fashioned mechanical typewriter, the argument for the low-level backrest was perhaps just plausible; with the modern electronic keyboard it is no longer valid. A medium-level backrest gives better back support and permits a more reclined (and thus more relaxed) working position (see below). Grandjean (1987) recommends a 500-mm backrest.

In order to give the user the greatest possible variety of working positions, the angle of the backrest should be adjustable (independently of the seat). The backrests of many modern office chairs are spring loaded such that they follow the user's changes of position. In theory this seems like a good idea. In practice, some users like it and some do not. Thus it is important that the user should also be able to lock the backrest in place if he or she wishes.

Finally, the backrest should be contoured to the form of the lumbar spine and adjustable in height (again relative to the seat) so that the user may match the mid-point of the seat's lumbar pad to the curve of his or her own back. As we noted earlier, it is important that the contour of the backrest not be excessive. Some modern office chairs are definitely over the top in this respect (see also the discussion of backrests in Sections 5.4.4 and 5.4.5).

7.3.3 Armrests

Traditionally, typists' chairs did not have arms, whereas 'executive' chairs commonly did. As with backrests (see above), this was in part a matter of ergonomics and in part the differentiation of status. Some keyboard users like to support their elbows on the arms of a chair as they work, and insomuch as it reduces the static loading on the muscles of the neck and shoulder girdle, this would seem no bad thing. An alternative which achieves the same end is to support the wrists (see Section 7.8 below). Armrests can be a mixed blessing, however, if they prevent the user from getting close to the desk.

7.3.4 The Usability of Adjustment Controls

The more different ways it is possible to adjust a chair, the more difficult it becomes to design adjustment mechanisms that are easy to locate and operate. The more difficult an adjustment mechanism is to operate, the less likely it is that the chair will be adjusted properly. As a general principle, each mode of adjustment should have its own dedicated control lever. Coupled adjustment mechanisms, in which the backrest angle and seat tilt are controlled by the same action, are especially undesirable (for reasons which have already been discussed in Section 5.6). All adjustment controls should be easy to reach and use from a seated position.

While adjustability is desirable, Helander et al. (2000) showed that this does not necessarily have to be continuous adjustment, and they suggested that incremental steps of about 2.5 cm in seat height and 3° in seat or backrest angles would be satisfactory. Their trials were, however, of a chair in isolation, and further study is needed to test whether these findings could be applied to chairs used for office tasks.

Office workers frequently do not know how to operate the adjustment mechanisms of their chairs. Training them how to do so may often improve their comfort dramatically.

7.4 VISUAL DEMANDS OF SCREEN-BASED WORK

7.4.1 Viewing Distance

For visual comfort in screen-based work a viewing distance of 50 cm may be regarded as an absolute minimum; 75 cm would probably be better (see Section 4.6). Jaschinski et al., (1999) found that operators reported more strain with the screen placed at 66 cm than at 98 cm. However, they also found that operators had widely different preferences, so that placement should be as flexible as possible, allowing up to 100 cm if at all possible. The screen should not be too close or it will cause visual fatigue. The material displayed on the screen should thus be designed in such a way as to be legible at an adequate distance. This is partly a matter of character size and partly one of image quality. A modern high-resolution screen has a degree of legibility that approaches that of printed text, but older screens may be much less legible (particularly if the image is unstable due to technical faults, etc.). There is evidence that dark characters on a light ground are superior in this respect to light characters on a dark ground (Radl, 1980; Bauer and Cavonius, 1980). Both the brightness and the contrast of the screen should be separately and easily adjustable by the user. One other factor to remember is the vision of the user. Neither reading nor distance viewing spectacles (or contact lenses) are entirely suitable for working at a computer screen, and a change to spectacles prescribed specifically for such work can greatly improve posture and comfort. If the computer user needs bifocal or multifocal lenses because of other task demands (for example communicating with customers or consulting display screens located at several distances), both lens prescriptions and workstation layout should be matched to the situation (Horgen et al., 1989).

Lighting is also a factor. In general, glare from light sources will be minimised if the screen is placed at right angles to a window and parallel to overhead fluorescent tubes. Diffused uplighting is better, however. (For a more detailed discussion see, for example, Pheasant, 1991a,b.) There may be other sources of glare within an office and care should be taken to check for these and remove or shield them. Glare on a display screen is a common cause of poor posture.

7.4.2 Display Screen Height

There is some controversy over the appropriate height for the display screen. This has to satisfy visual comfort (related to the relaxed direction of gaze) while minimising neck flexion, so that the optimal position for the screen would be expected to be influenced by the sitting posture adopted. Delleman (1999) conducted an experiment in which touch-typing VDU operators worked at various screen and seating arrangements spanning screen heights between 5 cm above and 40 cm below eye height with an upright and a 15° reclined seat backrest. Of these, the position preferred by the operators was a screen 10 cm below eye height with whichever seat they were using. Delleman suggests that a slightly flexed (approximately 15°) neck posture is preferred and recommends a screen centred on a gaze direction 6 to 9° below the horizontal.

On the basis of visual functions, Jaschinski and Heuer (2004) also recommend a gaze direction at an angle slightly below horizontal (around 10°) but note that there is considerable individual variation in preferences for both screen height and viewing distance. Some flexibility for both should therefore be provided in the workstation arrangement.

Other researchers have recommended very different vertical screen positions, from 15° above to 45° below the horizontal. In two reviews of the many papers published, Psihogios et al. (2001) and Delleman (2004) found that the general consensus was for a screen at a little below eye level, such that when looking at the most important part of the display (or centre of the screen for word processing), the user has a downward visual angle of between 0° and 15° (see also Section 4.5). Psihogios et al. confirmed the laboratory-based experimental results with a month-long field study of VDU operators who worked with their screens at eye level and at 17.5° below the horizontal (for two weeks in each condition). This screen position will eliminate the poked chin and rounded back of the 'yuppie hump' (Figure 7.3)

FIGURE 7.3 VDU user demonstrating the 'yuppie hump', from an original in Stephen Pheasant's collection. (From Pheasant, S. (1991). *Ergonomics, Work and Health,* London: Macmillan, Fig. 5.4, p. 111. Reproduced with kind permission.)

which comes from looking at a screen which is too low. (The legibility of the display may also be a factor in causing the poked chin.) Ankrum (1997) mentions an additional advantage of having a screen positioned below eye height in that it gives much greater flexibility to adopt a variety of head and neck postures, promoting desirable variation in posture. ISO 9241-5 recommends that the most important part of the visual display should be ±15° around the line of sight (ISO, 1998b).

When looking at the centre of the screen, the user's line of sight should be approximately perpendicular to the screen surface. To meet all these requirements, it will be necessary for the screen to be physically separate from the keyboard, to tilt and rotate on its base, and to be designed in such a way that it can if necessary be raised on a plinth or some other suitable support.

7.4.3 Document Holder

The provision of a holder for source documents (so that they can be read at screen level and without turning the head) will reduce the postural loading on the neck muscles considerably (particularly in data entry, copy typing tasks, etc.). It is also desirable to have some adjustment in the viewing distance between document holder and screen, since accommodation of the eyes may differ when reading from these two sources. The preferred location of the document holder, relative to the screen, depends upon the task. In general, the item that the user looks at most frequently should be placed directly in front. If screen and source documents are referred to equally often, then they should be placed on either side of the straight-ahead position and angled slightly inwards towards each other.

7.4.4 The Unskilled Keyboard User

The two-finger 'hunt and peck' typist presents a special set of problems, since he or she will inevitably spend much of the time looking down at the keyboard rather than at the screen or source documents, tending to work in a hunched position. The problem can be ameliorated to some extent by encouragement to adopt a more reclined posture and be as relaxed as possible. Beyond this, it is difficult to see what else can be done — other than giving training in touch typing (which for anyone using a keyboard on a habitual basis is clearly desirable).

7.4.5 Multiple Display Screens

Some jobs involve consulting several display screens, each with a different feed channel of information, or monitoring a bank of video screens. Where two or three VDUs have to be arranged on the desk, it severely limits writing space or room for placing input devices, other equipment or documents. Such cramped conditions are likely to cause poor postures of the head, neck, shoulders and back. When a bank of display screens is arranged on a wall, perhaps for monitoring security cameras covering a building or selecting views from various cameras sending live feeds from a broadcast sports event, there is difficulty in fitting all these within the 'acceptable display zone' (see Figure 4.13) while still maintaining an adequate viewing distance with the picture or text on the screen clearly visible. The upper rows of screens may also be placed too high, so that the viewer has to adopt an extended neck posture, which is very undesirable.

Both situations are difficult to resolve, but both are cases where McCormick's principles of rational workspace layout (Table 4.1) make a good starting point. The solution may be in redesigning the software for the displays in addition to rearranging the layout of the individual screens and other equipment. Thus, for example, a bank of display screens might be replaced with multiple views presented in tile formation on a single screen, with the facility to select any to view enlarged on a second screen.

7.5 THE PORTABLE (LAPTOP OR NOTEBOOK) COMPUTER

The increased use of portable computers for extended periods of work is currently becoming a cause for concern in some circles. The problems, however, seem to be mainly related to the environments in which portable computers are used and some design features resulting from the desire to make them compact and lightweight. A comparison of experiences of 2192 computer users (Heasman et al., 2000) showed that the health problems of portable computer users were little different from those of desktop computer users and in both cases were correlated with length of time for which they were used. Portable computers, however, tend to be used for shorter periods of time, probably because their users are engaged in many other tasks. One source of discomfort mentioned by the users of portable computers was the carrying of them (together with their peripheral devices) which, in the employment context, should be assessed in relation to manual handling risks.

The environments in which portables tend to be used are far from ideal and often induce poor work postures. They are frequently used while travelling (in trains or planes), in cramped conditions in cars, and in hotel bedrooms and airport lounges. The most obvious effect is the difficulty in finding a surface of suitable height on which to place the computer (even if it is not on the user's lap). Even when a suitable surface is available, users are unlikely to pay attention to adjusting the seat or their own postures, just because it is seen as a temporary situation. Many studies have shown that portable computer users have much more inclined head postures and neck flexion than when working at a desktop computer (e.g., Straker et al., 1997; Wilson, 2001; Jonai et al., 2002).

The problem is, first, the height of the screen and, second, that it is only clearly legible when viewed through a narrow range of angles. Together these factors fix the head and neck in an unsatisfactory position. The problem may not be quite as bad as it seems, insomuch as the screen of a laptop computer may be tilted back to any angle the user chooses. The typical postural adaptation that occurs therefore (when the laptop is used on an office desk) is a simple inclination of the head (Figure 7.4). Although this is by no means desirable (see Section 4.6), it is probably not as bad as the yuppie hump (which comes from looking at a *vertical* screen which is too low). Users of laptop computers can adopt very unconventional postures, particularly as compared to the two common postures adopted by desk-based VDU users illustrated in Figure 7.5. Harris and Straker (2000), for instance, discovered

FIGURE 7.4 The laptop computer. Note the posture of the neck.

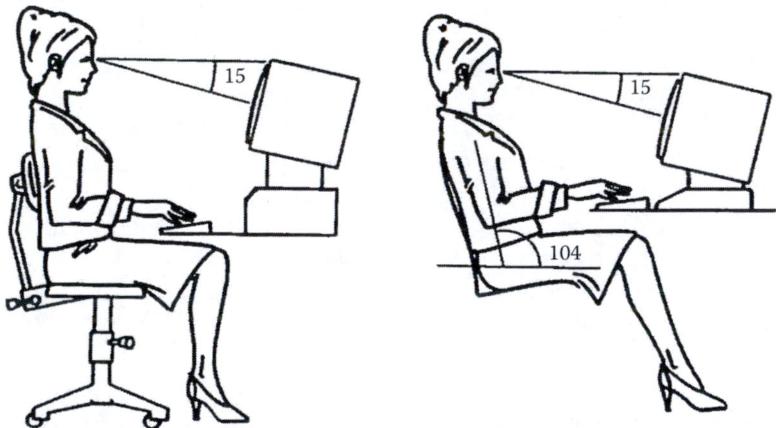

FIGURE 7.5 Working posture at the visual display terminal as recommended by Çakir et al. (1980) (left) and Grandjean and colleagues (Grandjean, 1987; Grandjean et al., 1984) (right).

that schoolchildren use their portable computers sitting cross-legged, lying prone propped up on their elbows, leaning against a wall with the computer on their flexed knees and even sitting on the floor with the laptop on a desk. Portable computer users are unlikely to use document holders, although, for many, their work involves the simultaneous use of paperwork and screen. This may be an additional factor leading to poor head and neck postures.

Among design issues are the position of the screen, the layout of the keyboard and the size and spacing of the keys on the keyboard. The viewing distance for the screen of a portable is likely to be shorter than for a desktop VDU. There has been little discussion of this issue, although several studies have noted reports of visual tiredness and discomfort (Straker et al., 1997; Heasman et al., 2000; Wilson, 2001; Jonai et al., 2002). Potentially, the screen could be too close for visual comfort (or poorly matched to the prescription of the spectacles being worn); this will also result in the screen subtending a wider visual angle and requiring larger glance angles. It would be preferable to have a detachable screen, both so that the placement of the computer to give a good height for keying is not to the detriment of the head and neck posture and so that the screen can be placed at a more comfortable viewing distance.

Keyboards of portable computers may be smaller than keyboards of desktop PCs. A small keyboard will constrain arm postures more and, inevitably, mean that keys are spaced closer together. Heasman et al. (2000) found that this could cause difficulty for people with large fingers but did not seem to be a problem for people with small fingers. However, closely placed keys probably demand more precision in typing. Keyboards should have an adequate surface in front of the keyboard to support the wrist and have tilt adjustment. ISO 9241-5 recommends a wrist support surface with a depth of at least 100 mm in front of all input devices (ISO, 1998b). The edge of the support should be radiused so that it does not cut into the wrist.

The device used to control the cursor on a portable computer is likely to be a fingertip control such as a touch pad (glide pad), a trackball or a trackpoint (a pointing stick which is smaller than fingertip width). The last of these was found to be the least preferred in Heasman et al.'s (2000) survey, and they recommend that users should be able to attach an external mouse. Sommerich et al. (2002) have shown that postural fixity is greater when such an external mouse is not available.

7.6 COMPUTERS IN SCHOOLS

While not strictly an office environment, computer workstations in schools have very similar requirements — with the added constraint on design imposed by the greater variation in anthropometry among the users. In other words, how well is display screen equipment and school furniture matched to children's needs, and how can the workstation best be designed to accommodate adult and child users at the same time?

This is a difficult design problem, and the second question at least has not been much studied. Whether sitting or standing, teachers will have difficulty in reading from a screen which is placed at a suitable height for children, and they will have to bend forward when using input devices, with the consequent musculoskeletal loading. In nursery and primary schools, these problems are of course even greater, and teachers have the options of kneeling, crouching or sitting on child-sized seats — all of which are undesirable. Coole and Haslegrave (2000) showed the extent of the resulting problems in a study of work in nursery schools (for children aged 3 to 5 years).

Berns and Klusell (2000) surveyed computer workplaces in schools and showed that the dimensional problems of computer workstations, as already present in offices, are accentuated in schools. For example, when children use a standard keyboard together with a mouse, their reach distance, relative to their anthropometry, will be much greater than that of an adult. In classroom work, several children tend to use a computer together, and desks should be developed to provide more adequate views of the screen and access to input devices. Both Coole and Haslegrave (2000) and Berns and Klusell (2000) suggest that children could work at raised workstations (with adequate foot support of course) to facilitate interaction with teachers.

Briggs et al. (2004) have illustrated the consequences of children of different ages using a typical Australian classroom computer workstation (where the desk was 720 mm high). These consequences were mainly in head tilt and neck flexion, rather than in trunk posture. Head tilt and neck flexion were greater among the older, and taller, children and less in younger children (4 to 8 years).

Harris and Straker (2000) have noted the increasing use of portable computers in schools, with their advantage of flexibility in use. Briggs et al. (2004) compared the postures adopted by children when working at VDU screens and at laptop computers on school desks. They found effects similar to those for adults, with more flexed postures when working at a laptop, probably due to both the screen being lower and the lack of separation between keyboard and screen. However, head tilt was less during laptop use than when using a desktop computer, but children up to 9 years old tended to extend their atlanto-occipital joint and flex their lower cervical spine, adopting an undesirable forward head posture. The effects on posture are complex, since they result from interactions between the geometry of the workstation and of the equipment, the size of the children and the tasks that are being performed. One conclusion from the study is that workstations should be better matched to the age group of the children.

Use of computers at poorly designed workstations has become a major source of musculoskeletal problems among children. Berns and Klusell (2000) drew attention to the need for training of teachers in the ergonomics of workstation arrangement, as well as improvement of workstation design, since teachers have an influence on both the selection of furniture and the ways in which desks, seats, VDUs and input devices are arranged and used (and also, very importantly, on whether children understand the need for a healthy arrangement of their workplaces).

7.7 INPUT DEVICES

Use of input devices can influence whole upper body posture as well as the localised hand and arm posture and can be a source of considerable static loading during work at a computer. Cook and Kothiyal (1998) showed that the position of the mouse on the desk had a great influence on muscular activity in the anterior and middle deltoid muscles. When keyboard and mouse must be used together at a workstation, it is advisable for right-handed people to use a keyboard without a numeric key pad, since Cook and Kothiyal found the resultant more distant position of the mouse caused an increase in abduction and flexion of the shoulder. Delisle et al. (2004) investigated the alternative solution of operating the mouse with the left hand and found that most right-handed users (60% of their experimental subjects) could convert to this within a month's practice after ergonomics training, although they still took longer to complete mouse tasks. On whichever side the mouse (or other input device) is placed, Woods et al. (2002) recommend the use of a (curved) L-shaped desk, in which the computer is placed within the curve, giving an additional surface for full forearm support when using any nonkeyboard input device to the right (or left) of the monitor.

The actions required differ when using different input devices, and many alternative devices are now available, those other than the keyboard often being called NKIDs (non-keyboard input devices). The effects of using NKID devices differ amongst themselves almost as much any differ from the use of a keyboard. With a keyboard, for example, touch typists perform very rapid movements with the fingers of both hands but hold their upper arms static, flexed and slightly abducted and their forearms inwardly rotated. Wrists are held in extension and ulnar deviation. In contrast, mousing involves only one arm, with a more static posture of the hand, wrist and fingers; most of the dynamic activity is of the two fingers for clicking, with slight ulnar/radial movements of the wrist for pointing and dragging.

Erdelyi et al. (1988) showed that there was continuous 'static' activity of muscles in the trunk, neck, shoulders and arms maintaining the stability of the body during typing. Cooper and Straker (1998) compared muscle loading while using a mouse with that while using a keyboard and found that there was a tendency for the anterior deltoid muscle to be more heavily loaded during mouse work and the trapezius muscle during keyboard work, possibly because of mouse users supporting their whole forearm. Karlqvist et al. (1994) showed that posture did indeed differ between mouse and keyboard use. During mouse use (in word-processing tasks) wrist deviation was more than 15° to the ulnar side for much of the working time, with the shoulder rotated outward by more than 30°. In contrast, those who used a keyboard without a mouse had their wrists in a more neutral position (and with slight radial rather than ulnar deviation), while their shoulders were in a generally neutral posture with slight inward rotation. From the results of this and other studies, it would seem advisable to use a short keyboard (without a number pad) whenever possible so that the mouse can be positioned more centrally when it is used in combination with a keyboard.

The general problems with current nonkeyboard input devices were found in Woods et al.'s (2002) survey to be inappropriate size (in relation to the user's hand anthropometry), awkward shape (in relation to the grip surfaces and other surfaces used to generate movements), discomfort in posture or grip and poor precision when using the device. The aim in the design of any of the input devices (combined with workstation arrangement) should be to keep the wrist within its neutral posture zone. Static loading on the arm, shoulder and neck should be minimised, with the arm supported if the upper arms cannot be relaxed.

7.7.1 The Keyboard

The keyboard should be as thin as possible. It is usually recommended that its slope be adjustable, although in biomechanical terms a very slight rake would seem preferable to a steeper one. Users often have strong views about the feel of different keyboards. One user once described the difference between working at two particular makes of keyboard as 'like the difference between walking on turf and walking on pavement'. At present, however, we know little or nothing about these matters on a formal scientific basis. The features of the keying action which give the keyboard a 'good feel' remain elusive. Different users seem to have different (and sometimes conflicting) views on the subject.

Different users also have different styles of typing, which can in turn lead to exertion of excessive force on the keys. Martin et al. (1996) measured the force used by different fingers during keying when typists were keying at a comfortable speed. They found that, on average, peak key reaction forces ranged from 3.33 N at the thumb to 1.84 N at the little finger, with an overall average of 2.54 N. The forces that they applied were therefore much higher than the force needed to depress the key switch on the keyboard tested (0.47 N), but there was considerable variability between the 'heavy touch' and 'light touch' typists. Both ISO 9241-4 and the U.S. standard ANSI/HFS 100-1988 recommend an upper limit of 1.5 N for key actuation force, key displacement between 2 and 4 mm and tactile feedback to confirm that the key stroke has registered (ISO, 1998a; ANSI/HFS, 1988), but Rempel et al. (1997) found a significant difference in force applied by typists between keys with actuation forces of 0.47 N and 1.02 N and recommended a lower force limit. ISO 9241-4 recommends 0.5 to 0.8N as the preferred force level for key actuation.

There is good evidence that neither the conventional QWERTY key layout nor the presentation of the alphabetical keys in a single rectangular key tray on a plane surface is an optimal ergonomic solution to the problem of keyboard design (see Pheasant, 1991a,b for a detailed discussion). The former issue is to all intents and purposes a lost cause, since existing keyboard users have invested so much time and effort in developing their skills in using the QWERTY layout. There are more realistic opportunities for improvement in terms of the keyboard itself, and various keyboard designs are now available (with negative tilt, split keyboard, and contoured key trays).

The basic problem with the plane keyboard and rectangular key tray is that a degree of ulnar deviation of the wrist will inevitably be required to maintain the alignment of the fingers at the keys. (This is accentuated if the keyboard is too high [see above] and if the user leans forward [see below].) A number of split keyboard designs have been produced to overcome this problem by allowing the half of the keyboard used by each hand (of touch-typists although not of hunt and peck, two or three finger typists) to be angled and sloped in natural orientations for the wrists of the right and left hands. Trials of these keyboards seem encouraging (see, for example, the early study of Grandjean [1987]). Studies of wrist and finger postures when using a range of nonplanar and split keyboards indicate that these new designs have potential for reducing biomechanical stress but that the degree of improvement varies between individuals, perhaps because of differences in hand anthropometry. Research is needed to fully understand the effects of the design parameters (Marklin et al., 2000; Treaster and Marras, 2000; Zecevic et al., 2000). Çakir (1995) advises that a split keyboard should have angle-adjustment, so that the user can select the best position for him- or herself. Hedge et al. (1999) and Simoneau and Marklin (2001) have shown that a negative tilt to the keyboard can also reduce wrist deviation, and Gerard et al. (1994) showed that a contoured keyboard can reduce the levels of finger and wrist muscle activity.

7.7.2 The Mouse

The second input device, the mouse, must be adequately positioned on the desk relative to both VDU and keyboard and be designed according to the general principles of hand-operated controls given in Chapter 6. For tasks which involve the use of both keyboard and mouse (and probably for the majority of office workers engaged in word-processing and data-entry tasks), it is not possible to have both input devices in their optimal positions at the same time. However, both keyboard and mouse mat are relatively easy to move provided that a large clear area is left on the desk to facilitate their positioning. McCormick's first principle (Table 4.1) applies when one or the other is used predominantly.

There has been concern that people who use the mouse extensively (in layout work, etc.) are prone to suffer from upper limb disorders similar to those of keyboard users — the causative factors presumably being the particular combinations of static muscle loading due to working posture and the repetitive motions of the wrist and fingers that the tasks in question entail. There is probably more potential for static loading of the fingers and hand when using a mouse than when using a keyboard, particularly for 'click, drag and drop'. Users can also be observed gripping the mouse continuously, even when it is not actually in use. There is a role here during training to help in raising computer users' awareness of the subtle influences of their task behaviour and postures on musculoskeletal loading. The mouse should be grasped with arm, wrist and fingers all relaxed and must be an appropriate size and shape for the user's hand, as well as having adequate surfaces for the fingers and palm of the hand controlling sensitive movements of the mouse (while leaving the first two fingers free to operate the buttons and for scrolling). From personal experience, both mouse design and handling technique seem to be important in minimising static muscle loading. In a 3-year field study, Aarås et al. (2002) showed that a mouse designed to be used in a more neutral wrist posture can significantly reduce musculoskeletal symptoms among VDU operators.

Various styles of mouse can be found, offering different handholds and hand postures (for example a flat 'whale' mouse or a joystick mouse) while still attempting to keep the wrist in a neutral posture. Wardell and Mrozowski (2001) measured differences in wrist flexion and extension between these but noted that the majority of movement in the use of a mouse is in the fingers rather than the wrist and that this will be influenced by software settings controlling sensitivity. Moreover, wrist posture for most mice is probably more affected by the position of the mouse on the desk than by the design of the mouse itself.

7.7.3 Other Input Devices

We may confidently predict that as time goes by, the keyboard or mouse as such will increasingly be replaced by other devices for entering information onto the computer and otherwise controlling its functions. The extent to which these will supplant the keyboard and mouse in everyday applications — or perhaps more realistically the rate at which they will do so — is impossible to estimate at present. Current input devices include joysticks, trackballs, trackpads (or touchpads), light pens, tablets and pens and touchscreens. The last three are best used for imprecise pointing tasks or item selection (menu or icon entry). The other devices are used for various tasks, including selecting, dragging, cutting, pasting and scrolling.

Woods et al. (2002) studied the patterns of use of nonkeyboard input devices (including the mouse). Apart from the mouse, relative importance of the different devices is probably as much a function of the individual user's work tasks and of the software used as of the choice of the user. Woods et al. found that users of joystick mice, trackballs and CAD tablets used these intensively and tended to keep their hand on the device for the majority of the time, in contrast to most mouse users who used them in conjunction with a keyboard. Another problem they identified was the lack of space available on workstations for such devices, since many have to incorporate a keyboard or other equipment.

Of the few studies that have been carried out on trackballs, Karlqvist et al. (1999) found the trackball they tested for a text editing task to produce lower shoulder elevation than a mouse but greater wrist extension. However, this cannot necessarily be generalised to other designs of trackball or to other input tasks.

Aarås et al. (1997) found that muscle effort was reduced when using a joystick (vertical) mouse, but Straker et al. (2000) reported that this type of mouse was less easy and slower to use than a traditional mouse. It is operated with a relaxed power grip (to some extent a hook grip) rather than a precision grip. This means that the control of the mouse movements is likely to be less precise and so may be less suitable for some tasks than others. There has, so far, been very little investigation of these matters or of the suitability of particular input devices for different tasks.

The ultimate alternative to the keyboard (short of psychokinesis) is voice input with speech recognition software. This is an option for word processing, data entry and control functions. It will doubtless be suspected of causing repetitive strain injury of the vocal chords! Nevertheless, it is an important development for computer users who at present have to use keyboards or other input devices intensively, particularly if they experience musculoskeletal problems in their use.

One further type of input device deserves mention: the web camera. For its informal use, the anthropometric requirements influencing its location are relatively simple (assuming that its orientation can be adjusted in all axes), and the effectiveness of its location can easily be assessed by monitoring its view on the VDU. However, in a more formal context, web cameras are an important element of the effectiveness of teleconferences; for these, the locations and fields of view can have a significant impact on the positioning of and interaction between the participants. This aspect seems to have been little studied so far.

7.8 WHAT MAKES A 'GOOD POSTURE' IN SCREEN-BASED WORK?

Concern over the increased reporting of work-related upper limb disorders in the white-collar population (see Section 9.6) initially led (and to some extent still does) to a demand that keyboard users should be taught 'the correct way to sit'. To the lay person this requirement may seem straightforward enough. It is not quite as simple as it seems, however, and a number of unresolved issues remain. There are two basic schools of thought concerning the matter (although, with dynamic seating, many other schools of thought are emerging). We could call those the orthodox or 'perpendicular' approach, and the alternative or 'laid-back' approach. The postures in question are illustrated in Figure 7.5. The views that people take on these matters are not always stated entirely explicitly, and indeed the whole issue is surrounded by something of an air of vagueness. Overall, the adherents of both schools would be in a broad measure of agreement concerning the anthropometric criteria proposed at the beginning of this chapter. They would also agree that working postures that entail *forward leaning* (and in particular the yuppie hump) are highly undesirable. The principal points at issue are, first, the desirable position of the trunk, and second, whether or not the wrists should be supported.

The *perpendicular position* for keyboard work (in which the trunk is as far as possible kept upright with the back principally supported in the lumbar region so as to maintain its 'normal' curve, as on the left of Figure 7.5) has, to the authors' knowledge, been taught in schools of typing since the 1930s at least. It may be regarded as representing the conventional wisdom on the subject, which until the mid-1980s or thereabouts would have commanded more or less universal consent. People who still teach the perpendicular position sometimes refer to it as 'sitting in balance'. This sounds very good, but it does not really get us much further forward in understanding the issues involved, in the absence of a formal explanation of what is meant by 'balance', framed in the language of physiology and biomechanics. Such an explanation has not been forthcoming.

The first significant challenge to the orthodox view was made by the late Professor Etienne Grandjean and his coworkers in Zurich in a series of papers published in the early 1980s, the findings of which are summarised in his book *The Ergonomics of Computerized Offices* (1987). Grandjean's views were based upon trials in which VDU users were provided with fully adjustable chairs and workstations to use in their own offices. The great majority of subjects preferred a 'laid-back' position in which the trunk was reclined by between 10° and 20° to the vertical. Only about 10% of subjects chose to sit upright. Where a padded wrist support was available, the great majority of subjects (80%) chose to use it; where it was not available, around half of the subjects chose to rest their wrists on the desk. On average, the subject's elbows were flexed to a little less than a right angle, so the forearms were inclined slightly upwards.

Grandjean (1987) argued forcefully that in biomechanical terms there is nothing whatsoever wrong with this position. He mainly based his views on the experimental evidence of Andersson et al. (1974, cited in Section 5.2), who showed that when the trunk is thus reclined, the loading on the lumbar spine is substantially less than it is when sitting upright. He did not, however, explicitly address the issue of upper limb disorders in this context.

On balance, a supported wrist when using the keyboard or mouse would seem to be desirable rather than otherwise, in that it will reduce the static loading on the muscles of the neck, shoulder and arm, as clearly shown by Aarås et al. (1997, 1998) for both keyboard and mouse work. There are two caveats to this general position. One is that supporting the wrist on the sharp edge of the desk (which you see quite commonly) can cause blunt trauma to the tissues of the front of the wrist (and in particular to the ulnar nerve). The second is that it may result in a 'cocked' (i.e., extended) wrist, which will in turn cause static loading of the muscles in the extensor compartment (i.e., back) of the forearm. The latter is particularly likely to be a problem if the keyboard is abnormally thick or if it is used in a steeply raked position. Both are highly undesirable. Both may be avoided by the use of a padded wrist support (while raising the seat height slightly to allow for this). In addition to reducing the mechanical loading on the lumbar spine, the laid-back sitting position (particularly when combined with the supported wrist) has the additional advantage of tending to increase the overall horizontal distance between the user's shoulders and the keyboard. (The disadvantage that would accrue in terms of static load is eliminated by the supported wrist.) It follows, as a matter of geometry, that the degree of wrist deviation required to maintain the alignment of the fingers on the keys will be correspondingly diminished. As the trunk moves from a reclined position to an upright position, and then from an upright position to a forward sitting position, the elbows must also move out sideways to accommodate the width of the lower part of the rib cage. This results in a further and progressively more pronounced ulnar deviation of the wrist.

Opponents of the laid-back approach argue that it tends to degenerate into a slumped position similar to that of the yuppie hump. By the same token, the upright position will tend to degenerate into a forward slump. With a well-designed seat that gives good back support, the former tendency will be minimal, whereas the latter will still be present to a more marked extent, particularly when the user is tired or under stress.

On the basis of these considerations, therefore, the laid-back position for keyboard work offers material advantages as compared with the perpendicular position. Having said this, one should not be unduly prescriptive in advising the individual keyboard user as to what constitutes a good posture. It is more important that he or she learn the importance of postural diversity in the workplace and the avoidance of unnecessary muscle tension. The laid-back position is only desirable in that it is materially more likely to achieve these aims.

Figure 7.6 (an unposed original photograph) shows a keyboard worker in a natural relaxed sitting position at a well-designed fully adjustable workstation. While her display screen appears to be lower than desirable, this may be acceptable for her specific task if she is only using the top portion of the screen for data entry (since this portion of the screen falls within the acceptable display zone.

One final aspect of office work is use of the telephone, which can induce poor head and neck postures as well as raised arms. Simpson and Buckle (1999) showed the advantage of substituting a headset for a traditional telephone when its use is frequent. Their study group of office workers, who used the telephone on average for 2.6 hours a day, found improved neck and trunk postures and reported significantly less neck discomfort after a month of head set use.

7.9 THE DESIGN OF SCREEN-BASED WORKING TASKS

It is widely recognised that prolonged periods of intensive screen-based keyboard or mouse work (particularly repetitive tasks such as data entry and copy-typing), unbroken by rest pauses or changes of working activity, are highly undesirable and have the potential to result in musculoskeletal injury. The greater the exposure to

FIGURE 7.6 Keyboard worker in a natural, relaxed position.

intensive keyboard or mouse use, the greater the risk, however good the ergonomics of the workstation and working posture; the worse the ergonomics, the greater the level of risk for a given level of exposure (see also Chapter 9). The risks of maintaining a seated posture for a prolonged period also apply.

In organizing office work, therefore (and in allocating the various tasks to be performed in the office to individual workers), periods of intensive screen-based keyboard use should, wherever possible, alternate on a frequent and repeated basis with other working tasks of a contrasting nature. Where this cannot reasonably be achieved (as regrettably is all too often the case in the electronic sweat shops of the data entry trade), a suitable daily schedule of rest pauses should be set up, but this is by way of being second best.

It is difficult to be precise about these matters. As an approximate rule of thumb, 'screen breaks' of around 5 minutes in each continuous period of keyboard work would seem about right — these being taken *in addition* to the person's normal lunch break and mid-morning and mid-afternoon coffee breaks. The office worker should be actively encouraged to get up, move around and stretch during these breaks. The habit of taking lunch at the desk is strongly to be discouraged. 'Micropauses', in which the user consciously stops work and relaxes against the backrest of the chair for a few seconds every few minutes, are also very valuable. Rearrangement of the workstation may also help to introduce occasional changes of posture and movement. Simply relocating a printer at some distance from the desk can help in this.

In other words, if an adequate degree of task diversity cannot reasonably be achieved, then the working period should be structured in such a way that it is broken up with micropauses, short pauses and long pauses, in which recovery from fatigue can occur.

8 Ergonomics in the Home

8.1 INTRODUCTION

The house and home may in general be divided into a number of more or less discrete spaces, each of which is specialised for the performance of a particular range of purposeful activities, which in a broad sense we could call 'working tasks'. Most typically the boundaries of these spaces will coincide with the rooms of the house. The spaces may overlap physically to some extent, and tasks may intrude to some extent from one space to another. In this chapter we shall consider the ergonomics of three such 'workspaces': the kitchen, the bathroom and the bedroom and an important circulation space: the staircase.

8.2 THE KITCHEN

Of the specialised spaces of the home, the kitchen is the one that can most obviously be treated as a functional working area, in the ordinary narrow meaning of the word '*work*', and it is perhaps because of this that it is the one that has been discussed most extensively in the ergonomics literature. We shall consider, first, the layout of the floor plan and, second, the heights of working surfaces and other related matters.

8.2.1 Layout

The literature on the layout of a kitchen reveals two basic design principles, which turn out to be variants of McCormick's sequence-of-use and frequency-of-use principles, respectively (see Table 4.1).

1. For a right-handed person, the sequence of activity proceeds from left to right thus: sink to main worksurface to cooker (or hob) to accessory worksurface for putting things down. It clearly makes sense for this sequence to be unbroken by tall cupboards, doors or passageways; but it need not be in a straight line — an L- or U-shaped configuration will serve just as well. Another accessory work surface to the left of the sink completes the layout (Figure 8.1). The reverse sequence would suit a left-handed person better, but both are workable. The proximities of the units are more important than the direction of the sequence.
2. The refrigerator (or other food store such as larder, freezer, etc.), sink and cooker constitute the much discussed 'work triangle' of frequently used elements. For reasons of safety, through circulation should not intersect this triangle — particularly the route from sink to cooker, which is used more than any other in the kitchen. The sum of the lengths of the sides of the triangle (drawn between the centre fronts of the appliances) should

TABLE 8.1
Optimum Heights (millimetres) of Kitchen Sinks and Worktops

	Men			Women		
	5th %ile	**50th %ile**	**95th %ile**	**5th %ile**	**50th %ile**	**95th %ile**
Worktop	930	1015	1105	855	930	1005
Sink (top edge)	1005	1090	1180	930	1005	1080

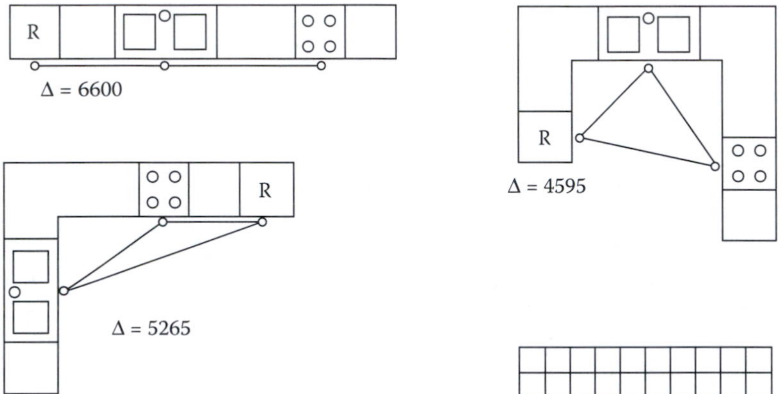

FIGURE 8.1 Three kitchen layouts designed, according to the principles discussed in the text, on a 300-mm modular grid. Each includes refrigerator (R), cooker and similar lengths of working surface. The work triangle is indicated, and the sum of the lengths of its sides, (Δmm) is given.

fall within certain prescribed limits. Grandjean (1973) quoted maxima of 7000 mm for small to medium-sized kitchens or 8000 mm for large kitchens. The Department of the Environment (1972) gave a minimum combined length of 3600 mm and a maximum of 6600 mm for the kitchen 'to leave adequate working space and yet be reasonably compact' and also specified that sink–cooker distance should be between 1200 and 1800 mm in length (Figure 8.1).

8.2.2 Worktop Height

In order to determine an optimal height for kitchen working surfaces, we must consider both the anthropometric diversity of the users and the diversity of tasks to be performed. If, as is commonly the case, a 175-mm deep sink is set into the worksurface, then the effective working level may range from perhaps 100 mm below the worktop height when washing up to a similar distance above it when operating machinery or mixing with a long-handled spoon. We should expect differences, even among tasks performed upon the surface itself, associated with varying requirements of downward force, for example between rolling pastry and spreading butter.

Ward and Kirk (1970) studied these matters by means of a fitting trial. The subjects, who were all women, performed three groups of tasks and selected the following worktop heights as optimal:

- Group A — tasks performed above the worktop (peeling vegetables, beating and whipping in a bowl, slicing bread): 119 [47] mm below elbow height
- Group B — tasks performed on the surface (spreading butter, chopping ingredients): 88 [42] mm below elbow height
- Group C — tasks involving downward pressure (rolling pastry, ironing): 122 [49] mm below elbow height

These results were subsequently confirmed using a variety of physiological measurements (Ward, 1971) in which it was also shown that the optimal height for the top edge of the sink was approximately 25 mm below the elbow.

The next stage in the analysis is to allocate priorities to these three groups of tasks. Ideally, this would best be done by means of an observational survey of user behaviour. By way of second best, we ask a sample of 'typical' kitchen users and find a general agreement that the group B tasks are more important and the group C tasks least important. Allocating a weighting of 4 to group B and a weighting of 1 to each of the others, we arrive at an overall recommendation of 100 mm below elbow height for the optimum height of the worktop. Combining this with anthropometric data for our standard population (Table 2.5) gives us the figures set out in Table 8.1.

The dimensional coordination of kitchen equipment is obviously desirable, and the purchaser should have every confidence that a new oven will fit into the existing range of units. The provision of adjustable worktops, or fittings in a range of heights, is not incompatible with this goal, but it does make things more difficult (and therefore more costly). In response (presumably) to Ward's studies, BS 3705 (BSI, 1972) read thus:

> Subject to the need for field research and solving technical problems, it is thought that a 50 mm incremental range of heights of working surfaces may be adopted in the future, the ranges being 900 to 1050 mm for sinks and 850 to 1050 mm for worktops. Because studies show that generally the worktop surface needs to be higher than the present 850 mm for the greater number of users, this standard is omitting the 850 mm worktop height, although this might be included in any subsequent range after the above research is completed. As an interim measure the standard will remain at the BS 3705 (imperial) height, rounded off metrically to 900 mm for sinks and worktops.

A decade later this interim measure had acquired a distinct air of permanence. BS 6222 (Part 1; BSI, 1982) read, 'The co-ordinating heights of all units and appliances shall be as follows ... top of worktop: either 900 or 850 mm (second preference)'. ISO 3055 (1985) also specifies a standard worktop height of 850 or

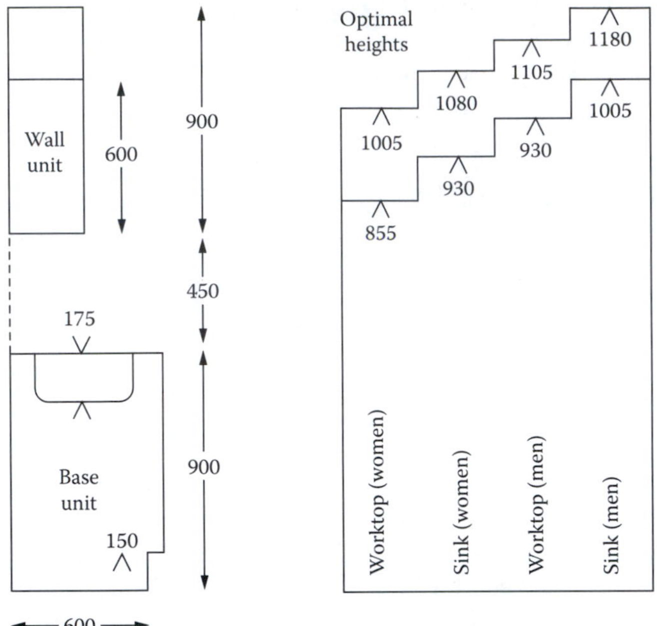

FIGURE 8.2 Standard kitchen unit: (left) base and wall units and (right) optimal ranges of worktop heights (5th to 95th %ile) (dimensions in millimetres). Analysis of storage space according to the criteria of Table 4.11.

900 mm but hedges its bets in an annex (which is not regarded as part of the standard proper) by referring to 'appropriate' heights of 850 to 1000 mm for food preparation and 900 to 1050 mm for washing up, suggesting that 'adjustments can be different plinth heights and other means'.

Figure 8.2 shows a set of standard kitchen units in side elevation as compared with the height optima from Table 8.1. The 900-mm worktop is lower than ideal for approximately half of women and virtually all men. The sink (with its rim at the same height as the worktop) is too low for just about everyone.

Compare this state of affairs with that of the standard office desk, which we discussed in the last chapter (section 7.2). Our analyses indicate that the 900-mm standard for the kitchen worktop is far from being the best single compromise height (and the 850-mm standard, which fortunately does not seem to be used in practice, would be that much worse).

How serious a problem is this in practice? This question (as is always the case with such matters) is more difficult to answer. The potentially deleterious effects on the tall user of working at a sink or worktop that is too low will depend to a considerable extent on how long he or she does it for at a stretch. (Compare this again with the matter of exposure to intensive keyboard use discussed in Section 7.9.) For those of us who are fortunate enough to have sound backs, a sink or worktop that is too low will be no more than a minor annoyance which will in all probability go more or less unnoticed. In other words, it is something that we are readily able to adapt to (the third fundamental fallacy; see Table 1.1). For those of us who are less fortunate in this respect — and low back trouble is very common indeed — the task of washing up or preparing a meal at a sink or worktop that is too low may be an intensely painful experience. At the point at which it becomes intolerably painful, the person is effectively 'disabled' in respect of this task.

Some elderly people, by contrast, may find higher worktops too high. Ageing is associated with a shrinking in height, starting perhaps in the 40s and becoming marked in the 70s and 80s. A comparison between the mean statures for British men aged 19 to 65 years and those aged 65 to 80 years (Tables 10.1 and 10.5) shows a difference of 55 mm, which is similar to Stoudt's (1981) findings for the U.S. population. The corresponding difference for British women is 40 mm. Since the 19- to 65-year-old population data is usually used for design purposes, the needs of elderly people may not be met. Although arm length does not change significantly, the elderly do find it more difficult to stretch to reach objects (Stoudt, 1981). So, as Kirvesoja et al. (2000) pointed out, the elderly will also find it more difficult to adapt to use worktops of an inappropriate height. The options of adjustability or of selecting from a range of worktop heights (as implied by the annex of ISO 3055) are more important for them. British standard BS 4467 (BSI, 1991) provides a useful guide to dimensions for many aspects of the layout of a kitchen and of the home more generally.

It might seem desirable to have worktops of, perhaps, two different heights within a kitchen to cope with the range of tasks in groups A–C and the various heights of cooking utensils, but this is not as satisfactory when the worktops form a continuous run. Steps in height create hazards, particularly in view of the hot and potentially flammable ingredients used in a kitchen. Slight differences in height can, however, be tolerated between clearly defined units, as at the edge of a draining surface of a sink.

8.2.3 Storage

It is instructive to compare side elevation views of standard kitchen units (Figure 8.3) with ergonomic recommendations for storage zones (discussed in Chapter 4; see Table 4.11). These recommendations should also be considered for the siting of eye level ovens and microwave ovens. The dishes and pots lifted into ovens can be very heavy. Noble (1982) recommends that shelves (300-mm deep) up to the height of 1400 mm can be used by 95% of the elderly population, but this should be restricted to a maximum height of 1350 mm if they have to reach over an obstruction such as a counter top. Both Noble (1982) and Kirvesoja et al. (2000) found that shelves below about 300 mm should be avoided for the elderly, who have difficulty in bending and kneeling. Deep shelves are particularly difficult and, in this case, the shelves should be at a height of at least 500 mm.

Cupboard space in the 800 to 1100 mm optimum height range being strictly limited, the most accessible storage space in the kitchen becomes the worktop, which disappears under a clutter of food mixers, spaghetti jars and other homeless objects, another example of Pheasant's principle of ergonomic decay (see also Section 7.2).

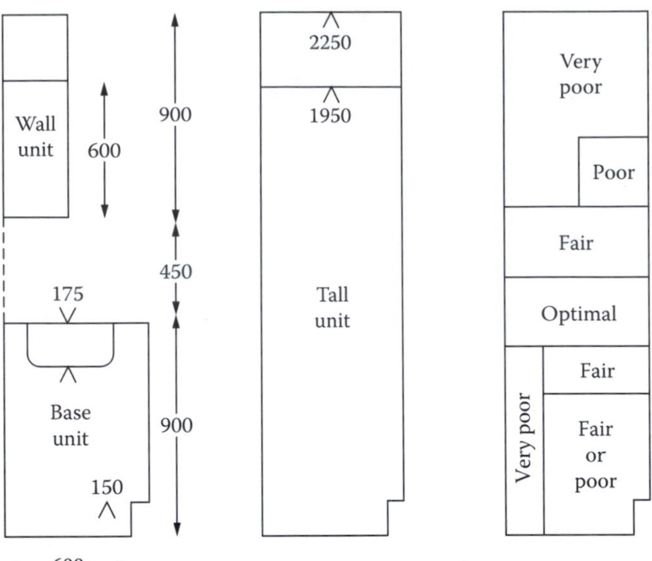

FIGURE 8.3 Standard kitchen tall unit (compared with the arrangement of base and wall unit [left]) (dimensions in millimetres). Analysis of storage space according to the criteria of Table 4.11.

8.3 THE BATHROOM

The bathroom should combine hedonistic luxury with functional efficiency. It is an environment in which to relax and unwind, soaking in a hot tub, but also a configuration of workstations for the practical activity of washing, grooming and excretion (assuming a special room is not set aside for the latter). *The Bathroom,* by Alexander Kira (1976), is a classic of user-centred design research, which every interested person should endeavour to read.

8.3.1 The Bathtub

The bathtub presents interesting problems of dimensional optimization. It must be large enough for comfortable use by one person (or perhaps two) but should not have needless volume, requiring filling with expensively heated water. It is also a notoriously hazardous environment for the frail and infirm.

Two principal postures are adopted in the bath: a reclined sitting position and a recumbent position (possibly with the knees flexed) in which the body is submerged to neck level. For comfort in the sitting position the horizontal bottom of the tub must be sufficient to accommodate buttock-heel length (95th %ile man: 1160 mm), and the end of the bath should provide a suitable backrest. Kira (1976) recommends a rake of 50 to 65° from the vertical and contouring to conform to the shape of the back. This seems excessive — a rake of 30° and a suitable radius where the base meets the end should be quite adequate. We are not particularly looking for postural support since the buoyancy of the water will both unload the spine and lift it away from the backrest. The more we increase the length of the horizontal base, the greater the possibility for total submersion. We may shorten our recumbent bodies by around 100 mm by flexing our knees, given that we wish to keep our heads above water and that, as the 95th %ile male shoulder height is 1535 mm, there seems little point in lengthening the horizontal part beyond around 1400 mm.

The width of the bath must at least accommodate the maximum body breadth of a single bather (95th %ile man: 580 mm). The ergonomist, who is usually a broadminded sort of person, should also consider the accommodation of couples. For couples wishing to sit side-by-side, the necessary clearance is given by their combined shoulder breadth (920 mm for a 95th %ile couple of opposite sex). For couples sitting at opposite ends (probably the more common arrangement) the clearance is given by combined breadth of the hips of one person and the feet of the other. This is greatest when the hips are female and the feet are male, in which case the 95th %ile combination is 625 mm. This arrangement does, however, demand that the taps should be in the centre of the bath to avoid arguments (Figure 8.4).

Consider a 95th %ile man (sitting shoulder height: 645 mm) reclining against the end of the bath. His shoulders will be 645 cos 30° = 558 mm above the base of the bath. He could not reasonably require more than 400 mm of water. If the backrest was raked further, say to 45°, then 300 mm of water would suffice. (These figures are mere speculation; it would be very interesting to perform a fitting trial to find out what depth of water people really do want.) Assuming a 30° rake, a bath depth of 500 mm would be required for an adequate quantity of water without too much danger of it splashing over the edge. In fact, a typical tub depth at the present is about 380 mm, although older models are often deeper. The outside height of the rim (above the floor) is, of course, generally greater than the tub depth (often by as much as 100 mm).

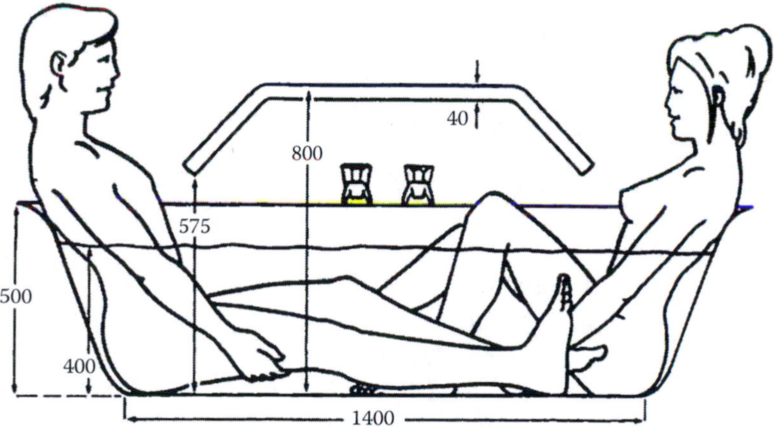

FIGURE 8.4 The ergonomically designed bath (dimensions in millimetres).

A deep bath or a high rim is generally deemed to make entering and leaving more difficult and hazardous, although Kira (1976) casts some doubt on this, arguing that the manoeuvres people use to enter and leave baths have been insufficiently analysed. Grandjean (1973) cites evidence that a height of 500 mm is acceptable to most elderly or frail people. Grab rails are usually advocated as an aid to stability. These could reasonably be a little above knuckle height at the point where you climb in (e.g., 800 mm above the bath base), around shoulder height (e.g., 575 mm) at the sitting end and about 40 mm in diameter. Vertical grab rails may be better for the infirm, provided they have a nonslip surface. Additional holds along the side of the bath are also desirable and, for the frail, a nonslip mat inside the bath is essential.

8.3.2 The Handbasin

The handbasin will be used for washing the hands and face and sometimes the hair. The criteria are relatively simple: it should be possible to wet the hands without water running down the forearms, and bending should be minimised. Hence, a basin rim that is at about the elbow height of a short user would be appropriate (5th %ile woman: 930 mm unshod). Kira (1976) studied the above activities experimentally by observing subjects first miming the actions without the constraints of an appliance and then using an adjustable rig. On the basis of these fitting trials, he concluded that, for washing the hands, the water source should be located some 100 mm above the rim of the basin, which should be set at 915 to 965 mm. Conventional handbasins are much too low (commonly less than 800 mm), except perhaps for use by children. The present practice of placing the taps at or below the level of the rim seems based on the assumption that people will fill the bowl and wash in the water therein. In fact, according to Kira, 94% of people prefer to wash under a running stream of water.

8.3.3 The Toilet (or Water Closet)

The toilet (or water closet or WC) poses some surprisingly complex design problems, if sanitary, functional and comfort criteria are all considered, and there have been relatively few studies on this. The two main types of toilet design are intended for use in either a sitting or a squatting posture. One or the other design tends to predominate in any one country, but both types may occur in public areas outside the home.

McClelland and Ward (1976, 1982) provide a thorough discussion of the ergonomics of the seated toilet, designed for a preferred posture for defecation in which the body is supported under the buttocks and thighs and with the thighs almost horizontal. They also note that the elderly, who tend to have more restricted joint movement, prefer a relatively high seat. In their study they found that, among the U.K. population, the preferred height for a toilet seat was 404 mm for women and 430 mm for men, although there was considerable variability (standard deviation: 30 to 33 mm). There were no significant differences across age groups and, interestingly, there was no correlation between seat height preference and stature. McClelland and

Ward suggested that a seat height of 450 mm would be the best compromise between the preferred heights for adult men and women, but that this would significantly disadvantage small women (and presumably children also). Their recommendation was therefore a seat height of 400 mm.

There is a strong body of opinion that takes the view that the sitting posture, which in western society we use when emptying our bowels, is physiologically unsound. Proponents of this view — most notably Hornibrook (1934) — argue that a squatting position, in which the thighs are pressed against the abdominal wall, encourages an easy and physiologically more efficient bowel movement, which in the long run will help prevent a variety of nasty diseases (to which we are prone as a result of our diet and sedentary habits). We are not aware of this theory having been tested experimentally (it would be difficult to do so), but experience suggests that it is basically correct. The physiology remains unclear. It is not solely a matter of increased intra-abdominal pressure, as has been suggested, since this is the act of 'straining at stool' which is allegedly deleterious.

The plinth of the conventional WC is typically around 380 to 400 mm in height. To facilitate the adoption of a squatting posture, this would have to be halved. One consequence of reducing the height is that the buttocks would take a much greater proportion of body weight. So the contouring of the toilet seat itself would become much more critical for comfort. However as you would not have to sit there for so long, it probably would not matter so much.

Cai and You (1998) studied the anthropometric design of the traditional (i.e., unsupported) squatting-type toilets. When they interviewed 100 people in railway stations, parks and department stores in Taiwan, their respondents listed difficulties they encountered when using a squatting-type toilet as numbness in the legs (72.0%), difficulty in maintaining balance (26.9%), difficulty in standing up after use (20.4%) and difficulty in squatting down (12.9%). The maximum age in the sample was 55 years, so the survey did not include any elderly users. Thus even people familiar with using this style of toilet find problems with it, although 86.0% preferred it to sitting-style toilets in public facilities. One of the design parameters that Cai and You considered important was the footstep slope (in addition of course to the footstep dimensions, separation distance and angle of separation). From experimental tests, they judged a 15° slope to be the least fatiguing.

One major constraint on toilet design is that space is usually limited — the toilet often being referred to as the 'smallest room in the house' — and outside the home such constraints are particularly severe in the various types of transport (and particularly in aircraft). The layout therefore needs careful consideration to optimise the use of space — care which is conspicuously absent in all too many public facilities. Often, access clearance is poor because a toilet paper dispenser mounted on the wall impedes access or because there is no standing room to turn to close the door once you are inside. A second factor is the location of the toilet paper dispenser or water tap. This should be considered in relation to the method of dispensation so that it is easily accessible in the posture imposed by the type of toilet. It seems that little attention is given to the functional aspects of access and tasks performed in toilets.

The space requirements are even more critical for wheelchair access. The Good Loo Guide published by the Centre for Accessible Environments and RIBA Enterprises (Lacey, 2004b) gives helpful design recommendations with layouts and dimensions. This indicates that a clear area 1.5 m in diameter is needed for wheelchair turning space, while the room length and width for a fully accessible toilet should be 2.2 m by 2 m. A study to develop design guidelines for accessibility of aircraft toilets by people who use wheelchairs was sponsored by the Paralyzed Veterans of America, National Easter Seal Society, Multiple Sclerosis Society and United Cerebral Palsy Association (Warren and Valois, 1991). This study discusses the main principles and shows the absolute minimum space required when the wheelchair user is assisted in a lateral or 90° transfer between the wheelchair and toilet seat.

8.4 THE BEDROOM

Considering the amount of time we spend in bed and the importance of sound sleep to our overall well-being, it seems remarkable how little formal scientific study has been devoted to the ergonomics of bed design. Tall people commonly complain about beds being too short. Noble (1982) cites the results of a survey of beds on the market in the United Kingdom. Both single and double beds ranged in length from 1900 mm to 2360 mm. The length of the recumbent body is somewhat greater than stature, and the bed should be somewhat longer still, since people sometimes like to sleep with their hands above their heads. Assuming that a person will require a bed length of at least 150 mm greater than his or her stature for comfort, we may calculate that a bed length of:

- 1980 mm will be too short for 1 man in 10
- 2055 mm will be too short for 1 man in 100
- 2105 mm will be too short for 1 man in 1000
- 2150 mm will be too short for 1 man in 10,000 and so on (see Section 2.2)

Bed width is more complicated, being not solely a matter of anthropometrics. A sound sleeper may make up to 60 gross changes in posture during the course of a night. Physiologically, these are the equivalent of fidgets. They serve to preserve sleep, by relieving muscle tension, preventing the build-up of pressure hot-spots, and so on — these being potential sources of neural signals of discomfort that might wake us (see Section 4.4.1). The bed should be wide enough to allow these changes in posture to proceed unimpeded. In practice this tends to mean the wider the better. There must logically be a point beyond which further increases in width would carry no further benefit. The determination of this point would have to be the subject of empirical studies.

In Nelson's day (c. 1800) the poles of a sailor's hammock were a standard 18 in. (450 mm) long. This, being less than a 95th %ile male shoulder breadth (510 mm), would have permitted many men side-lying postures only. A modern navy bunk is at least half as wide again at 27 in. (685 mm). By way of comparison, a typical beach mat is 24 in. (610 mm) wide and the standard hospital bed is 910 mm. The single beds in the survey cited by Noble (1982) ranged from 750 to 1000 mm

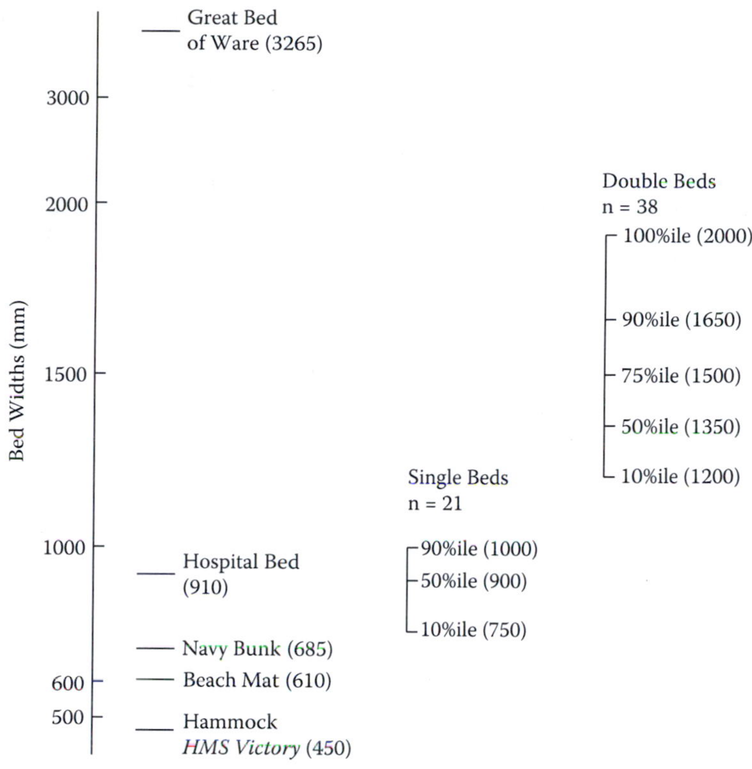

FIGURE 8.5 A historical review of bed widths (dimensions in millimetres).

in width (n = 21), and the double beds from 1200 to 2000 mm (n = 38) (see Figure 8.5).

According to *Brewer's Dictionary of Phrase and Fable,* the legendary Great Bed of Ware (*Twelfth Night*, iii, 2), which was said to have belonged to Warwick the Kingmaker, was 12 feet square and capable of holding 12 people. The historical Great Bed of Ware (c. 1580) was a little smaller than this, measuring 10 ft, 8.5 in. wide by 11 ft, 1 in. long (3265 × 3380 mm). It came from the Crown Inn, in the village of Ware in Hertfordshire (on the old road from London to Cambridge), where it was something of a tourist attraction. It is now in the Victoria & Albert Museum.

People who suffer from back trouble are often advised to sleep in 'a hard bed'. Experience indicates that, not uncommonly, this advice turns out to be incorrect, and in some cases an excessively hard bed can make things worse rather than better. Norfolk (1993) reports a questionnaire survey of the advice that osteopaths give their patients concerning these matters. Of the osteopaths in his sample, 93% said they offered their patients advice about choosing a bed, although, interestingly enough, 83% also said that they would welcome more technical information on the ergonomics of bed design. Many were very critical of what Norfolk refers to as the 'current vogue' for excessively hard mattresses, with 98% of the sample saying that (presumably in their experiences and those of their patients) beds can be too firm for comfort. This has been confirmed in a user trial reported by Nicholson et al. (1985).

Part of the problem seems to be based upon a confounding of two different physical properties of the bed which we could call 'conformability' and 'sag'. Conformability is the ability of the bed to adapt to the contours of the body and to support it in a diversity of positions with a minimal build-up of pressure hot-spots. Hänel et al. (1997) have pointed out that thermal comfort is important, as well as minimising local pressure, and that 'sinking' into a bed (with a large body contact area) could affect thermal balance, so designing appropriate conformability is a complex issue. Conformability is principally a property of the mattress itself. There are a variety of ways of achieving this technically, in terms of the design of the bed springs, etc. Unless the mattress is very soft indeed, however, the tendency of a bed to sag into a hammock shape will be more a property of the construction (or state of wear) of the supporting surface upon which the mattress is placed. For both sleeping comfort and postural support, it would seem desirable that the combination of mattress and bedstead (or supporting surface) should provide conformability without sag.

8.5 THE STAIRCASE

Of the circulation spaces within a house, the staircase is the most hazardous. Apart from very young children who do not recognise the dangers at all, toddlers and the elderly may lack adequate motor control and balance to negotiate stairs safely. Other users face the risks of slipping or tripping on stairs. In Britain, every year there are nearly a quarter of a million injuries in falls on stairs or steps inside the home (Department of Trade and Industry, 2000) and more than 22,000 serious injuries and nearly 1,000 deaths a year among older people falling on the stairs (Haslam et al., 2001). Home interviews carried out by Haslam et al. with people between the ages of 65 and 96 revealed that they often hurried on the stairs (to answer the door bell or to go to the bathroom) despite knowing the risks and that they had to carry loads, such as laundry or even a walking stick, that could impede their movements. It is generally agreed that the most serious accidents occur when descending stairs, because it is so difficult to recover from loss of balance (at the same time as attempting to halt body momentum if it is possible to grasp the handrail). When ascending a staircase, the steps in front can help to recover from slipping or tripping, and there is additional support from the handrail.

Some of the main design considerations are the step height and depth (which should not, of course, vary over the course of the staircase) and the surface finish, which should be slip-resistant. According to Grandjean (1988), stairs with a gradient of 25 to 30° result in both the lowest physiological cost when climbing and the fewest accidents. The dimensions he recommends for a flight of stairs are a riser (step height) of 170 mm and a tread depth (step depth) of 290 mm, although he adds the following 'stair formula' to allow for some variation while maintaining an acceptable gradient (within the range tested between 16 and 38°).

$$2 \times \text{riser} + \text{tread depth} = 63 \text{ cm} \tag{8.1}$$

Other studies have used different measures to judge the acceptability of stairs, and there is some variation between the various recommendations. In a study of lower-limb joint moments during stair climbing, Mital et al. (1987) found that these were minimised with a riser of 102 mm and tread depth of 305 mm and that, if riser height were increased, ankle and knee joint moments would increase by 20% at 152 mm, with a further 10% increase by 203 mm. Irvine et al. (1990) used psychophysical methodology to find the stair dimensions that users considered acceptable. The optimum they recommend is a riser of 183 mm and tread depth of 279 or 300 mm, but they report that larger dimensions were not acceptable to shorter people, and a riser as low as 102 mm was judged unacceptable by almost everyone. Grandjean's (1988) recommendations fall within the ranges of dimensions that Fitch et al. (1974) indicated were least likely to lead to mis-steps when ascending or descending stairs, except that slightly longer tread depths (at least 312 mm) had the least mis-steps when descending stairs. This is understandable when you consider that weight is largely taken on the ball of the foot as you step, so that, when descending, the step needs to be at least as deep as the full length of the foot. The ball of the foot is also used in locating the next step through proprioceptive feedback, which again is more difficult on a narrow step. Thus those who feel unsure of their balance tend to place their feet sideways as they descend the staircase.

Taking the results of all four studies, the optimal dimensions for stairs will differ according to the criterion. Overall, Grandjean's (1988) recommendation of a riser of 170 mm and a tread depth of 290 mm seems to be a fair compromise.

For readers concerned with staircase design, Archea (1985) and Cohen (2000) give valuable insights into behaviour when descending stairs and the design parameters which are relevant to this. Archea, for example, has observed the 'cautious foot wiggle' we use as we take our first couple of steps on to a staircase to 'get the feel of the treads' and of the geometry and condition of the particular staircase, coordinating this with a prior visual check. It is clear that this behaviour is closely linked to placing the ball of the foot on a point that is both safe to support the weight of the body and sufficiently rearward of the edge of the step to ensure that the 'foot base' is within the step — so that balance can be maintained for the whole body. Archea also points out the changes in vision which may occur with ageing and which may lead to difficulty in reacting to visual cues that are absorbed unconsciously by younger people. He cites studies by Palastan which showed that the degradations to vision which occur during ageing lead to difficulties in detecting stair edges and thus to a much more cautious descent.

If the staircase design does not adequately provide the visual and proprioceptive feedback, there will be a greater risk of mis-steps. Staircase design involves a complex compromise between the many influencing factors which have been mentioned (anthropometric, biomechanical, physiological, proprioceptive, visual and behavioural). That is not to say that it is difficult to achieve a good design, as witnessed by the staircases in most homes and public spaces. It does, however, provide a very clear illustration of the need for the designer to make an initial task analysis (as discussed in Section 4.1) and to identify the user population and their capabilities and needs (Sections 2.3 and 4.9).

However, limitations of space will sometimes require a steeper than desirable staircase. In such circumstances, Ward and Beadling (1970) recommended a riser of 217 mm and a tread depth of 245 mm, which gives a gradient of 41°, although this falls outside Grandjean's (1988) range and will not be physiologically desirable. The recommended ranges in British standard BS 5395 (Part 1; BSI, 2000) (for residential buildings) are somewhat greater:

- Riser: minimum, 100 mm; maximum, 220 mm (preferably less)
- Tread depth: minimum, 225 mm; maximum, 350 mm
- Pitch: maximum, 41.5°

The U.S. military standard MIL-STD-1472F (Department of Defense, 1999) recommends risers of 165 to 180 mm and tread depths of 280 to 300 mm, which is very similar to Grandjean's (1988) recommendations, but MIL-STD-1472F does permit ranges of 125 to 200 mm and 240 to 300 mm for risers and tread depths, respectively.

Informal observation suggests that risers of greater height than those recommended by either Grandjean (1988) or Ward and Beadling (1970) will require some effort to climb and that narrower tread depths will be more difficult to negotiate (either because of poor balance or through lack of space to place the whole of the foot), especially when descending the stairs. These difficulties are likely to increase with age. It would seem sensible then to have steps at a 'comfortable' height which will facilitate natural movement. Voorbij and Steenbekkers (1998) surveyed elderly people's opinions of the step heights that they found to be comfortable and that were the maximum they felt to be safe with one hand on a handrail. (Those who wished to were allowed to step down backwards, and 7.6% of men and 21% of women aged 50 years or older chose to do so at the maximum step height.) There was considerable variation in the comfortable step height with anthropometry and between men and women, but interestingly no large or systematic trend with increasing age. However, all the subjects in their sample were healthy, so the dimensions recorded do not take account of the effect of disabilities other than the use of a walking aid. Taking men and women together, the 5th %ile and 95th %ile step heights judged comfortable were 140 to 230 mm at age 50 to 54 years and 100 to 230 mm at age 80 years and older. The values were virtually the same for ascending and descending.

The measurements of maximum step height, by contrast, indicated that people's abilities become more variable with age, with the 5th %ile value decreasing, while the 95th %ile value remained fairly constant. The 5th %ile and 95th %ile maximum step heights judged safe were 320 to 500 mm at age 50 to 54 years and 230 to 500 mm at age 80 years and older. Again, there was no obvious difference between capabilities when ascending and descending. The tests were performed on a single step and thus do not necessarily reflect the riser height that would be judged comfortable over a whole staircase, but Grandjean's (1988) recommended riser height of 170 mm does fall within 5th %ile to 95th %ile range of comfortable step heights and is well below the 5th %ile maximum capability of all but one of the age and gender subgroups of Voorbij and Steenbekker's (1998) sample.

Handrails are helpful in assisting climbing up a staircase and even more in maintaining balance and reducing the risk of slipping when descending. Staircases should have at least one and preferably two easily gripped handrails. As with step dimensions, the choice of handrail height involves a compromise between several factors, as well as between the needs of younger and older users. However, there is some degree of flexibility in the effective height because the user can adjust the position of his or her hand along the rail. The handrail is used more by older than younger people, but in the event of tripping both groups will use it to prevent or ameliorate a fall. Cohen's (2000) study showed that most people stay within easy reach of the handrail when walking down stairs. The height of the handrail should assist the user in generating stabilising forces and moments at the hand, and for this Maki et al. (1984) recommend a height of 910 to 1002 mm above the front edges of the treads on the basis of experimental trials with able-bodied subjects. The mean preferred height among their subjects was 910 mm. British standard BS 5395 (Part 1; BSI, 2000) also recommends handrail height to be within 900 to 1000 mm above the pitch line of the staircase and banisters (guards) to be at least 900 mm on the staircase and landings (preferably not less than 1100 mm on the latter).

9 Health and Safety at Work

9.1 INTRODUCTION

About 226 people are killed each year in the United Kingdom in accidents that take place at work; that is 0.8 per 100,000 workers (2002/2003 statistics from HSC, 2003). The exact figure fluctuates a little from year to year. A further 28,426 are seriously injured, and nearly four and a half times this number sustain injuries that, although they are of a less severe nature, are none the less serious enough to keep them off work for three days or more (and thus find their way into the official statistics). Added to this, we have an unknown (but doubtless very great) number who sustain minor injuries requiring first aid treatment only and an unknown (but again large) number who develop diseases or ill health, of one sort or another, as a result of their work. An estimate of the *prevalence* of work-related ill-health (as distinct from the *incidenc*e of new cases in any one year) is that in 2001–2002, 2.3 million workers in the United Kingdom were suffering from an illness that they believed had been caused or made worse by their current or past work, which resulted in 33 million working days lost in that year (HSC, 2003).

If we take fatalities as an index of safety, however — and there seems to be good reason to do so, since they are likely to have been recorded more carefully than mishaps with less severe consequences — then the available evidence indicates that work is getting steadily safer. The current U.K. annual fatality rate of 0.8 deaths per 100,000 workers is dramatically decreased from the rate in the first decade of the twentieth century, when it was 17.5 per 100,000, and from the rate of 5.6 per 100,000 in 1961 (as shown in Figure 9.1, compiled from data from various national statistics reports). The downward trend is thought to be due in part to better regulation of working practices and in part to changes in the nature of work such that fewer people are engaged in its more hazardous varieties. However, major and less severe injuries have not reduced as significantly as fatalities and, in particular, musculo-skeletal disorders are still a major problem to tackle in workplaces (HSE, 2002). In the United Kingdom they account for more than half of all self-reported occupational ill-health (Jones and Hodgson, 1998).

The causes of fatal and nonfatal accidents appear to differ, as illustrated by the comparison in Table 9.1. The figures for fatalities have been averaged over several years in order to smooth the annual fluctuations which arise in the data because of the relatively small numbers involved. The table is set out by rank order for the nonfatal accidents. The data have a number of striking features. In the case of the nonfatal accidents, the top seven statistical categories account for about 90% of all the accidents. For the fatalities, the grouping is slightly less marked, with the top

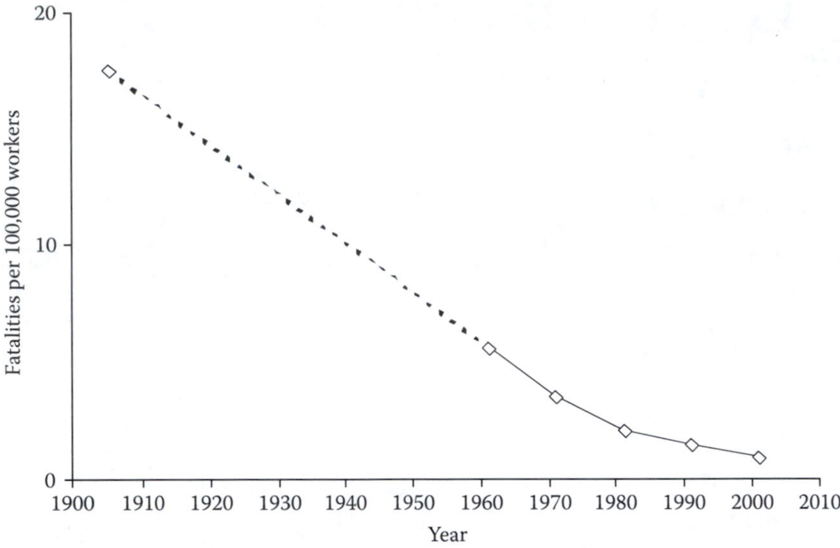

FIGURE 9.1 Fatal accidents at work, 1900–2001.

TABLE 9.1
Accidents at Work, Classified by Principal Cause

Principal Cause	Nonfatal		Fatal	
	% Total	**Rank**	**% Total**	**Rank**
Lifting, handling, carrying	32.0	1	0.3	14
Slip, trip, fall on same level	21.4	2	1.2	11
Struck by moving object (including falling object)	13.8	3	13.2	3
Fall from a height	9.5	4	25.0	1
Striking against stationary object	6.0	5	0.6	12
Contact with moving machinery	4.9	6	7.5	5
Contact with harmful substance	2.9	7	2.4	8
Struck by moving vehicle	2.5	8	15.8	2
Electricity	0.5	9	6.2	6
Animal	0.5	10	0.3	13
Trapped by collapsing or overturning object	0.4	11	8.9	4
Fire	0.4	12	1.8	10
Explosion	<0.1	13	2.3	9
Drowning and asphyxiation	<0.1	14	4.0	7

Source: All figures based on U.K. Health and Safety Executive Annual Reports. Nonfatal accidents are for the year 1992. Fatal accidents are for the period 1986–1992 (excluding those resulting from the Piper Alpha oil rig disaster on the 6th July 1988).

seven categories accounting for just 80% of the total. More importantly, the rank orderings of causes for the fatal and nonfatal accidents are quite different. The largest difference is for lifting and handling accidents, which take first place in the case of the nonfatal accidents but are in last place in the case of the fatalities. You would find similar differences if you were to compare nonfatal accidents having different degrees of severity. The differences can in many cases be predicted on a common sense basis, in that they reflect the relative propensity for causing serious injury of different types of mishap. Thus the relative positions of 'fall on the level' and 'fall from a height' are different for the fatal and nonfatal accidents in Table 9.1. Overall, you are much more likely to fall on the level than to fall from a height; but if you fall from a height, the injuries are more likely to be severe.

The relative frequencies with which accidents having consequences of varying degrees of severity occur are often summarised in the form of an '*accident pyramid*'. Figure 9.2 is an example, based on the U.K. figures for 2002–2003 summarised at the beginning of this chapter together with additional data from various sources. We note, however, that the shape of the pyramid will be very different for different types of accidents. Thus for 'fall from a height', which accounted for 22% of all fatalities and where we had about 79 major injuries per fatality in 2002–2003, the pyramid is sharply peaked, whereas for 'lifting and handling' accidents (accounting for 12% of major injuries and 39% of >3 day lost time injuries but no fatal injuries) the pyramid is almost flat (HSC, 2003).

The accident pyramid is a reflection of the two factors (or sets of factors) that determine the risks inherent in an activity or working practice: the probability of a particular event (i.e., accident) occurring and the probability of particular consequences (i.e., injury) resulting from such an event (should it occur). The distinction can be an important one, insomuch as the steps required to control these two components of risk may in some cases be different. Thus the distinction is sometimes

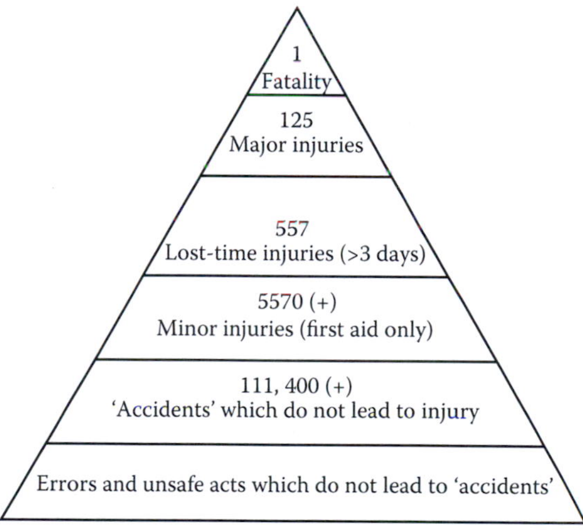

FIGURE 9.2 The accident pyramid.

drawn between *primary safety*, the prevention of accidents per se and *secondary safety*, the protection of the person in the accident situation (for example, designing safer roads and more 'crashworthy' vehicles, respectively; or making loads easier to handle, as against providing safety boots in case you drop them on your feet). An equivalent distinction may be drawn for preventive medicine in general, where it is customary to speak of *primary prevention* (of the precursors of disease), *secondary prevention* (of the disease itself) and *tertiary prevention* (of its long-term consequences).

A closely related distinction is the one that is nowadays drawn between *risk* and *hazard*, a hazard being the potential to cause harm and a risk being the likelihood that this harm will be realized. The distinction is an important one. The terminology is confusing, however, since in everyday language the words *risk* and *hazard* are used interchangeably.

Within the European Union, national regulations under the Safety and Health of Workers at Work Directive 89/391/EEC require all employers to assess the risks in their employee's workplaces and to remove or reduce these as far as is reasonably practicable. A similar *risk assessment* approach is taken in the Liability for Defective Products Directive 85/374/EEC and the Machinery Directive 89/392/EEC, which lay the onus on designers and manufacturers to perform risk assessments as a part of the design process for consumer products and machinery. (It should be noted that subsequent amendments have been made to all these directives and the latest versions should be consulted when making formal risk assessments, and that, in the case of product liability, retailers may also be held liable for faults in the products they sell.) This has emphasised the ergonomic approach in the core of *design for safety* and the various regulations, taken together, mean that designers, manufacturers and managers must consider risks to *both* safety and health for *both* employees and consumers (or end-users) of products, equipment, machinery and workplaces.

The following sections discuss the factors which influence risk, firstly in relation to safety and secondly in relation to musculoskeletal health, concentrating mainly on those factors which are influenced by anthropometric characteristics and workplace layout. Many other factors influence risk (to a greater or lesser extent depending on the task performed, the way the work is organised and the psychosocial environment) but these are outside the context of *Bodyspace*. Guidance on risk assessment procedures can be found, for example, in European standard EN 292 Parts 1 and 2 (CEN, 1991a,b) and in the U.K. Health and Safety Commission's Management of Health and Safety at Work — Approved Code of Practice (HSC, 2000).

9.2 ACCIDENTS AND HUMAN ERROR

An *accident* is an unanticipated, unplanned or uncontrolled event — generally one that has unhappy consequences. There are two main theories as to the prevention of accidents:

- *Theory A*: Accidents are caused by unsafe behaviour; they may therefore be prevented by modifying the ways in which people behave

- *Theory B*: Accidents are caused by unsafe systems of work; they may therefore be prevented by redesigning the working system

The former approach could be characterised as 'fitting the person to the job', the latter as 'fitting the job to the person'. The two theories may be regarded as complementary. Neither provides a complete explanation of the ways in which accidents occur, but both tell us an important part of the story, and the prevention of accidents generally involves considering both working system and behaviour while working (and of course, and most importantly, the interaction between these).

Unsafe behaviour may stem from:

- Difficulties imposed by poor work and workplace design
- Time and output pressure imposed by organisational and psychosocial factors
- Lack of awareness of the risks of the work
- Lack of adequate instruction and training in safe working methods
- Lack of supervision in the working situation
- A foolhardy attitude towards the risks

Not all unsafe acts lead to accidents; likewise, not all accidents result in injuries (see Figure 9.2). In general, unsafe behaviour is common and accidents are rare (and injuries more so). Unsafe behaviour is thus reinforced.

The basic elements of a *safe working system* are:

- A safe working environment
- Safe plant, equipment and personal protective equipment (PPE; where necessary)
- Safe procedures and working practices
- Competent personnel
- Competent management

The employer has a responsibility at law (in the United Kingdom, under the Health and Safety at Work Act 1974 and the regulations implemented under the European Union's Management of Health and Safety at Work Directive and its associated Directives, as well as in terms of the common-law duty of care) *not only* to put the basic elements of a safe working system in place, to warn employees about the risks of work and to provide adequate and sufficient instruction and training in safe working methods, *but also* to take such steps as are necessary to ensure that the system of work continues to operate safely on an ongoing basis. The latter (which is in many ways the most difficult) is contingent on what is often called the *safety culture* of the organization: a set of factors, deeply embedded in its social ethos, that colour the attitudes of its members and influence their actions and risk-taking behaviour (at all levels in the organizational hierarchy). This is underpinned with the requirement to assess whether risks are present, to remove or reduce the risks as far as practicable, and to 'manage' (monitor and control) any residual risks which remain.

Accidents commonly have multiple causes, in that they stem from the conjunction (i.e., coming together) of a number of adverse circumstances. It is widely accepted that *human error* makes a significant contribution in the causation of many accidents at work — probably the overwhelming majority. The contribution may very well be a decisive one, in that *but for* the error in question, the accident would not have occurred. Errors do not arise in isolation, however. Their occurrence is often contingent upon other adverse circumstances or features of the working system that lie outside the jurisdiction and control of the person concerned.

Psychologists have studied and produced various classifications of human error. For practical purposes it is important to recognise two particular categories:

- Errors of judgement in the perception or appraisal of risk
- Errors of execution in the performance of the working task

True errors of judgement in the perception or appraisal of risk stand at one end of a continuum which stretches through violations of safe working practice (having greater or lesser degrees of conscious intent) to the deliberate and premeditated criminal act (of vandalism, sabotage, assault, etc.). Errors arising both in perception and appraisal of risk and in task performance are commonly *system-induced*, in that there may be deficiencies in the design of the working system (most typically at the operator–machine interface, but also in provision or design of training) which make the person's working task more difficult and thus render him or her more error prone.

9.2.1 The Catastrophic Failure of Complex Systems

When an accident has particularly serious consequences (e.g., multiple loss of life or large-scale environmental contamination), we are likely to refer to it as a disaster or a catastrophe. The role of human error in the catastrophic failure of large-scale human-made systems — nuclear incidents, plane crashes, etc. — has been discussed at length elsewhere (Pheasant, 1988a, 1988b, 1991a). For present purposes we shall limit ourselves to just one example, the loss of the *Herald of Free Enterprise*, which illustrates the main points at issue particularly well.

The reader will recall that on the 6th March 1989, the cross-Channel car ferry *Herald of Free Enterprise* put to sea from Zeebrugge in Belgium with her bow doors open. An inrush of water flooded the large unobstructed spaces of her lower car deck, causing her to capsize, and 188 lives were lost. The direct responsibility for ensuring that the bow doors were closed lay with the Second Mate who, as it transpired, was asleep in his cabin (where he remained until he was awakened by the ship rolling over). The overall responsibility for the safety of the ship lay with the Captain, who was in his normal place on the bridge. From a human factors standpoint, the most striking feature of the accident was that from his customary position on the bridge, the Captain had no direct means of knowing whether the bow doors were open or not. There was no visual display — such as a simple warning light, for example — to provide him with this critical information. Neither was it anyone's particular duty to tell him (although this latter point remains surrounded by a certain air of vagueness). The Captain's fatal decision to put to sea may thus be construed as a classic system-induced error.

A number of other adverse circumstances were contributory factors. The tide that day was particularly high, making the loading of the ship difficult (particularly since the ramp at Zeebrugge had been designed for a different type of vessel); because of the fierce economic competition of the cross-Channel ferry routes, the crew were under considerable pressure to achieve the fastest possible turnaround time in port; because of vandalism, life jackets were stored in inaccessible lockers; because of the open-space design of the car deck, a layer of water could develop great momentum as the ship rolled with the waves and thus overturn the vessel very rapidly; and so on. The relationships between these various contributory factors, in the chain of causation which led to the catastrophe, are summarised in Figure 9.3. Pheasant (1988a) referred to this overall process as the Zeebrugge–Harrisburg Syndrome. (The reader will recall that the ill-fated Three Mile Island nuclear reactor was in Harrisburg, Pennsylvania.)

It became apparent at the subsequent Court of Inquiry that it was by no means unknown for ferries of this type to go to sea with their bow doors open. Other captains had commented on the problem, and the suggestion that warning lights should be installed had been passed up to the board level of the ferry company. The suggestion met with derision. At the Court of Inquiry, the honourable Mr. Justice Sheen concluded that 'from top to bottom the body corporate was infected with the disease of sloppiness'. The Captain, the First Officer and the Second Mate were

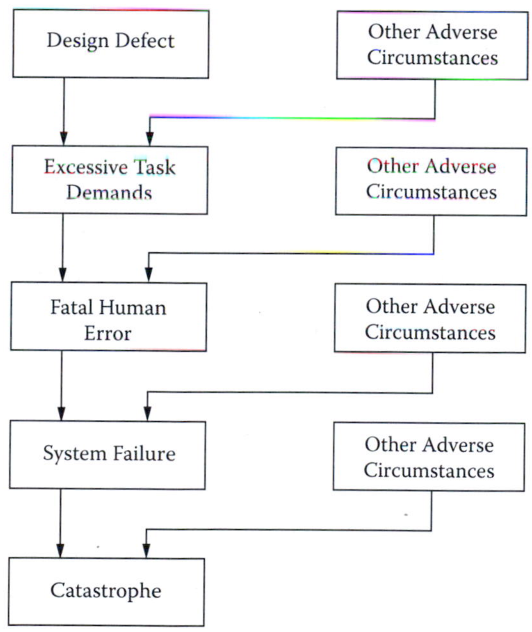

FIGURE 9.3 The Zeebrugge–Harrisburg syndrome. (*New Scientist,* 21 January, 55–58 from S. Pheasant, 1988a)

found to be negligent. The Captain lost his operating certificate for one year and the First Officer, for two years.

The episode has two interesting legal postscripts. The Captain appealed against the loss of his certificate on the grounds that going to sea with the bow doors open was a common practice. The appeal failed on the grounds that the fact that a particular form of negligence was rife in the world of car ferries did not condone it in any individual case. This finding is in some measure unusual in that the defence of 'normal custom and practice' is often successful in personal injury claims. An attempt to bring a criminal prosecution against the ferry operators for 'corporate manslaughter' also failed, it being ruled that the risks attendant on going to sea with the bow doors open were not sufficiently obvious to warrant such a charge. The common-law test of 'reasonable foreseeability' does not apply here, the criterion applied in the criminal charge of manslaughter being a more demanding one. The risk would have to 'stare you in the face'.

9.2.2 Everyday Accidents

We now return to some of the most frequent types of workplace accidents ('everyday' accidents on the 'shop floor') as categorised in Table 9.1. Each of these presents its own set of ergonomics (or human factors) issues.

'*Slip, trip and fall*' accidents (for example) very often result from a lack of good housekeeping — the failure to mop up spillages, keep walkways free from obstructions or trailing cables, highlight or guard changes in floor level and so on. Much of this stems in turn from a defective safety culture and 'the disease of sloppiness'. Issues of environmental design may also be involved, however, for example the layout and lighting of the working area, the slip resistance of flooring materials, etc.

'Contact with moving machinery' accidents remain a common cause of serious injury at work and are, traditionally, perhaps the immediate focus of safety inspections. In U.K. legislation, at least, these are unique in that there is an absolute requirement to safeguard employees (and any others present) from contact with moving machinery. No defence of reasonable practicality is allowed. The accidents arising from contact with moving machinery are often horrific and must be prevented, but this should not obscure their relative importance and the need to address the causes of apparently less immediate but severe long-term musculoskeletal injuries (as discussed later).

The safeguarding of machinery raises some interesting points in the application of anthropometrics. The concept of a *safety distance* is based upon a reversal of the normal criteria for reach and clearance. A safety guard or barrier will fulfil its function of separating people from the hazardous parts of machines, *either* if apertures in the guard are sufficiently small to prevent access by a particular body part (finger, hand, arm, etc.), *or* if the distance between the aperture and the hazard is sufficiently great for the latter to be out of reach (by the body part in question) (see Section 4.2.3). The limiting user is thus one with a small finger (hand, arm, etc.) in the case of aperture size or a long finger (hand, arm, etc.) in the case of distance. (In theory we also need to allow for the correlation between the length and girth of the body parts in question, but in practice this is likely to be small, and if we assume it to be zero we will err on the side of caution.) Safety distances are the subject of a series of British and European standards to which the reader is referred for further information, in particular to European Standard EN 294 (CEN, 1992).

Regrettably, it is all too common for people to seek ways of 'defeating' the safety mechanisms of machinery in the interests of increased output — and for serious or fatal injury to result. There can be many motivations for this, both organisationally induced and personal. Some of the most common are related to payment systems, time pressures (sometimes related to reducing down-time for the machinery during maintenance) and poor design of the guard itself.

The reader will thus note that for both the classes of accident discussed above, theories A and B are both applicable to some extent to accident causation. Overall, this is true for most other classes of accident too, up to and including the catastrophic failure of complex human-made systems.

We note also the very general applicability of the ergonomic approach to accident prevention. We turn now to a large and important class of work injuries in which ergonomic issues are of decisive causative significance and in which theory B will in general be much more applicable than theory A.

9.3 MUSCULOSKELETAL DISORDERS

Many health-related injuries, diseases and disorders occur as a direct or indirect consequence of the nature and demands of the person's working task, rather than as a result of some hazard to which the person is exposed during the course of his or her work but which is not intrinsically part of the working task itself. These include (for example):

- Lifting and handling injuries
- Work-related upper limb disorders
- Musculoskeletal pain and dysfunction resulting from unsatisfactory working posture, etc.

In *Bodyspace,* we are not concerned with those resulting from toxic or environmental hazards to which the person is exposed at work, although environmental factors (e.g., heat, cold, etc.) may play a contributory role in the causation of *musculoskeletal disorders*.

In other words, musculoskeletal disorders result from a *mismatch* between the *demands* of the working task and the *capacity* of the working person to meet those demands, generally when the former exceeds the latter and the person is placed in a situation of *overload.* Such disorders may occur as discrete events which take place at a particular point in time as a result of a single episode of *overexertion.* They may occur insidiously over a period of time as a result of *cumulative overuse*, or they may result from a combination of both, in that the effects of cumulative overuse may render the person susceptible to subsequent injury by overexertion.

Over-exertion injury occurs when an anatomical structure fails under peak loading, because its mechanical strength (usually its tensile strength) is exceeded. This most characteristically occurs in the execution of some voluntary action. *Over-use injury* occurs when the rate of damage to an anatomical structure exceeds the rate of repair. The repair of damaged tissue is a natural ongoing biological process. The injury process usually involves repeated micro-trauma. The timescale may be one of hours, days, months or years.

Cumulative trauma to soft tissues (and other anatomical structures) resulting from prolonged overuse may lead to a progressive diminution in their mechanical strength, thus rendering the structure more susceptible to injury by overexertion at some subsequent point of peak loading (perhaps at a level of loading that, under other circumstances, could be tolerated with impunity). This 'history dependence' of tissue strength when exposed to repeated cycles of loading is sometimes referred to as a 'creep effect'; it may be likened to the phenomenon of metal fatigue. The physiological fatigue of muscles may also be a factor in that it may lead to a breakdown in the normal coordination and control of voluntary movement, with the attendant risk that an articulation will be driven beyond the limits of its normal range of motion.

Some musculoskeletal injuries are accidents in the technical sense of the word as defined above. For example, a person may lose his or her balance when handling a load that is beyond the individual's safe capacity, causing an overstrain in trying to regain control of the situation. However, the great majority of overexertion injuries are not accidents in this sense of the word — in that they do not entail any identifiable intervening event that interrupts the normal execution of the action in question — other than the direct manifestation of the injury itself. Thus the person may feel a sudden sharp pain (in back, shoulder, wrist, etc.) whilst executing a familiar action or procedure in what appears to be the normal way. The only thing that is unexpected is the pain itself.

There is also another and more radical sense in which these injuries are not accidents — they are often entirely foreseeable. Consider, for example, a working population such as nurses, who are called upon to handle the awkward, unstable and excessively heavy load of the human body on a frequent and repeated basis and very often under circumstances that are adverse in other respects. It is wholly predictable that, as a population, nurses will suffer a high incidence of back injuries. We can also predict (both on a theoretical biomechanical basis and on the basis of experience) the kinds of patient-handling manoeuvres in which such injuries are most likely to occur, although we may not be able to predict (with any degree of accuracy) whether a particular nurse will be injured on a particular occasion (see Pheasant and Stubbs, 1992a).

We could say much the same thing for overuse injuries to the hand, wrist and arm, which are endemic in people who do repetitive hand-intensive work on industrial assembly lines. It has been known for over 5 years that people who work in assembly plants are liable to suffer from these sorts of conditions (Thompson et al., 1951). The risk is equally well recognised in the poultry processing industry, to the extent that in these industries at least, such injuries must be regarded as foreseeable in any conceivable meaning of the word.

The back injuries of the nurse or the upper limb disorders of the assembly line worker are inherent in the nature of the systems of work in question. The employer is under a legal obligation to take such steps as are 'reasonably practicable' to set up a safe system of work.

9.4 BACK INJURY AT WORK

Something in the order of 4.8 million working days are lost each year in the United Kingdom due to back pain (HSE, 1995). Back pain is a condition of complex multifactorial aetiology in which many risk factors may play a part. These may be grouped under two main headings:

- Occupational risk factors associated with the mechanical stresses to which the person is exposed in the course of his or her working life
- Personal risk factors particular to the individual concerned, which may be further subdivided into those factors associated with the person's lifestyle and those that stem from his or her constitutional make-up

The epidemiological literature dealing with these matters is vast. Taken as a whole, it indicates that (in a statistical sense at least) occupational risk factors have a greater causative significance than personal risk factors and that of the personal risk factors, the lifestyle factors are of greater causative significance than the constitutional factors (see Pheasant, 1991a). One should note, however, that since these factors must almost certainly act interactively and in complex combinations, the picture in any individual case may be very much less clear-cut.

Although now more than 30 years old, the classic epidemiological studies of the Israeli researcher Magora (1972, 1973a,b) remain in many ways definitive. These were based on a large sample of men and women drawn from many different walks of life. To summarise the results of what was an extensive investigation, Magora's findings indicated that two quite distinct groups of people were particularly at risk: those whose jobs were physically demanding (entailing frequent heavy lifting, forceful exertion, etc.) and those whose jobs were fully sedentary. However, those falling into a middle category, whose jobs were moderately physically demanding and who spent some of their time sitting and some of their time standing, fell into a particularly low-risk category. Psychological factors were also important, in that people who reported low levels of job satisfaction or found their work mentally demanding (in the sense of the demands it placed on their concentration) were more likely to suffer with their backs.

The most striking feature of Magora's findings was the magnitude of the difference between the categories. About 20% of people in the physically demanding and fully sedentary categories suffered with their backs, as against around 2% in the middle low-risk category. In other words, there was a ten-to-one difference in risk level. The inference to be drawn is that the relationship between back pain risk and physical workload is U-shaped (or possibly J-shaped) (see Figure 9.4). Let us suppose that the relatively low prevalence in the middle category represents a baseline level of risk attributable in a general sense to 'the human condition' (or to the summative effects of all the various personal risk factors). It would follow that any excess prevalence, over and above this level, found in the other categories,

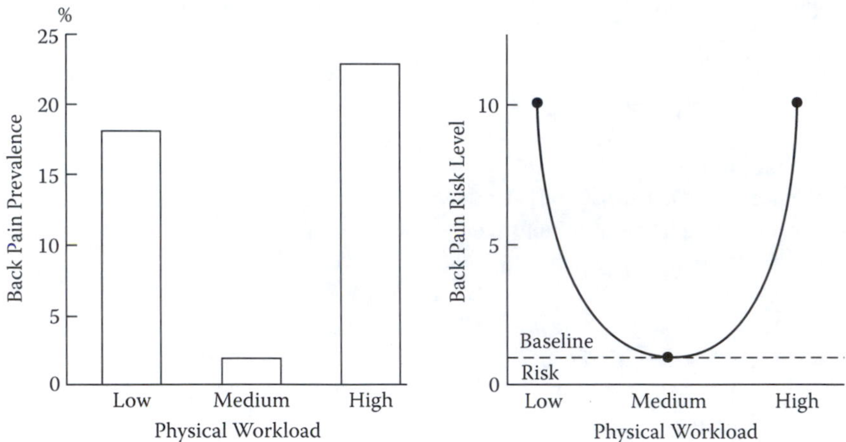

FIGURE 9.4 Back pain risk as a function of physical workload. (Based on data from Magora, A. (1972). *Industrial Medicine,* 41, 5–9.)

represents back trouble in which the person's work was a significant causative factor. In the great majority of cases, therefore, back trouble must be regarded as a work-related condition.

Occupational risk factors for low back trouble are summarised in Table 9.2 (see Pheasant, 1991a for a summary of the literature upon which this table is based). Thirty percent of workers in the European Union report that their work causes low back pain, and the prevalence of low back pain is particularly high in agriculture, construction, driving and health and social care (Op De Beeck and Hermans, 2000). Reviews of the epidemiological evidence for work-related causes of low back pain provide strong evidence of association with manual materials handling, whole-body vibration, lack of social support and low job satisfaction, and evidence of association with heavy manual labour and awkward work postures (Bernard, 1997; Op De Beeck and Hermans, 2000).

The probability is that these various risk factors have an additive effect. There is surprisingly little direct evidence for this proposition — although it certainly

TABLE 9.2
Back Pain: Occupational Risk Factors

- Heavy work: lifting, pushing, pulling, sudden maximal force exertion, bending, twisting, etc.
- Stooped working posture
- Prolonged sedentary work
- Lack of task diversity
- Unaccustomed physical activity
- Vibration and shock
- Psychosocial factors

makes physiological sense — and given that the different risk factors are not associated with different sorts of clinical problems (and we have no particular evidence that this might be the case), it is difficult to see how things could be otherwise.

The probability also is that these factors act cumulatively over a period of time (although the process of cumulative injury will doubtless be offset to some extent by the body's natural mechanisms of repair). There is some striking epidemiological evidence for this — at least over the very long time scale. Degenerative disc disease is generally thought of as being part of the natural ageing process. The physiological changes in the properties of the disc which underlie the process of degeneration commence at around age 25; past middle age we are all affected to a greater or lesser extent. Superimposed over these physiological changes are the effects on the discs of normal wear and tear. A constitutional predisposition is also thought to be involved, although the principal evidence cited for this proposition is that people who show severe signs of degenerative changes in one part of the spine are likely to show them in other parts as well, and this evidence could be interpreted in other ways.

As part of a much larger epidemiological study of back and neck trouble, Hult (1954) compared the prevalence rates of radiological signs indicative of advanced disc degeneration in men who were in light and heavy occupations, respectively. The results are shown in Figure 9.5. In the youngest age group there is no detectable difference in prevalence. With the passing of the years the prevalence for the two occupational categories diverge, until in the oldest (>50) age group there is a difference in prevalence of around two to one. In other words, in addition to the normal wear and tear of everyday life, the backs of men in heavy jobs showed clear and objective signs of a further degree of abnormal wear and tear resulting from the nature of their work.

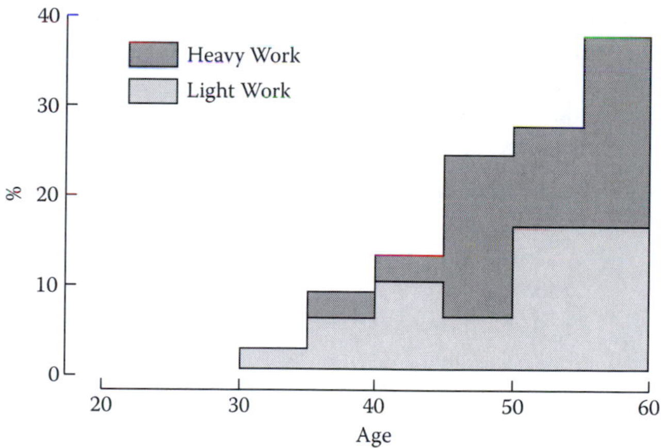

FIGURE 9.5 Prevalence of pronounced disc degeneration. (After Hult, L. (1954). *Acta Orthopaedica Scandinavia,* Supplement 17.)

Other structures in the spine and its associated soft tissues are liable to wear and tear as well as the intervertebral discs and are also likely to be involved in causes of back injury and back pain.

We shall turn now to the particular problems associated with lifting and handling heavy loads at work, which are a major component of the heavy work which is a risk factor for back pain.

9.5 LIFTING AND HANDLING

Recent decades have seen major changes in the nature of industrial work as human muscle power has been increasingly replaced by machines. Overall, work is not as heavy as it was 40 years ago. Lifting and handling injuries continue to be a major problem, however. The percentage of all reported work injuries attributed to lifting and handling has not shown much change since the early 1950s. This is something of a paradox.

The percentage of injuries attributable to lifting and handling in different sectors of the economy are likewise surprisingly similar. The figure is a little lower than the average of 38% in the pulp and paper products industry and a little higher in the electrical machinery industry, for example, but the differences generally only amount to a few percent either way (HSE, 2004).

The two areas of working life which stand out as having the highest proportions of lifting and handling injuries are air transport (51.6%) and human health activities (52.3%), where these account for over half of all reported injuries leading to 3 days' absence from work. In the latter case, the difference is attributable to the particular difficulties attendant on lifting and handling the human load. Even within this high-risk sector, however, there are striking differences between occupational groups. Ambulance personnel have a much higher incidence of patient-handling injuries than nursing staff; nursing auxiliaries, student nurses and community nurses all have a higher injury rate than qualified nurses working on the wards (HSE, 1982; Pheasant and Stubbs, 1992a). In general, these differences reflect differences in the overall amount of lifting which the occupational groups in question are called upon to do and the difficulty of the circumstances under which they are called upon to do.

Discussions of the prevention of lifting injuries have tended to revolve around two seemingly simple questions:

- What is the safest way of lifting heavy weights?
- What is the maximum safe weight a person can lift?

The first question stems from theory A as to accident causation, the second from theory B. Neither has a simple answer.

The available evidence points to the conclusion that training people in 'safe' lifting techniques alone is unlikely to have a sustained impact on injury rates. This is equally so for lifting training in general and for the special case of patient handling. Training is necessary but not sufficient. Pheasant has discussed the matter of lifting training at length elsewhere (Pheasant, 1991a; Pheasant and Stubbs, 1992b). The following discussion will concentrate on the design of safe systems of work.

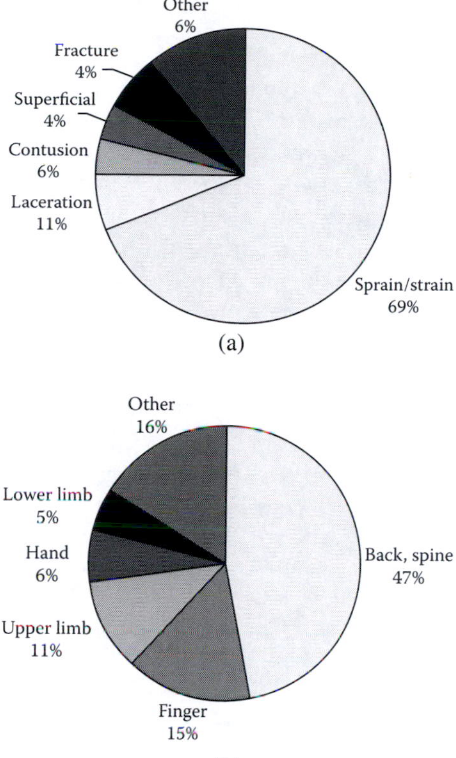

FIGURE 9.6 Lifting and handling injuries classified (a) by type and (b) by site of injury. (Based on U.K. statistics for 2001 [HSE, 2004].)

Not all lifting injuries are overexertion injuries: a significant minority are accidents in the narrow sense of the word (as defined above) and result in lacerations, contusions, fractures, and so on. Neither do all lifting injuries affect the back, although the low back is the part of the body most commonly affected (see Figure 9.6). It is worth noting that lifting tasks not infrequently result in overexertion injuries to the neck, shoulder and wrist.

We have mentioned already that heavy manual work may accelerate the degenerative processes that occur in the spine with age. There is also good epidemiological evidence for an association between work that entails heavy lifting in a squatting or kneeling position and osteoarthritis of the knees (Cooper et al., 1994).

In summary, lifting and handling tasks entail three distinct classes of risk:

- The risk of accidental injury
- The risk of injury due to overexertion
- The risk of injury due to cumulative overuse

In practice, however, the measures that are required to control these three classes of risk will tend to be similar ones.

9.5.1 Workspace Layout

The first and most fundamental principle of safe lifting is that the load should at all times be as close as possible to the body. There are two reasons for this. First, the closer the load, the less is its leverage about the various articulations of the body; hence less muscular effort is required, and there is less mechanical stress on potentially vulnerable structures (e.g., those of the back). Second, the closer the load, the more easily it is counterbalanced by the weight of the body so it is less likely to get out of control. Thus the strength of the lifting action falls off rapidly as a function of the distance of the load from the body (see Figure 9.7), and the weight that can be handled safely becomes correspondingly less.

A second important principle is that symmetrical lifting actions are in general safer than asymmetrical lifting actions, particularly if the latter involve turning actions which impose a rotational twist on the spine. This is partly because the lumbar spine is anatomically vulnerable to injury under a combination of flexion and torsional loading and partly because in turning actions we naturally tend to lead with the hips, thus exposing the lumbar spine and its musculature to particularly high levels of loading.

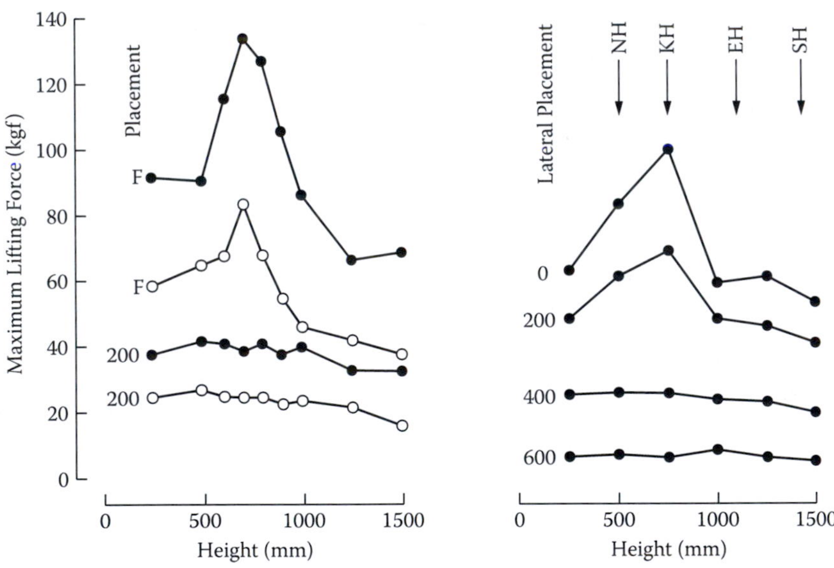

FIGURE 9.7 Strength of a static lifting action as a function of height above ground and foot placement. Left: freestyle placement (F) and feet placed behind the axis of lift. Data kindly supplied by Anne-Marie Potts (●, group of 16 men; ○, group of 14 women). Right: feet placed 20 mm behind the axis of lift 400 mm apart, various distances to the left. The placement figures are for the mid-line of the body. NH, knee height; KH, knuckle height; EH, elbow height; SH, shoulder height of the 21 male subjects. Data kindly supplied by Jane Dillon of the Furniture Industry Research Association.

FIGURE 9.8 Lifting at a distance: palletization task. From an original in Stephen Pheasant's collection. (From Pheasant, S. (1991). *Ergonomics, Work and Health,* London: Macmillan, Fig. 15.17, p. 302. Reproduced with kind permission.)

In practice, the distance of the load from the body and the symmetry of the lifting action will be largely determined by foot placement, and this in turn is determined by the presence or absence of obstacles that prevent the person from getting his or her feet beneath or around the load (as illustrated in Figure 9.8). Lifting and turning actions likewise very often stem from deficiencies in workstation layout.

Given good foot placement, the strength of the lifting action is greatest at around knuckle height (around 700 to 800 mm) and falls off rapidly above and below this level. When exerting a vertical lifting force at knuckle height or thereabouts, the upper limbs are vertical and almost straight, and the hips and knees are slightly flexed. The muscles of the lower limb thus exert a powerful extensor thrust along the line of the almost straight limb at their best possible mechanical advantage. When the force is exerted at a distance from the body, however, this peak in lifting strength disappears (see Figure 9.7).

If the lift commences at much below knuckle height, the person will either have to incline his or her trunk and therefore increase the loading on the spine (which will tend also to be flexed upon itself and thus anatomically vulnerable to injury) or else strongly flex the knees, thus reducing the mechanical advantage and also rendering the knees anatomically vulnerable to injury. In either case the power of the lifting action is diminished and the weight that can be handled safely is likewise reduced.

A lift that is commenced at knuckle height or lower can (provided the load is not excessive) be continued comfortably to elbow height or a little more. If the load is a box or carton which is held by its lower edges, or a crate with handholds on the sides, the lifter will then begin to encounter difficulties as wrists reach the limit if their range of abduction. The grip will thus either have to be changed or else the lifter must make awkward compensatory movements of upper limbs and trunk,

FIGURE 9.9 Lifting outside the normal height range. Note the hyperextension of the lumbar spine.

neither of which is at all desirable (see Figure 9.9). The wrist is also anatomically vulnerable in this position.

Lifts that commence at elbow height may be continued to shoulder height or thereabouts without too much difficulty, but beyond that point the reduction in strength really begins to tell. There is a particular danger that loads that must be handled at shoulder height and above will get out of control. In the authors' experience, lifting tasks that entail the handling of loads outside the comfortable height range of the person in question are a common cause of injury.

On the basis of these considerations, we may divide the reach envelope of the standing person into *lifting zones*, as shown in Figure 9.10 (after Pheasant, 1991a; Pheasant and Stubbs, 1992b). The heights given for the various landmarks are based upon anthropometric data for the standard reference population (see Section 2.8) but rounded up to convenient whole numbers. The verbal categories describing each zone may be regarded as giving a general indication of the weight of load that might be considered acceptable in each zone (see also Section 9.5.2 below).

When carrying a load such as a box or carton, a person will generally hold it by the lower edges at hip height or above (800 to 1100 mm) so as not to impede walking. The effort required to lift loads from conveyor belts, etc., may often be reduced, therefore, by setting the belt at a level that allows the worker to pull the

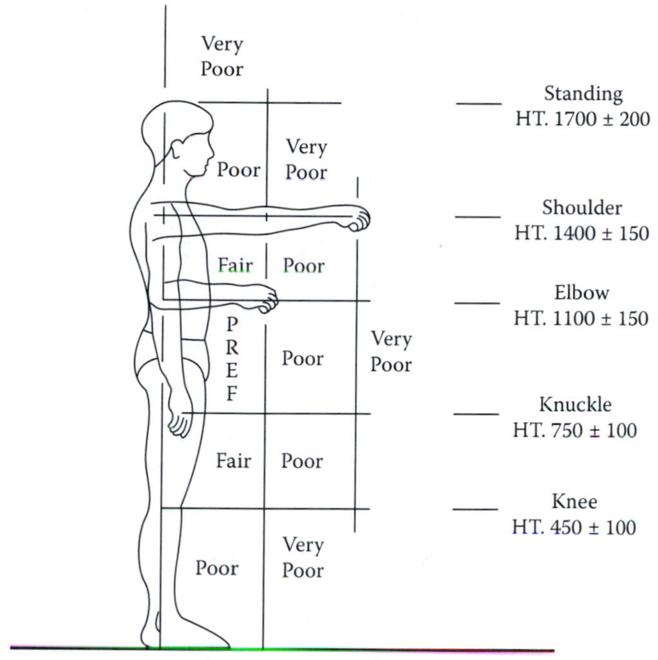

FIGURE 9.10 Height ranges for lifting actions. (From Pheasant, S. (1991). *Ergonomics, Work and Health,* London: Macmillan, Fig. 15.20, p. 305. Reproduced with kind permission.)

load forwards and take its weight at a suitable height for carrying. (This will depend in some measure on the nature of the load.)

9.5.2 The Load

In general, a compact load is safer to lift than a bulky load of the same weight because its centre of gravity will be closer to the body; so the moment about the spine will be less and the body posture more stable. This is especially important if the load is to be lifted from the ground, since a compact load (<300 mm in width or from back to front) can be lifted between the knees rather than in front or to the side of the knees.

Unstable loads and loads with unexpected inertial characteristics are a particular hazard. The centre of gravity should be as close as possible to the geometrical centre; if offset, its position should be marked. Contents should be securely packed to prevent unexpected shifts in the centre of gravity — a common cause of lifting injuries. Secure handholds are an advantage (see Chapter 6), but their shape and orientation need careful consideration (Drury et al., 1985; Deeb et al., 1985).

There remains the question, What is the maximum weight that a person of a particular age and sex, and of 'normal fitness', may be expected to lift under a given set of circumstances, without undue risk of injury? This is a very difficult question indeed. First, there is no weight of load, however small, that guarantees safety. You can injure your back by stooping down to pick up a pencil. Biomechanical calculations show that when the trunk is inclined forward to a horizontal position, the loading on the base of the spine is the same as the loading that results from lifting a compact 30-kg weight close to the body (Pheasant, 1991a).

If for one reason or another the back is at all vulnerable to injury, there may be no spare capacity for lifting an external load, over and above the weight of the body itself. Unfortunately, the *precursors of lifting injury*, which determine a person's level of vulnerability, are not always easy to recognise. A person may be vulnerable to injury without knowing it. Overall we are not particularly good judges of what we can handle safely, and we commonly injure ourselves lifting loads that we believe to be within our capacity. Conversely, experience shows that if a load *feels* as if it is too heavy to handle safely, then it probably is too heavy. In other words, our subjective appraisals of our safe limits are systematically biased in the direction of risk (although there are doubtless important individual differences in this respect).

In any given set of circumstances (as defined by load characteristics, lifting position, and so on), we should in general expect the risk of injury due to overexertion, *for a particular individual*, to increase steadily with the weight of the load. The relationship may or may not be linear (but biological systems being what they are, a nonlinear relationship seems more probable). Given the nature of human variability, the rate at which the risk of injury increases with the weight of the load will necessarily vary greatly between individuals. Thus if there are important *threshold effects* (where the level of risk takes a sharp upswing), the location of these will likewise vary greatly. Taking one consideration with another, one would expect the risk of injury by overexertion, *for a given working population*, to increase with load weight in a smooth upwardly accelerating curve, in which such threshold effects as might be present for given individuals are masked.

We also need to take into account *both* the immediate risks of overexertion injury *and* the long-term risks of cumulative overuse — and probably also the interactions between the two. The risk of accidental injury presents an even more difficult set of problems. One would expect the probability of some unexpected mischance that leads to injury to increase with the weight of the load, but the connection is not as clear as it is for overexertion injury or cumulative overuse, except insomuch as the heavier the load, the more likely you are to injure yourself seriously if it gets out of control.

Even if we fully understood these matters, to the extent of being able to plot out a curve relating overall population risk to load weight, we should still be faced with the question of where to set the limit. What is the cut-off point beyond which the risk of injury becomes unacceptable? One possibility is to set the limit at a level of loading that would result in a *just noticeable risk of injury*, that is, a level at which the risk of injury due to work would be just measurably greater than the background level of risk associated with life as a whole. This is, broadly speaking, what we attempt to do for chemical hazards, radiation hazards, and so on, since in these cases the risk in question may be controlled by a more effective containment of the hazard so that the working person does not come into contact with it. In the case of lifting and handling, however, we cannot do this, and to set the limit at the level of just noticeable risk would to all intents and purposes be equivalent to calling for the abolition of all useful manual work. This would be pointless — not least because guidelines that cannot be met in practice are ignored and fall into disrepute.

Neither does it make much sense to set the limit at a level of loading at which injury becomes 'probable' to the extent of being 'more likely than not', because in practice this would mean that we were constantly having to replace our workforce. There are industries where this happens. They are recognizable by the age distributions of their workforces.

The problem thus becomes one of reaching a *reasonably practicable compromise* position which allows working life to continue without incurring excessive risk. This roughly approximates the legal concept of a risk that is *reasonably foreseeable*, or, in this context, the level of risk which a reasonable person would agree to accept in the course of his or her working life (were he or she fully appraised of the facts of the matter).

In reaching this point we are faced with two principal difficulties: the first is the inadequacy of our scientific knowledge; the second is that legal conceptions of probability are not altogether the same as scientific ones, in that moral certainty does not equate easily with statistical certainty. Some countries have seen fit to impose limits on the weights that people may lift at work; others have not. The International Labour Office has published a compilation of such weight limits (ILO, 1990). These can be summarised in the form of a cumulative distribution showing the percentage of countries in which a load exceeding the weight in question would be considered unacceptable. For adult men these cluster around a median figure of 50 kg, with 50% of values falling within the 45 to 55 kg range; for women they cluster around a median value of 25 kg, with 50% falling within the 20 to 25 kg range.

The U.S. National Institute of Occupational Safety and Health published an influential set of guidelines in 1981 which dealt specifically with symmetrical two-handed lifting actions performed directly in front of the body (NIOSH, 1981). For any such action it was considered possible to calculate an *action limit* (AL), beyond which there was deemed to be a moderate increase in risk and a *maximum permissible limit* (MPL), beyond which the risk was considered unacceptable. These guidelines were based upon biomechanical, physiological, psychophysical and epidemiological criteria. The resulting equation that defines the AL and MPL takes into account the horizontal and vertical positions of the load, the distance it is lifted, the frequency of lift and the duration of the task. The original guidelines have subsequently been revised (Waters et al., 1993). The equation defining the limits has been modified, and two new elements, dealing with asymmetry and ease of grasp, have been added. The action limit and maximum permissible limit have been dropped in favour of a *recommended weight limit* (RWL), which is set at a level that is approximately equivalent to the old action limit. The concept of a *lifting index* (LI) has been introduced. This is the ratio of the load on the job to the RWL. It thus represents a relative measure of the severity of risk. It is essentially seen as a tool in job redesign, and no specific cut-off point is proposed. Pheasant has discussed the theoretical and practical strengths and weaknesses of the NIOSH guidelines at length elsewhere (Pheasant, 1991a). In his view both the old and new NIOSH equations greatly underestimate the importance of the vertical height range of the lift.

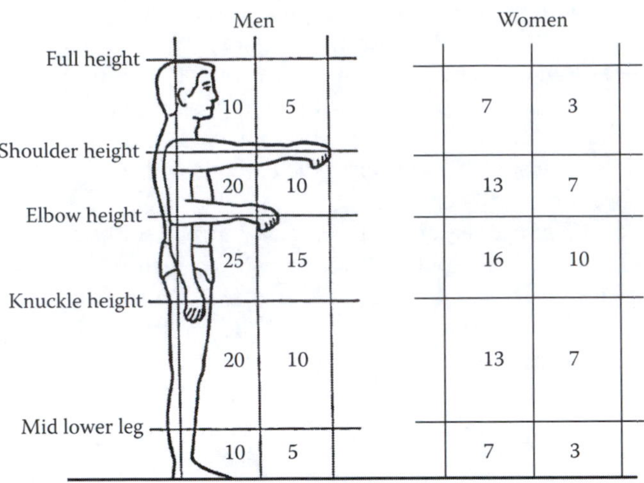

FIGURE 9.11 Lifting zones (when standing) with the HSE (2004) risk filter guidelines data (load weights in kilograms). Note that the guideline figures are much lower when seated — 5 kg for men and 3 kg for women — and only apply to a zone at waist level close to the body.

'Guideline figures' for lifting and carrying have been published by the U.K. Health and Safety Executive (HSE, 2004), as shown in Figure 9.11. These were based in part upon a series of studies at the Robens Institute, in which intra-abdominal pressure (IAP) was used as an indirect index of spinal loading (Davis and Stubbs, 1977a,b, 1978; David, 1987), in part upon the concept of lifting zones as set out above (and illustrated in Figure 9.10). The Health and Safety Executive document in which the guideline weight values are set out stresses that they are not 'limits' as such, but are for guidance purposes only — to be used in the context of a broader approach to the assessment of risk based upon ergonomic principles. The guideline figures are said to be such as to afford 'a reasonable level of protection to around 95% of working men and women'. Correction factors are given for twisting, repetition rates and team lifting. In addition to this, the HSE has developed a set of 'MAC' manual handling assessment charts as a tool to help health and safety inspectors assess the most common risk factors in lifting, carrying and team handling tasks (HSE, 2003b). The MAC charts indicate four levels of increasing risk by colour bands and generate total scores to help to prioritise remedial actions.

In essence, the guideline figures in Figure 9.11 define a boundary level beyond which there is a potentially significant risk of injury and a detailed risk assessment should be made. The HSE (2004) document in question goes on to say that operations exceeding the guideline figure by a factor of more than about two may represent a serious risk of injury 'even for a minority of fit, well-trained individuals working under favourable conditions'. We take this to be a way of saying that beyond this point the risk of injury may be unacceptably high.

In summary, therefore, there are several tools which can be used for assessing risk in handling tasks, the two mentioned here being the NIOSH (1991) lifting formula and the HSE risk filter diagram (2004) and MAC charts (2003b). Standards

ISO 11228 and EN 1005 parts 2 and 3 also provide guidance on manual handling activities (ISO, 2002a; CEN, 2003a, 2002). In each case the values of load weights are indicated on the assumption that all other factors are optimal. In other words, the risks of injuries when handling loads are multifactorial and load weight is only one of the factors influencing the risk of injury. The assessment tools, moreover, are in agreement in showing clearly that the best position for holding a load is close to the body and between knuckle and elbow height; if the load is lifted outside this zone, the risk increases.

9.6 WORK-RELATED UPPER LIMB DISORDERS

The terms *work-related upper limb disorder* (WRULD) and *repetitive strain injury* (RSI) are, to all intents and purposes, synonymous. Both terms are used generically to refer to a diverse class of conditions affecting various anatomical sites in the hand, arm, shoulder and neck, which occur in people doing a wide variety of types of work involving intensive use of the hands (not all of which are necessarily repetitive in the strict and narrow sense of the word). A national household survey carried out in 1995 estimated that over half a million people in the United Kingdom had suffered during the previous 12 months from an upper limb or neck musculoskeletal disorder which they attributed to their work or which was exacerbated by their work (Jones and Hodgson, 1998). Over 80% of these said that their condition had resulted in a limitation of movement or loss of strength. This experience seems to be reflected in many countries within the European Union, and overall, neck and shoulder problems appear to be more prevalent than wrist, hand or elbow problems (Buckle and Devereux, 2002).

Occupational groups most notably affected include:

- Industrial assembly line workers, for example in the automotive, electronics, pottery and food processing (especially meat and poultry) industries
- Workers at supermarket checkouts
- Garment sewing machinists
- Musicians (particularly those playing string instruments and the piano, but others as well)
- Keyboard users (particularly data entry workers, copy typists, legal secretaries and journalists)

Table 9.3 summarises the results of a number of epidemiological studies of disorders falling into this category, which were gathered together by Armstrong et al. (1993). Note that the 'relative risk' is the ratio of the frequency with which the condition occurs in a sample of people drawn from the occupational group in question to the frequency with which it occurs in a control sample, who are not exposed to the same sort of risk.

One feature of Table 9.3 which is particularly worth noting is that data entry workers (the only group of keyboard users included) appear fairly low on the list. In other words, although data entry workers (and by inference other intensive keyboard users) are at an elevated risk level (compared with other people), the level

TABLE 9.3
Work-Related Upper Limb Disorders: Relative Risk in Selected Occupational Groups

Job	Risk
Industrial workers (high force/high repetition)	29.4
Sausage makers	24.0
Shipyard welders	13.0
Industrial workers (hands above shoulders)	11.0
Frozen-food factory workers	9.4
Assembly line packers	8.1
Meat cutters	7.4
Shoe assembly workers	7.3
Packers	6.4
Data entry workers	4.9
Packaging and folding workers	3.9
Scissor makers	1.4

Source: Armstrong, T. J., Buckle, P., Fine, L. J., Hagberg, M., Jonsson, B., Kilbom, A., Kuorinka, I. A. A., Silverstein, B. A., Sjøgaard, G. and Viikari-Juntura, E. R. A. (1993). *Scandinavian Journal of Work, Environment and Health,* 19, 73–84.

of risk is by no means as high as it is for some of the industrial groups that have been studied. This runs contrary to the popular impression of RSI being 'the keyboard users' disease'. Nevertheless, the total number of intensive keyboard users within the working population is high and increasing.

The risk level for data entry workers was borne out in the great RSI 'epidemic' which swept Australia in the 1980s (see Figure 9.12). Even when the epidemic was at its height (in 1985–1986), the annual incidence of new cases remained highest in the blue-collar jobs which have been traditionally associated with injuries of this kind. The epidemic itself, however, was caused to a very great extent by an increased reporting of such injuries in keyboard users. Because there are nowadays a very large number of keyboard users, the problem of keyboard injury must be treated as a very important one (see below).

The term *RSI* is widely deprecated on the grounds that it is inexact and misleading. This is broadly the case. Discussions of the matter generally revolve around fairly arid issues of semantics which need not concern us here. The only point we need bother to note in this respect is that *repetition* is but one risk factor amongst many in the aetiology of these conditions, and it need not necessarily be a particularly important one in any individual case. The likelihood is that static muscle loading is of greater causative significance in the overuse injuries sustained by keyboard users; likewise, in the repetitive, short-cycle time tasks of the industrial assembly line,

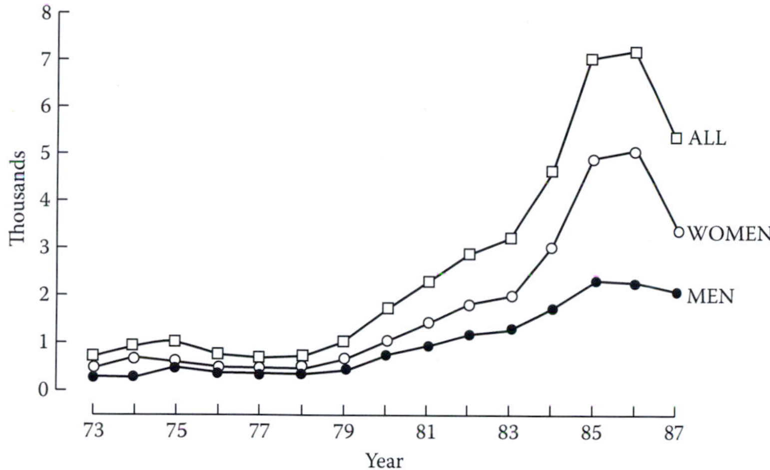

FIGURE 9.12 Annual incidence of repetitive strain injury (RSI) from New South Wales. (From Pheasant, S. (1991). *Ergonomics, Work and Health,* London: Macmillan, Fig 4.1, p.79. Reproduced with kind permission.)

force requirements and the nature and extent of the movements in question may be more important than repetition rates.

In the United Kingdom, 'informed opinion' (as embodied, for example, by the Health and Safety Executive) currently prefers the term *work-related upper limb disorder*, although *repetitive strain injury* remains in more general use. In Australia, where the term *RSI* originated (and likewise in New Zealand), it has been replaced by *occupational overuse syndrome* (OOS). In North America, a broadly similar range of conditions are referred to as *cumulative trauma disorders* (CTDs); in the Japanese literature the term *occupational cervicobrachial disorder* (OCD) is used. It must be stressed that all of these terms are generic ones, which encompass a range of different clinical entities (i.e., specific conditions).

9.6.1 On the Varieties of RSI/WRULD

The chief difficulty with generic terms like *RSI* and *WRULD* is the diversity of clinical conditions to which they are applied. Two main subdivisions are generally recognised, which are sometimes referred to as Type I and Type II RSI.

The former (Type I) are relatively discrete and localised overuse injuries to specific anatomical structures. This subdivision includes conditions resulting from traumatic inflammation of soft-tissue structures, such as the various forms of *peritendinitis* and *tenosynovitis* which affect the tendons of the muscles of the forearm and their soft-tissue coverings (and also muscles at other sites in the upper limb and shoulder region), as well as lateral *epicondylitis* and medial *epicondylitis* (otherwise known as tennis elbow and golfer's elbow), which affect the points of origin of the extensor and flexor muscles, respectively.

The Type I subdivision also includes the so-called *entrapment neuropathies*, although the term is not a particularly good one, as the probability is that the symptoms of the condition may arise from irritation of the nerve as well as entrapment. The best known of these is *carpal tunnel syndrome* (which affects the median nerve as it passes through the confined space of the carpal tunnel at the front of the wrist). The median nerve may also be affected at other sites, as may the other nerves of the limb.

A meeting of relevant healthcare professionals was convened by the U.K. Health and Safety Executive in 1998 to establish case definitions, as far as possible, for work-related upper limb pain syndromes — a first stage in establishing diagnostic criteria. Consensus was reached and definitions established for carpal tunnel syndrome, tenosynovitis of the wrist, de Quervain's disease of the wrist, epicondylitis, shoulder capsulitis and shoulder tendonitis (Harrington et al., 1998). The group also identified a condition of nonspecific diffuse forearm pain, although the criteria were largely exclusive rather than inclusive.

The underlying pathologies of the Type I conditions are relatively well understood (although some grey areas remain), and they are thus relatively uncontentious, except that in the medico-legal context there may be an entirely legitimate dispute as to whether the condition is caused by work in any individual case, as against being the consequence of normal wear and tear, degenerative changes or a constitutional predisposition, etc. At law, the decision must be made in each individual case on the balance of probabilities, although for some conditions (e.g., peritendinitis and tenosynovitis) there is a stronger a priori assumption of work relatedness than for others (e.g., carpal tunnel syndrome), in that the available scientific evidence points to occupational or constitutional risk factors as being of greater or lesser relative importance in the (multifactorial) aetiology of the conditions.

It stands to reason that for almost any occupational or work-related condition one could name, some people will be more at risk than others — otherwise all the members of a particular work force would be affected rather than only some of them. This being so, it would be fallacious to draw a distinction between a condition that is caused by constitutional factors and merely aggravated by work, as against being caused by work in a person who is at risk for constitutional reasons.

More recently it has become increasingly clear, however, that many people with RSI/WRULD cannot easily be allocated to any of the traditionally recognised clinical categories. These people — who are described as suffering from Type II RSI — typically report symptoms of pain and dysfunction at multiple sites in the upper limb (or limbs), shoulder region and neck. These symptoms are often described as 'diffuse'. This is an unfortunate choice of word, in that it tends to imply that the symptoms are vague and insubstantial. They are not — at least, not always. In some cases they are crippling. A better description is 'disseminated'. It is likewise often said that these people have no objective clinical signs. (In medical parlance, a symptom is something reported by the patient; a sign is something that the physician observes.) This distinction is only partly true at best, in that the principal signs that may be observed are ones in which the patient reports pain, either on the palpation of tender structures (mainly muscles) or on the performance of certain diagnostic manoeuvres (the details of which need not concern us here).

The experienced examiner will also be able to detect palpable changes in the physical quality of the muscles, which may feel hard or compacted, etc. In some cases there will be a change in the temperature of the affected limb, indicative of a disturbance of blood flow.

The term *RSI* (or alternatively, 'repetitive strain syndrome' [RSS]) is sometimes applied to these disseminated conditions by default and for want of a better alternative, as if it were a diagnosis. This use of the same term in both the generic and quasi-diagnostic senses has been a source of much avoidable confusion. The practice is to be discouraged. This leaves us, however, with the problem of what else to call these conditions. Our own preference is for the term *disseminated overuse syndrome* (DOS).

Although, from experience, the disseminated forms of RSI/WRULD are by no means unknown in manual workers on industrial assembly lines, they occur most prominently and characteristically in keyboard users. They are in fact the classic *keyboard injury* (see below). In contrast, assembly workers seem to be most commonly affected by the localised varieties of RSI/WRULD. This suggests that different causative mechanisms are involved.

The disseminated overuse syndrome of the keyboard user has a characteristic natural history. The first symptoms are typically minor ones: most often a tingling in the hands or aching at the wrist; less often there is a dull ache in the neck or shoulder area. At first this comes on towards the end of the working day and subsides in the evenings and at weekends. The symptoms gradually become more severe, more unremitting and come to affect the person's activities away from the workplace, interfere with sleep, and so on. As they do so the symptoms spread, either proximally (up the limb) or distally (down the limb), as the case may be. In due course they may cross over to the opposite limb or they may affect the upper back, the breast, or the side of the face and even occasionally the low back and lower limbs. From our experience, the typical characteristics of disseminated overuse syndrome are those summarised in Table 9.4.

The underlying pathology of the syndrome remains both obscure and contentious. There are those who take the view (often in the medico-legal context) that what is unknown is unreal. 'If I wasn't taught about this disease at medical school, and I don't know how to treat it, then how can it possibly exist?' They therefore say that people who claim to suffer from this syndrome are deluded or they argue that the symptoms these people report are the product of 'conversion hysteria' or

TABLE 9.4
Typical Characteristics of Disseminated Overuse Syndrome of the Keyboard User

- Insidious in onset
- Difficult to reverse
- Multiple sites of tenderness in upper limb(s), shoulder girdle and neck
- Signs of 'adverse neural tension' often present
- Biochemical changes in tissues and pain sensitisation
- Secondary or tertiary psychological effects very common

'somatization' or other kinds of psychobabble. They are wrong. The syndrome undoubtedly has a basis in organic pathology, and, complex as they are, the underlying mechanisms are slowly being unravelled (as discussed in the reviews of Buckle and Devereux, [1999, 2002]). Greening, Lynn and their colleagues have shown changes in both sensory nervous function and kinematics of the anatomical structures in the wrist for people reporting nonspecific arm pain (Greening and Lynn, 1998; Greening et al., 1999, 2003). They first found that the threshold for perception of vibration in the hand was raised (an early sign of peripheral neuropathy) among patients with RSI but also among office workers who used a keyboard regularly in their work (compared with a control group who only used a computer keyboard occasionally). A period of keyboard work was found to further exacerbate the change in function for the RSI patients. In later studies using magnetic resonance scanning, they showed that patients with nonspecific arm pain have considerably reduced movement of the median nerve in the carpal tunnel, indicating nerve entrapment, and they suggest that diffuse symptoms associated with RSI could be due to nerve entrapment at other sites.

The condition progresses from one of ordinary muscle fatigue which, when opportunities for recovery are inadequate, becomes chronic. At some point (and the mechanism is not well understood), a self-sustaining cycle of inflammation, pain and muscle spasm supervenes, involving the activation in the muscle of what are sometimes called 'trigger points' (see Wigley, 1990; Pheasant, 1991a,b). At the same time a cascade of changes is initiated in the central nervous system (CNS) mechanisms which mediate the experience of pain. This is sometimes called *pain amplification* or *neurological sensitization* (Pheasant, 1991a, 1992, 1994a,b; Cohen et al., 1992; Gibson et al., 1991; Helme et al., 1992). Via various feedback loops in the CNS, disturbances of motor control and blood flow may also ensue. The entrapment or irritation of peripheral nerves (which physiotherapists call 'adverse neural tension') may also be part of the picture, although how this relates to the other mechanisms is not clear.

Psychological factors may play a role in this process, but they are certainly not the sole causative factors involved, and there is good evidence that in many cases psychological symptoms that RSI victims report (anxiety, depression) are secondary or tertiary developments from the initial physical disorder (see below). In other words, they are consequences, not causes. Sleep disturbance may be an intervening causative link.

The classification of the conditions falling into the overall RSI/WRULD category into Types I and II is a useful one, but it is something of an oversimplification. Some of the possibilities are summarised in Table 9.5 and Figure 9.13 (see also Pheasant, 1994b).

9.6.2 Overuse Injuries to Process Workers

Overuse injuries to the hand, wrist and forearm, which are endemic in manual workers on industrial assembly lines, etc., are the product of a number of risk factors. The most prominent is very often a lack of task diversity, in that if you repeat the same wrist and hand movement, or short-cycle manipulative operation, endlessly

TABLE 9.5
A Provisional Clinical Classification of Work-Related Upper Limb Disorders

Type	Signs and Symptoms
Type I	Localised condition, generally involving soft-tissue injury, e.g., tenosynovitis, peritendinitis, epicondylitis, carpal tunnel syndrome or another entrapment neuropathy. Physical signs specific to the condition in question are present.
Type II	Diffuse condition involving pain and dysfunction at multiple sites. Tenderness on palpation in multiple muscle groups, often with signs of 'adverse neural tension' and sometimes with signs of neurovascular disturbance. Absence of physical signs of specific (i.e., Type I) conditions. The classic disseminated overuse syndrome (which occurs particularly in keyboard users).
Type III	Type I and II conditions are present concurrently.
Type IV	More than one of the Type I conditions are present concurrently (e.g., carpal tunnel with epicondylitis).
Type V	Localised condition involving chronic pain and dysfunction. No evidence of any specific (Type I) condition, either now or at any prior stage in the history.
Type VI	Localised condition involving chronic pain and dysfunction with clear (?) evidence of a Type I condition at some prior stage in the history but no ongoing physical signs thereof. Symptoms are very often activity related. May be regarded as a postinjury syndrome resulting from a process of sensitization.
Type VII	Disseminated (Type II) condition with clear (?) evidence of a Type I condition at some prior stage in the history but no ongoing physical signs thereof. May be regarded as a postinjury syndrome.

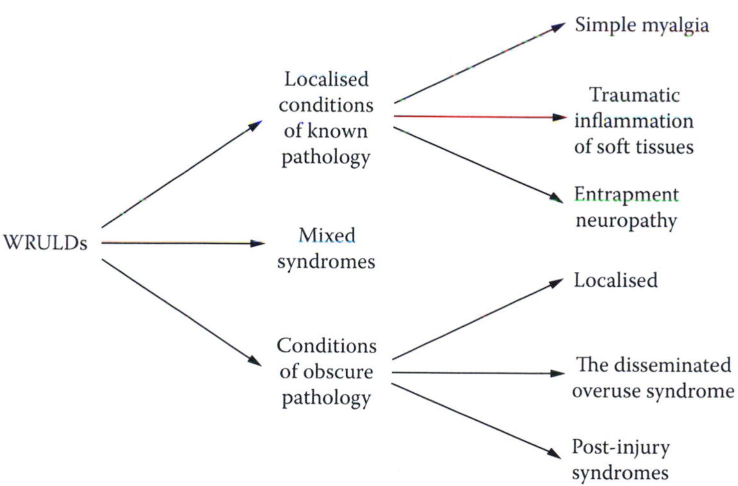

FIGURE 9.13 Classification of work-related upper limb disorders (WRULDs).

throughout every working day, then it stands to reason that some anatomical structure or another in the hand or arm is simply going to wear out. This must surely be a matter of common sense. We could call this 'abnormal wear and tear' as against the 'normal wear and tear' of everyday life. A closely related risk factor is time pressure, whether this is imposed by the pace of the machine or by production targets and incentive schemes in self-paced work.

Overuse injuries are not solely restricted to short-cycle time tasks, which are repetitive in the narrow sense of the word. Some quite complex tasks may carry a high level of risk if the demands they impose are excessive in other ways. Likewise, although job rotation is always in theory to be encouraged, it may be of minimal benefit if the jobs between which the person rotates all have *similar biomechanical profiles* — that is, if they impose similar patterns of loading on the muscles and soft tissues.

The magnitude of the forces that it is necessary for the person to exert in the task in question is also relevant. In an elegant study of workers from a number of industrial plants, Silverstein et al. (1986) compared the prevalence of WRULDs in those whose jobs entailed high and low levels of force and repetitiveness. Both force and repetition were statistically significant risk factors on their own, and the combination of the two carried a particularly high level of risk. This basic pattern of association was confirmed in a subsequent study of the specific disorder, carpal tunnel syndrome, which was reported by the same authors, although in this later study, force on its own was not a significant risk factor (Silverstein et al., 1987). The data from both studies suggest that when force and repetition are combined, their effects are at least multiplicative. Stenlund et al. (1992) demonstrated the cumulative effect of heavy work in a study of the shoulder disorder acromioclavicular osteoarthrosis, where risk was found to increase significantly with total load lifted during working life. Overall, these findings are just about what we should expect, on the assumption that the pathologies of these conditions are the consequence of abnormal wear and tear which results from hand-intensive work.

The specific risk factors involved in hand actions performed during manual tasks or in the use of hand tools have been discussed in some detail in Section 6.9.2. There is also epidemiological evidence that carpal tunnel syndrome may be caused by the use of vibrating tools (Cannon et al., 1981).

It is widely recognised that people become accustomed to repetitive (and otherwise hand-intensive) work over a period of time. This process of adaptation is sometimes known as 'work hardening'. It probably has a number of physiological components. A simple muscle training effect is part of the story, as in all probability is the greater economy of movement that comes with increasing skill. However, the physiology may be more complicated than this, and the process of adaptation may well be confounded with 'survivor effects' (in other words, those people who are unable to adapt leave the job).

Whatever the underlying physiology, however, it stands to reason that, given that such processes of adaptation do indeed occur, then newcomers to a particular job will be at an elevated level of risk as compared with 'old hands'. This has particularly been shown to be so for peritendinitis crepitans (Thompson et al., 1951). By the same token, we should expect any change in working practices that entails an increase in task demands to result in an increase in injury rate. Again this is borne out in practice. It also seems that people lose some of their physiological adaptation during holidays, periods of sickness absence and other lay-offs. (This is sometimes called de-adaptation.) When they return to work they are more at risk (Thompson et al., 1951).

It must be stressed, however, that these conditions are by no means to be regarded solely as training injuries, in that, although unaccustomed work is clearly an important risk factor, the process of injury may also occur insidiously, leading to the onset of the condition in a seasoned worker and without any change in working practices being involved. It may be that when the condition arises in this way, it is because the person's capacity to tolerate hand-intensive work is diminishing with time (due to the cumulative effects of the work itself or to normal ageing) until the point is reached when the demands of the task come to exceed that capacity. This is well illustrated in Armstrong et al.'s (1993) conceptual model of the development of work-related neck and upper limb musculoskeletal disorders (Figure 9.14), which shows how a 'dose' (internal forces acting on body tissue) arising from the work demands leads to a response within the body which may, in turn, diminish the person's capacity (or indeed enhance it, in the case of adaptation). This process continues over time in cascading sets of exposure, dose, capacity and response.

We are now in a position to summarise the principal ergonomic risk factors that are associated with overuse injuries to the hand, wrist and forearm in process workers. These are set out in Table 9.6.

Workers on assembly lines are also prone to overuse injuries to the muscles and soft tissues of the neck and shoulder region. These are most commonly caused by working for lengthy periods with the arms in a raised position or from making

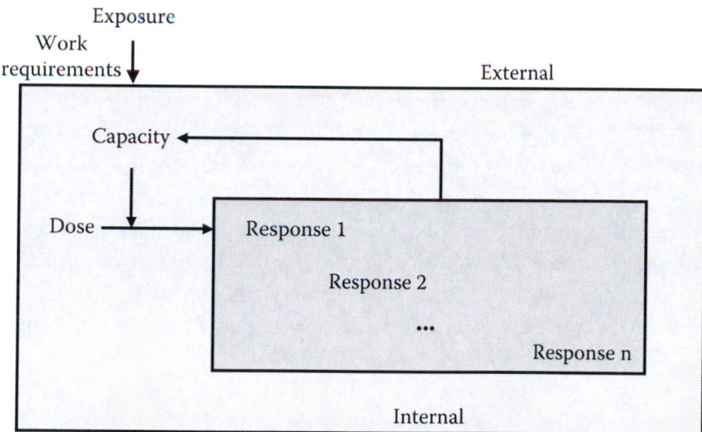

FIGURE 9.14 Conceptual model of the development of work-related neck and upper limb musculoskeletal disorders. (From Armstrong, T. J., Buckle, P., Fine, L. J., Hagberg, M., Jonsson, B., Kilbom, A., Kuorinka, I. A. A., Silverstein, B. A., Sjøgaard, G. and Viikari-Juntura, E. R. A. (1993). *Scandinavian Journal of Work, Environment and Health,* 19, 73–84. Reproduced with kind permission.)

TABLE 9.6
Overuse Injuries to the Forearm, Wrist and Hand in Process Workers: Occupational Risk Factors

- Lack of task diversity
- Time pressure
- Forceful exertion
- Frequent or repeated gripping actions, particularly if these are combined with turning actions or are made with a deviated wrist
- Pinch grip, claw grip, overspreading
- Vibration, impact, blunt trauma
- Unaccustomed work
- Psychosocial factors

frequent or repeated reaching actions (particularly overhead reaching or reaching behind the body).

Of the various WRULDs that affect the process worker, tenosynovitis has traditionally been regarded as the most important in the United Kingdom, whereas in the United States carpal tunnel syndrome has the greater prominence, and in the Scandinavian countries ergonomic and epidemiological studies have tended to focus on conditions affecting the neck and shoulder region (such as ‘tension neck’ or trapezius myalgia). It is difficult to account for these differences in emphasis. It seems wholly unlikely that they would reflect underlying differences in the true prevalence of these conditions. Nor does it seem likely that the conditions (at least as they are normally defined) can be mistaken for each other by anyone with more than a rudimentary knowledge of such matters.

9.6.3 Keyboard Injuries

There is nothing at all new about keyboard injuries. The earliest reports of ‘occupational cramps’ in typists date back to not long after the invention of the typewriter (see Quintner, 1991 for a historical review). In the first edition of his classic *Diseases of the Occupations*, Hunter (1955) listed the occupational groups that were prone to suffer from tenosynovitis, and along with various groups of industrial manual workers and agricultural craft workers he included ‘typists and comptometer operators’. (A comptometer was a primitive mechanical or electromechanical calculating machine, the keys of which sometimes had to be depressed in combinations like chords on a piano).

The rapid increase in the computerization of clerical and office work and introduction of the electronic keyboard, which occurred in the 1980s, led to a dramatic upswing in the reporting of these conditions. It is perhaps debatable whether the increase in reporting reflects a true increase in the underlying incidence of these conditions — we simply do not have the data to be certain either way — but assuming that it does (which on balance seems likely), then we must ask ourselves why this should be the case.

The force required to depress the keys on a modern electronic keyboard is only about one-twentieth of that which is required on a mechanical typewriter; and the 'travel' of the keys is likewise very much less. There are also more subtle differences in keying action, in that the old-fashioned mechanical typewriter required its user to modulate carefully the force with which each individual key was struck if the letters were to appear evenly on the page. There is no such requirement on an electromechanical or electronic keyboard. Modern keyboards are much thinner, furthermore, which in principle ought to reduce problems of anthropometric fit (see Section 7.2).

On the basis of these sorts of consideration, we might expect the introduction of the electronic keyboard to result in a reduction in injury rate rather than otherwise. Writing as long ago as 1980, Çakir et al. (1980) predicted that the introduction of the VDU would result in a *reduction* in the incidence of *tenosynovitis* in typists, although interestingly enough they predicted an *increase* in the incidence of what they called 'shoulder-arm syndrome' — what we might call the disseminated overuse syndrome or diffuse RSI, etc. Broadly speaking, this seems to have turned out to be correct.

The changeover from mechanical to electromechanical typewriters of increasing degrees of complexity — and then from electromechanical typewriters to screen-based word-processing systems — has led over the past few decades to a steady upward trend in typing speeds. Thus an average-to-good typist, who could reach 50 words per minute (wpm) on a mechanical machine, might reach 60 wpm on a basic electric typewriter and 70 wpm on a modern screen-based system, and nowadays, speeds of 90 wpm are not that unusual.

Common sense would perhaps tell us that the higher the keystroke rate (i.e., typing speed), the higher the risk. Overall, there is little or no hard evidence that this is the case, although the absence of such evidence does not of itself rule out the possibility, and there is little evidence to the contrary either. Physiologically, you could argue it either way. It could, for example, be argued that what matters is not so much the actual number of key depressions as such (per minute or per hour) as the proportion of that person's physiological capacity which the rate in question represents. This being so, if a person is required (for whatever reason) to work for lengthy periods at close to the limit of his or her individual capacity, there will be a potential risk, especially if circumstances are disadvantageous in other respects (for example, because of an unsatisfactory working posture). Thus a 50-wpm typist working flat out to meet a deadline would be exposed to much the same risk as a 90 wpm typist working under the same degree of pressure for the same length of time.

On balance, it seems probable that the direct agent of injury in most cases of (disseminated) RSI/WRULD that we encounter in keyboard users is the static muscle loading which is contingent on a fixed working posture, rather than the repetitive motions of striking the keys. Electromyographic studies by Onishi et al. (1982) have shown that (in a poorly designed workstation) this static loading may reach as much as 30% of the muscle's maximum capacity, which is more than enough to result in significant degrees of local muscle fatigue if there are not relatively frequent pauses for rest and recovery (see section 7.9). Early electromyographic studies of typists (Lundervold, 1958) showed that, with the onset of fatigue, muscle activation spreads to groups of muscles that were initially quiet, so the scene is set for a more generalised muscle fatigue to set in.

There are those who take the view that the light action of the electronic keyboard leads the user to hold back, particularly if they have been brought up on machines that have a more positive feel. Journalists transferring from their old portables onto screen-based systems often say this. If it is so, then physiologically this would equate to an increase in static muscle loading. As a hypothesis, it would lend itself to testing by electromyography, but the experiments have not been done. What is clearly the case, however, is that with the increased computerization of office work there has been a progressive diminution in the task diversity of the keyboard user — and an increase in the extent to which the work involves the inputting of streams of data or text onto the machine *and nothing else* — not even the manual operation of the carriage return and the winding of paper onto the platen.

There is also good evidence for an 'exposure effect'. In a study that has not been quoted anywhere nearly as widely as it deserves, Maurice Oxenburgh (1984) compared the prevalence of RSI in members of a particular organization who used the word processor keyboard for greater or lesser periods of time each day. The results are shown in Figure 9.15. The overall prevalence in this organization was 27%, but the prevalence varied from 9% in those who used a word processor for less than 3 hours per day to 70% in those who used it for more than 6 hours, which is equivalent to relative risks (compared with the overall prevalence) of 2.6 and 0.3 in the highest and lowest exposure categories, respectively. The most striking feature of these data is the sharp upswing in prevalence which occurs past the 6-hour point,

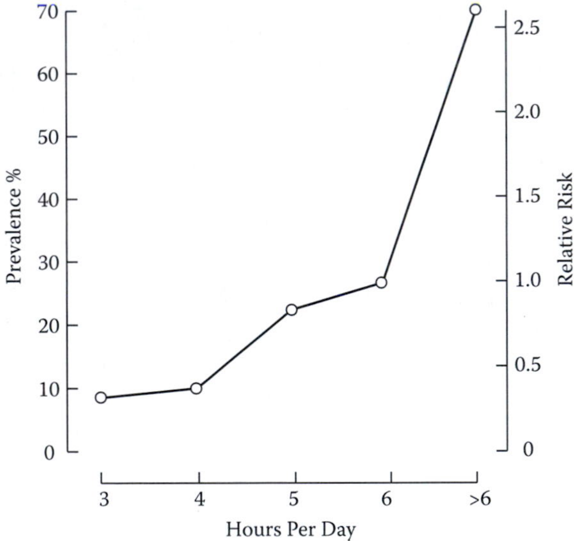

FIGURE 9.15 Repetitive strain injury (RSI) risk as a function of keyboard hours. (From a study by Oxenburgh, M. (1984). *Proceedings of the 21st Annual Conference of the Ergonomics Society of Australia and New Zealand,* 1984, 137–43.)

which is sufficiently marked that we should be justified in regarding it a threshold exposure effect. The moral is simple. Nobody should use the word processor for more than 6 hours per day.

To summarise, keyboard injury would appear to be associated with two main groups of risk factors in the working situation:

- Prolonged periods of intensive keyboard use in a constrained working posture, uninterrupted by rest pauses or changes of working activity, which may stem from lack of task diversity, pressure of work and so on
- The overall degree of musculoskeletal loading inherent in the keying action, of which static loading rather than dynamic loading is probably the more important contributory factor. The most important determinant of the static loading that the task entails will in most cases be the person's working posture, which will often be contingent upon deficiencies in the design or layout of the workstation and keyboard or other input device

Either or both classes of factors may be present in any particular case. Where both are present, they may be presumed to act at least additively and more probably multiplicatively. Having said this, however, one does at times come across cases where neither of these risk factors seem to be present. This presents us with something of a difficulty. Two possibilities suggest themselves: either there are other occupational or environmental risk factors involved that we have not so far taken into account, or perhaps these people are exceptionally constitutionally vulnerable for reasons as yet unidentified.

Returning now to the matter of static loading, we may distinguish between *generalised* and *localised* sources of muscle tension in keyboard work. The electromyographic studies of Lundervold (1958) showed, for example, that working with a seat that is too high (relative to the floor) results in an increased tension in a number of muscle groups. We could regard this as a generalised muscle tension contingent on discomfort, postural instability, and so on. There may also be a more localised tension, for example in the forearm or shoulder girdle muscles as a result of working with a deviated wrist and abducted shoulder — because of a keyboard that is too high (relative to the seat), a long reach distance to a mouse, or a forward leaning position, etc. Alternatively, the user may rest his or her wrists on the edge of the desk (perhaps again because it is too high relative to the seat) and thus work with wrists in extension, again leading to an increase in the static loading on the extensor muscles of the forearm. The studies of Duncan and Ferguson (1974) showed clear associations between postures of this kind and conditions that we would nowadays call RSI/WRULD; there was a particularly strong association between working with the wrists in extension or ulnar deviation and disorders affecting the muscles of the forearm. The need to rotate and incline the head and neck in order to read source documents placed flat on the table will likewise result in a static loading on the trapezius, sternomastoid and paraspinal neck muscles, leading to neck trouble. (For a further discussion of the ergonomics of keyboard workstations see Chapter 7).

Green and Briggs (1989) report an interesting study in which the anthropometric characteristics of keyboard users with and without symptoms of RSI/WRULD are compared. There was a tendency for those with symptoms to be more overweight, to have greater body breadths and lesser limb lengths. The authors discuss these findings in terms of lack of exercise and anthropometric mismatches with furniture and equipment. However, it is worth noting that the combination of greater body breadth with lesser relative limb length will inevitably result in a greater degree of ulnar deviation of the wrist when using the keyboard and thus a greater static loading on the forearm muscles.

In principle, we should expect both generalised and localised muscle tension to be of causative significance in the aetiology of keyboard injury. Of the two, we should probably expect the localised tension to have the more decisive effect, but this by no means need be infallibly the case.

Psychological stress is a potent generator of (generalised) muscle tension. Could psychological stress be the decisive causative factor therefore in some cases of keyboard injury? This is a difficult question indeed. The epidemiology certainly points to an association — at the statistical level, that is. In a study of data entry workers, Ryan and Bampton (1988) found that those who had symptoms in their necks, shoulders, arms and hands (and tenderness on palpation in relevant muscle groups) were more likely to report low levels of autonomy, peer group cohesion and role clarity and to say that they found their work boring and stressful. (They also reported missing rest breaks more often.) These associations could of course be explained in a number of ways. The possibilities have been narrowed down somewhat by a subsequent study reported by Hopkins (1990), who compared *symptom-free* workers in two sets of organisations doing similar sorts of keyboard work but having high and low prevalences of RSI within the organisations. Those in the high-prevalence workplaces reported lower levels of autonomy, peer group cohesion, task diversity, and job satisfaction and higher levels of stress and boredom. Thus the same psychological parameters that distinguish sufferers from nonsufferers (in Ryan and Bampton's study) also distinguish nonsufferers in high- and low-risk working environments (in Hopkins' study). By inference then, insomuch as psychological factors are of causative significance in the aetiology of keyboard injury, this influence stems from the external psychosocial features of the working situation rather than from the personal idiosyncrasies or inner mental turmoils of the individual in question. The medico-legal consequences of this finding are far-reaching.

To summarise, overall, the likelihood is that the proximate cause of keyboard injury is static muscle tension. This tension can in principle stem from a number of possible sources as shown in Figure 9.16.

The studies of Ryan and Bampton (1988) and Hopkins (1990) were, however, ones of people reporting relatively minor symptoms rather than the fully developed syndrome. This leaves the important question of whether such symptoms are more likely to progress to severe and disabling problems in people who have vulnerable personalities. An elegant study reported by Spence (1990) indicates that this is not the case. She compared five samples of subjects. People suffering from RSI of recent onset did not differ psychologically from people who had recently suffered other sorts of injuries to their upper limbs. Neither did either of these groups differ from

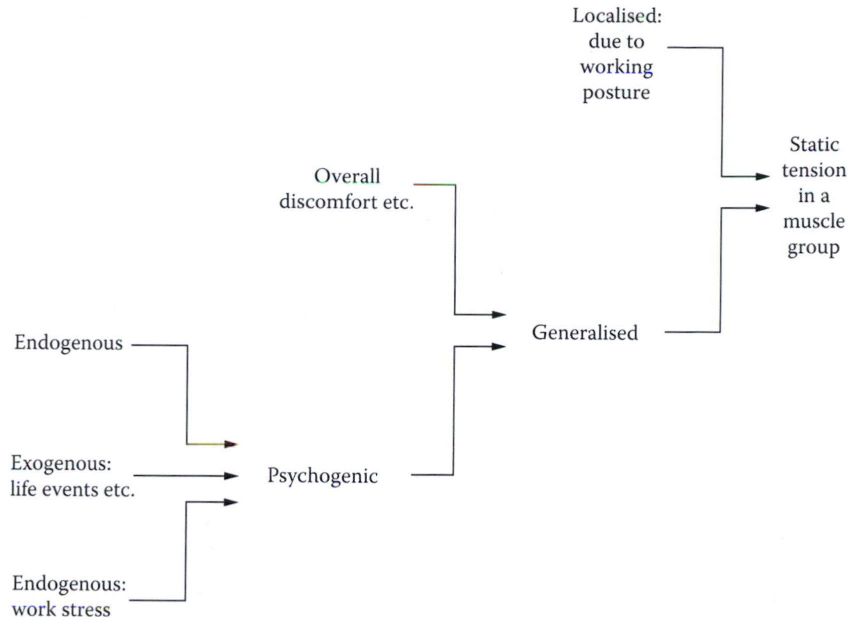

FIGURE 9.16 Static muscle loading in keyboard work: summation of contributory factors.

normal controls, whereas people with longstanding RSI and people with chronic problems resulting from injuries showed similar patterns of deviation from the psychological norm. This shows quite clearly that the anxiety and depression that are so characteristic of the RSI victim are the consequences of that person's physical condition, not its causes. We have little reason to find this surprising.

9.6.4 Assessment of Risk Factors for WRULDs

Epidemiological studies have shown the association of certain risk factors for the development of work-related upper limb disorders. Three extensive reviews of the evidence were made by Hagberg et al. (1995), Bernard (1997) and the National Research Council (1999). These showed that the work factors which have been most consistently shown to be associated with WRULDS are 'monotonous' (i.e., repetitive) hand work, high force, static muscle load and vibrating tools. Other factors may also be implicated, but the epidemiological studies performed to date have not demonstrated statistical significance for this. Organisational and psychosocial work factors that may be involved are high perceived workload, low job control and low social support from colleagues or supervisors (Smith and Carayon, 1996; Bernard, 1997; Devereux et al., 1999; Macfarlane and Hunt, 2000). Psychosocial factors can influence both the biomechanical load and the reactions to workplace conditions in a number of ways (Bongers et al., 1993). The psychological responses can increase static muscle tension, change the perception of symptoms or induce biochemical and physiological changes. The psychosocial stressors may also change work behaviour, such as working through rest breaks or adoption of poor work techniques (a simple example of this being the continuous holding of a pen, pair of scissors or other work tool even when it is not being used).

Many techniques have been developed for making WRULD risk assessments. Descriptions of these may be found in Li and Buckle (1999) and Colombini and Occhipinti (2004). The U.K. Health and Safety Executive (HSE, 2002) have produced a risk filter with an extended checklist to assist companies in making risk assessments, and ISO standard ISO/CD 11228-3 (ISO, 2003a) gives guidance relevant to process and assembly work.

Part III

The Bodyspace Tables – Anthropometric Database

10 Anthropometric Data

10.1 COMPILATION OF THE ANTHROPOMETRIC DATABASE

As we noted in an earlier chapter (Section 2.7), there are two possible approaches to the business of assembling an anthropometric database: the purist and the pragmatic. In compiling the tables that follow, the latter has been adopted. The technique used most extensively is that of ratio scaling, a more detailed discussion of which will be found in the Appendix Section A8.3 and in Pheasant (1982). Dimensions marked with an asterisk (*) are quoted from the original sources. The remainder have been estimated. Details of the sources are given below. Definitions of the dimensions which are tabulated together with some notes on application will be found in Chapter 2.

Each table is derived from a combination of two or more sources:

- A size source which gives us the mean and standard deviation for stature (or crown–heel length for infants) in the relevant target population
- One or more shape sources from which the coefficients E1 and E2 (as defined by equations A41 and A42 in the Appendix) are calculated. The shape sources must be of the same age and sex as the target population and have a similar ethnic admixture

10.2 POPULATIONS INCLUDED IN THE DATABASE

Table 10.1 is for the adult population of the United Kingdom. It is the same as Table 2.5, which is that of our 'standard reference population' used throughout the book and is repeated here for ease of reference. This is followed by tables for three subsets of the British adult population, broken down by age, who are somewhat taller or shorter as the case may be. The first of these (Table 10.2 for 19- to 25-year-olds) may be regarded as a best estimate of the anthropometric characteristics of the British adult population in the early part of the twenty-first century. The reader will note that the actual figures in these tables do not vary greatly; in practical terms, the differences between them are of only marginal significance. They have been included to enable the reader to decide for himself or herself the extent to which it may be necessary to take such differences into account in practical design work.

Tables 10.5 and 10.6 are two different estimates for the elderly population of the United Kingdom. The former is a more conservative estimate than the latter.

There follow four representative datasets for various European countries. Swedish men (Table 10.7) are much the same as British men, whereas Swedish women are a fair bit taller than British women and somewhat lighter in weight and more slender in their build. The Dutch (Table 10.8) are the tallest people in Europe (and the tallest included in the present database), in the case of the men by quite a fair margin. Table 10.9 is based upon a particularly thorough and well-conducted survey of French drivers reported by Rebiffé et al. (1983). On the assumption that these drivers were a representative sample of the adult population as a whole, the French are a little shorter than the British. Taken together, these various populations may be assumed to bracket the adult population of northern Europe. The populations of southern Europe will in general tend to be a little shorter than northern European populations. The data given for Polish industrial workers in Table 10.10 probably represent the shorter limit of European populations (or something very close to it).

U.S. adults (Table 10.11) are a little taller and a little heavier than their U.K. counterparts, although again the difference will be of relatively little significance in practical terms.

The data given in Table 10.12 for male Brazilian industrial workers place them close to the lower end of the stature range of European populations. Asian populations (Tables 10.13 to 10.16) are shorter still, the shortest population in the present collection being the sample of male Indian industrial workers reported by Gite and Yadav (1989), which is given in Table 10.14.

The tables conclude with data for British children from birth to maturity.

10.3 BRITISH ADULTS (TABLES 10.1 TO 10.6)

Stature data for British adults were taken from a survey of a nationwide stratified sample of households conducted in 1981 by the Office of Population Censuses and Surveys (OPCS, 1981; Knight, 1984). We may have considerable confidence in the validity and reliability of these data. Reference sources for E1 and E2 were U.S. civilians (Stoudt et al., 1965, 1970) for dimensions 8, 11 and 13–19; French drivers (Rebiffé et al., 1983) for dimensions 22–25 and 36; British drivers (Haslegrave, 1979) for dimensions 12, 20 and 21. The remaining dimensions were calculated from a variety of U.S. military surveys published by NASA (Webb Associates, 1978). Separate E coefficients were established for the different age bands.

The over-65-year-olds presented a problem. The OPCS stature data only extend to 65 years. An alternative source would be the survey by the Institute for Consumer Ergonomics (1983) of the inmates of geriatric institutions. The latter were measured shod, but if we subtract a nominal 20 mm for heels, we still see that their stature is much less than that of the 45- to 65-year-olds in the OPCS sample. In light of this discrepancy, two tables have been prepared. Table 10.5 is based on stature data estimated on the assumption that the decline in stature after 65 years is of a similar magnitude in Great Britain and the United States (as documented by Stoudt et al., 1965). Table 10.6 is based on the Institute for Consumer Ergonomics (1983) survey of 'elderly people', to which the reader should turn for further information.

10.4 ADULT POPULATIONS OF OTHER COUNTRIES (TABLES 10.7 TO 10.16)

Table 10.7 is based on surveys of Swedish male and female workers by Lewin (1969) and of Swedish women by Inglemark and Lewin (1968), Table 10.8 on a set of estimates for the Dutch population kindly provided by Johan Molenbroek of Delft University of Technology, Table 10.9 on the survey of French drivers by Rebiffé et al. (1983), and Table 10.10 on a survey of Polish industrial workers reported by Batogowska and Slowikowski (1974). Table 10.11 is based upon a major survey of U.S. adults conducted in 1971–1974 and deemed to be the most up to date available (Abraham, 1979). E coefficients were in all cases the same as for Table 10.1.

Table 10.12 is based upon a publication of the Instituto Nacional de Tecnologia (1989) in Rio de Janeiro, Table 10.13 on Abeysekera and Shanavaz (1987), and Table 10.14 on Gite and Yadav (1989). In each case missing data were, wherever possible, estimated by scaling up from the nearest available dimension and, failing that, the E coefficients of Table 10.1 were used.

Stature data and E coefficients used in Table 10.16 for men are based on sources cited in Webb Associates (1978); E coefficients for women are based on the assumption that differences in proportion between Japanese men and women are similar to those between European men and women. (In the absence of suitable reference data, estimates were made by scaling from the nearest available dimension.)

Table 10.15 is based on data from a hitherto unpublished survey of Chinese industrial workers in Hong Kong, kindly supplied by Bill Evans of the Department of Industrial Engineering at the University of Hong Kong. Dimensions marked with an asterisk (*) are quoted directly; the remainder are estimated as for Table 10.16.

10.5 INFANTS (TABLES 10.17 TO 10.21)

The crown–heel length data for infants were taken from Tanner et al. (1966). Male and female data were combined using equations A23 and A24 from the Appendix, and the 'point in time' values of the original data source were converted to 'period of time' values using the equation of Healy (1962). Values of E1 and E2 were taken from the survey of U.S. infants published by Snyder et al. (1977). Linear interpolation was required to adjust the data to appropriate mid-sample ages.

10.6 CHILDREN AND YOUTHS (TABLES 10.22 TO 10.38)

In the tables for British children and youths, a 5-year-old, for example, is anyone between their fifth and sixth birthdays. The stature data employed were taken from a major survey of British schoolchildren published by the Department of Education and Science (1972). Shape data were from Martin (1960) and Snyder et al. (1977). The predictions were edited to ensure steady unidirectional growth in all percentiles and compatibility with young adult data. The data are the same as those of the Department of Education and Science (1985) except that a different selection of anthropometric variables is included: additional details of the editing may be found in their report.

The 2-year-old data presented a particular problem since no suitable sources existed. It was therefore assumed that 2- and 3-year-olds differed by the same amount as 3- and 4-year-olds. This assumption may, however, be incorrect (as shown by Smith and Norris [2001]) and the dimensions overestimated, at least for male 2-year-olds, and the data in Table 10.23 should be used with caution.

No suitable sources exist for the chest or abdominal depths of the under 18-year-olds. E coefficients were calculated from Snyder et al. (1977) for equivalent circumferences; they were then scaled down according to young adult depth data. Relative chest depth for girls was assumed to be the same as that for boys until 11 years and then to approach the adult female proportions by steady annual increments.

Subsequent to the preparation of these estimates, a British Standard BS 7231 was published dealing with the body measurements of children (BSI, 1990b,c). Most of the dimensions given in this standard are appropriate for clothing design rather than for other areas of application. The stature data of this standard differ somewhat from those of the Department of Education and Science (1972) survey on which the present estimates are based. We have no means of knowing which is the more representative survey. The reader is invited to scale up the present estimates using the stature data given in BS 7231, if it seems appropriate to do so.

Similar scaling will be needed to update estimates as secular changes occur in the population. Smith and Norris (2001) have reviewed the secular changes in body sizes of U.K. children over the three decades between the surveys carried out by the Department of Education and Science (1972) in 1970–1971 (as used in the *Bodyspace* tables) and a survey by the Department of Health (1999) in 1995–1997 and found that there have been significant changes. Stature has increased by 1% for boys and 1.5% for girls, and weight has increased much more, by 6% and 7.5%, respectively. The greatest increases were found for children between the ages of 10 and 14 years. As Smith and Norris (2001) point out, the practical effect for design will depend on the dimension concerned and on the accuracy required, but limb dimensions and reach distances have increased significantly. The increase in weight will have had a corresponding effect on soft tissue dimensions. Just taking one example of the effects of these secular changes, they estimated that, for forward grip reach of an 11-year-old, the 5th %ile value in the *Bodyspace* tables and in Childata (Norris and Wilson, 1995) now represents the 2nd %ile, while the original 95th %ile value now represents the 86th %ile value. This type of underestimation will be especially critical for clearances and for safety-critical reach distances.

10.7 THE ANTHROPOMETRIC TABLES

The *Bodyspace* tables on the following pages give the anthropometric estimates for various populations of adults, youths, children and infants. Definitions of the dimensions which are tabulated are given in Table 2.4 and Figure 2.11.

- Table 10.1 British adults aged 19–65 years
- Table 10.2 British adults aged 19–25 years
- Table 10.3 British adults aged 19–45 years
- Table 10.4 British adults aged 45–65 years

TABLE 10.1
Anthropometric Estimates for British Adults Aged 19 to 65 Years (all dimensions in millimetres, except for body weight, given in kilograms)

	Men				Women			
Dimension	5th %ile	50th %ile	95th %ile	SD	5th %ile	50th %ile	95th %ile	SD
1. Stature	1625	1740	1855	70	1505	1610	1710	62*
2. Eye height	1515	1630	1745	69	1405	1505	1610	61
3. Shoulder height	1315	1425	1535	66	1215	1310	1405	58
4. Elbow height	1005	1090	1180	52	930	1005	1085	46
5. Hip height	840	920	1000	50	740	810	885	43
6. Knuckle height	690	755	825	41	660	720	780	36
7. Fingertip height	590	655	720	38	560	625	685	38
8. Sitting height	850	910	965	36	795	850	910	35
9. Sitting eye height	735	790	845	35	685	740	795	33
10. Sitting shoulder height	540	595	645	32	505	555	610	31
11. Sitting elbow height	195	245	295	31	185	235	280	29
12. Thigh thickness	135	160	185	15	125	155	180	17
13. Buttock–knee length	540	595	645	31	520	570	620	30
14. Buttock–popliteal length	440	495	550	32	435	480	530	30
15. Knee height	490	545	595	32	455	500	540	27
16. Popliteal height	395	440	490	29	355	400	445	27
17. Shoulder breadth (bideltoid)	420	465	510	28	355	395	435	24
18. Shoulder breadth (biacromial)	365	400	430	20	325	355	385	18
19. Hip breadth	310	360	405	29	310	370	435	38
20. Chest (bust) depth	215	250	285	22	210	250	295	27
21. Abdominal depth	220	270	325	32	205	255	305	30
22. Shoulder–elbow length	330	365	395	20	300	330	360	17
23. Elbow–fingertip length	440	475	510	21	400	430	460	19
24. Upper limb length	720	780	840	36	655	705	760	32
25. Shoulder–grip length	610	665	715	32	555	600	650	29
26. Head length	180	195	205	8	165	180	190	7
27. Head breadth	145	155	165	6	135	145	150	6
28. Hand length	175	190	205	10	160	175	190	9
29. Hand breadth	80	85	95	5	70	75	85	4
30. Foot length	240	265	285	14	215	235	255	12
31. Foot breadth	85	95	110	6	80	90	100	6
32. Span	1655	1790	1925	83	1490	1605	1725	71
33. Elbow span	865	945	1020	47	780	850	920	43
34. Vertical grip reach (standing)	1925	2060	2190	80	1790	1905	2020	71
35. Vertical grip reach (sitting)	1145	1245	1340	60	1060	1150	1235	53
36. Forward grip reach	720	780	835	34	650	705	755	31
Body weight	55	75	94	12	44	63	81	11*

Notes: See notes in Sections 10.1 to 10.3.

Dimensions marked with an asterisk (*) are quoted from the original source. The remainder have been estimated.

TABLE 10.2
Anthropometric Estimates for British Adults Aged 19 to 25 Years (all dimensions in millimetres)

	Dimension	Men 5th %ile	Men 50th %ile	Men 95th %ile	Men SD	Women 5th %ile	Women 50th %ile	Women 95th %ile	Women SD
1.	Stature	1640	1760	1880	73	1520	1620	1720	61*
2.	Eye height	1530	1650	1770	72	1415	1515	1615	60
3.	Shoulder height	1330	1445	1555	69	1225	1320	1410	57
4.	Elbow height	1020	1105	1195	54	940	1015	1090	45
5.	Hip height	850	935	1020	52	745	815	885	43
6.	Knuckle height	695	765	835	42	665	725	785	35
7.	Fingertip height	595	665	730	40	565	630	690	38
8.	Sitting height	855	915	980	37	800	855	915	35
9.	Sitting eye height	740	795	855	36	690	745	800	33
10.	Sitting shoulder height	545	600	655	33	510	560	610	31
11.	Sitting elbow height	195	245	300	32	180	230	275	28
12.	Thigh thickness	130	160	185	16	120	150	175	16
13.	Buttock–knee length	545	595	650	32	520	565	615	29
14.	Buttock–popliteal length	445	500	555	34	430	475	525	29
15.	Knee height	495	550	605	33	460	500	545	26
16.	Popliteal height	400	445	495	30	355	400	445	27
17.	Shoulder breadth (bideltoid)	415	465	510	29	355	395	435	24
18.	Shoulder breadth (biacromial)	370	405	440	21	330	360	390	18
19.	Hip breadth	300	350	400	31	300	350	400	29
20.	Chest (bust) depth	185	225	270	26	190	235	275	26
21.	Abdominal depth	195	240	280	26	185	220	260	22
22.	Shoulder–elbow length	335	370	405	21	305	330	360	17
23.	Elbow–fingertip length	445	480	515	22	400	430	465	19
24.	Upper limb length	730	790	850	37	660	710	760	32
25.	Shoulder–grip length	615	670	730	34	560	605	650	29
26.	Head length	185	195	210	8	170	180	190	7
27.	Head breadth	145	155	165	7	135	145	155	5
28.	Hand length	175	190	210	10	160	175	190	9
29.	Hand breadth	80	90	95	5	70	75	85	4
30.	Foot length	245	270	290	15	220	240	260	12
31.	Foot breadth	90	100	110	7	80	90	100	5
32.	Span	1670	1815	1955	86	1500	1615	1730	70
33.	Elbow span	875	955	1035	49	785	855	925	42
34.	Vertical grip reach (standing)	1950	2085	2220	83	1805	1915	2030	70
35.	Vertical grip reach (sitting)	1155	1260	1360	63	1070	1155	1245	52
36.	Forward grip reach	730	790	845	36	655	705	755	31

Notes: Best estimates for overall British population in the year 2000.

See notes in Sections 10.1 to 10.3.

Dimensions marked with an asterisk (*) are quoted from the original source. The remainder have been estimated.

TABLE 10.3
Anthropometric Estimates for British Adults Aged 19 to 45 Years (all dimensions in millimetres)

	Men				Women			
Dimension	5th %ile	50th %ile	95th %ile	SD	5th %ile	50th %ile	95th %ile	SD
1. Stature	1635	1745	1860	69	1515	1615	1715	61*
2. Eye height	1525	1635	1750	68	1415	1515	1615	60
3. Shoulder height	1325	1435	1540	65	1225	1315	1410	57
4. Elbow height	1015	1100	1185	51	940	1015	1085	45
5. Hip height	845	925	1005	49	745	815	885	43
6. Knuckle height	695	760	825	40	665	725	780	35
7. Fingertip height	595	660	720	38	565	625	690	38
8. Sitting height	855	915	970	35	800	855	915	35
9. Sitting eye height	740	795	850	34	690	745	800	33
10. Sitting shoulder height	545	595	650	31	510	560	610	31
11. Sitting elbow height	195	245	295	30	190	235	280	28
12. Thigh thickness	135	160	185	15	125	155	180	16
13. Buttock–knee length	545	595	645	30	520	570	610	29
14. Buttock–popliteal length	445	495	550	32	435	480	530	29
15. Knee height	495	545	600	31	460	500	545	26
16. Popliteal height	395	445	490	28	355	400	445	27
17. Shoulder breadth (bideltoid)	420	465	510	28	355	395	435	24
18. Shoulder breadth (biacromial)	365	400	435	20	330	360	390	18
19. Hip breadth	310	355	405	29	300	365	425	37
20. Chest (bust) depth	200	240	275	23	195	240	285	26
21. Abdominal depth	210	255	300	28	195	245	290	29
22. Shoulder–elbow length	330	365	400	20	305	330	360	17
23. Elbow–fingertip length	445	475	510	21	400	430	460	19
24. Upper limb length	725	785	840	35	655	710	760	32
25. Shoulder–grip length	615	665	720	32	555	605	650	29
26. Head length	185	195	210	8	165	180	190	7
27. Head breadth	145	155	165	6	135	145	155	5
28. Hand length	175	190	205	10	160	175	190	9
29. Hand breadth	80	85	95	5	70	75	85	4
30. Foot length	245	265	290	14	215	235	255	12
31. Foot breadth	90	100	110	6	80	90	100	5
32. Span	1665	1800	1935	81	1500	1615	1730	70
33. Elbow span	875	950	1025	46	785	855	920	42
34. Vertical grip reach (standing)	1940	2070	2200	79	1800	1915	2030	70
35. Vertical grip reach (sitting)	1150	1250	1345	59	1070	1155	1240	52
36. Forward grip reach	725	785	840	34	655	705	755	31

Notes: See notes in Sections 10.1 to 10.3.

Dimensions marked with an asterisk (*) are quoted from the original source. The remainder have been estimated.

TABLE 10.4
Anthropometric Estimates for British Adults Aged 45 to 65 Years (all dimensions in millimetres)

	Dimension	Men 5th %ile	Men 50th %ile	Men 95th %ile	Men SD	Women 5th %ile	Women 50th %ile	Women 95th %ile	Women SD
1.	Stature	1610	1720	1830	67	1495	1595	1695	61*
2.	Eye height	1505	1610	1720	66	1395	1495	1595	60
3.	Shoulder height	1305	1410	1515	63	1205	1300	1395	57
4.	Elbow height	1000	1080	1160	50	925	1000	1075	45
5.	Hip height	835	910	990	48	735	805	875	43
6.	Knuckle height	685	750	810	39	655	715	775	35
7.	Fingertip height	590	650	710	37	555	620	680	38
8.	Sitting height	840	900	955	34	785	845	900	35
9.	Sitting eye height	725	780	835	34	680	735	790	33
10.	Sitting shoulder height	535	585	635	30	500	550	600	31
11.	Sitting elbow height	190	240	290	29	185	230	280	28
12.	Thigh thickness	135	160	185	15	130	155	180	16
13.	Buttock–knee length	540	585	635	29	520	570	620	29
14.	Buttock–popliteal length	440	490	540	31	435	480	530	29
15.	Knee height	490	540	590	30	450	495	540	26
16.	Popliteal height	390	435	480	27	350	395	440	27
17.	Shoulder breadth (bideltoid)	415	460	505	27	350	390	430	24
18.	Shoulder breadth (biacromial)	360	395	425	19	325	355	385	18
19.	Hip breadth	310	360	405	28	315	375	440	37
20.	Chest (bust) depth	225	260	295	20	220	265	305	26
21.	Abdominal depth	230	285	340	34	220	270	320	31
22.	Shoulder–elbow length	330	360	390	19	300	325	355	17
23.	Elbow–fingertip length	435	470	505	20	395	425	455	19
24.	Upper limb length	715	770	830	34	650	700	755	32
25.	Shoulder–grip length	605	655	710	31	550	595	645	29
26.	Head length	180	195	205	7	165	175	190	7
27.	Head breadth	140	150	160	6	135	140	150	5
28.	Hand length	170	185	205	9	155	170	185	9
29.	Hand breadth	80	85	95	5	70	75	80	4
30.	Foot length	240	260	285	13	215	235	255	12
31.	Foot breadth	85	95	105	6	80	90	95	5
32.	Span	1640	1770	1900	79	1480	1595	1710	70
33.	Elbow span	860	935	1010	45	775	845	910	42
34.	Vertical grip reach (standing)	1910	2035	2160	76	1775	1890	2000	70
35.	Vertical grip reach (sitting)	1135	1230	1325	58	1055	1140	1225	52
36.	Forward grip reach	715	770	825	33	645	695	750	31

Notes: See notes in Sections 10.1 to 10.3.

Dimensions marked with an asterisk (*) are quoted from the original source. The remainder have been estimated.

TABLE 10.5
Anthropometric Estimates for British Adults Aged 65 to 80 Years (all dimensions in millimetres)

	Dimension	Men				Women			
		5th %ile	50th %ile	95th %ile	SD	5th %ile	50th %ile	95th %ile	SD
1.	Stature	1575	1685	1790	66	1475	1570	1670	60
2.	Eye height	1470	1575	1685	65	1375	1475	1570	59
3.	Shoulder height	1280	1380	1480	62	1190	1280	1375	56
4.	Elbow height	975	1055	1135	49	910	985	1055	44
5.	Hip height	820	895	975	47	740	810	875	42
6.	Knuckle height	670	730	795	38	645	705	760	35
7.	Fingertip height	575	635	695	36	550	610	670	37
8.	Sitting height	815	875	930	36	750	815	885	41
9.	Sitting eye height	705	760	815	34	645	710	770	38
10.	Sitting shoulder height	520	570	625	32	475	535	590	36
11.	Sitting elbow height	175	220	270	29	165	210	260	28
12.	Thigh thickness	125	150	175	15	115	145	170	16
13.	Buttock–knee length	530	580	625	29	520	565	615	29
14.	Buttock–popliteal length	430	485	535	31	430	480	525	29
15.	Knee height	480	525	575	30	455	500	540	26
16.	Popliteal height	385	425	470	27	355	395	440	26
17.	Shoulder breadth (bideltoid)	400	445	485	26	345	385	425	23
18.	Shoulder breadth (biacromial)	350	375	405	17	320	350	380	17
19.	Hip breadth	305	350	395	28	310	370	430	37
20.	Chest (bust) depth	225	260	290	20	220	265	305	26
21.	Abdominal depth	245	300	355	33	225	270	320	30
22.	Shoulder–elbow length	320	350	385	19	295	320	350	17
23.	Elbow–fingertip length	425	460	490	20	390	420	450	19
24.	Upper limb length	700	755	810	34	640	690	740	31
25.	Shoulder–grip length	595	645	695	30	540	590	635	28
26.	Head length	175	190	200	7	165	175	185	7
27.	Head breadth	140	150	160	6	130	140	150	5
28.	Hand length	170	185	200	9	155	170	185	9
29.	Hand breadth	75	85	90	5	65	75	80	4
30.	Foot length	235	255	280	13	210	230	250	12
31.	Foot breadth	85	95	105	6	80	85	95	5
32.	Span	1605	1735	1860	78	1460	1570	1685	68
33.	Elbow span	840	915	985	44	760	830	900	41
34.	Vertical grip reach (standing)	1840	1965	2090	75	1725	1835	1950	68
35.	Vertical grip reach (sitting)	1110	1205	1295	57	1040	1125	1210	52
36.	Forward grip reach	700	755	805	32	640	685	735	30

Notes: See notes in Sections 10.1 to 10.3.

Dimensions marked with an asterisk (*) are quoted from the original source. The remainder have been estimated.

TABLE 10.6
Anthropometric Estimates for 'Elderly People' (all dimensions in millimetres)

Dimension	Men 5th %ile	Men 50th %ile	Men 95th %ile	Men SD	Women 5th %ile	Women 50th %ile	Women 95th %ile	Women SD
1. Stature	1515	1640	1765	77	1400	1515	1630	70*
2. Eye height	1410	1535	1660	76	1305	1420	1535	69
3. Shoulder height	1225	1345	1465	72	1130	1235	1340	65
4. Elbow height	935	1025	1120	57	860	945	1030	52
5. Hip height	785	875	965	55	700	780	860	49
6. Knuckle height	640	715	785	45	610	680	745	41
7. Fingertip height	550	620	690	42	515	590	660	43
8. Sitting height	785	850	920	42	710	785	865	48
9. Sitting eye height	675	740	805	40	610	685	755	45
10. Sitting shoulder height	495	555	615	37	445	515	585	42
11. Sitting elbow height	160	215	270	34	150	205	255	32
12. Thigh thickness	120	145	175	17	105	140	170	19
13. Buttock–knee length	510	565	620	34	490	545	600	34
14. Buttock–popliteal length	410	470	530	36	405	460	515	34
15. Knee height	455	515	570	35	430	480	530	30
16. Popliteal height	365	415	470	32	330	380	430	31
17. Shoulder breadth (bideltoid)	380	430	480	31	325	370	415	27
18. Shoulder breadth (biacromial)	335	365	400	20	305	335	370	20
19. Hip breadth	290	340	395	32	285	355	425	43
20. Chest (bust) depth	215	255	290	23	205	255	305	30
21. Abdominal depth	230	290	355	39	205	260	320	35
22. Shoulder–elbow length	305	345	380	22	280	310	345	20
23. Elbow–fingertip length	410	450	485	23	370	405	440	22
24. Upper limb length	670	735	800	39	605	665	725	36
25. Shoulder–grip length	570	625	685	35	510	565	620	33
26. Head length	170	185	200	8	155	170	180	8
27. Head breadth	135	145	155	7	125	135	145	6
28. Hand length	160	180	195	11	145	165	180	10
29. Hand breadth	75	80	90	5	65	70	80	5
30. Foot length	225	250	275	15	200	225	245	14
31. Foot breadth	80	90	105	7	75	85	95	6
32. Span	1540	1690	1840	91	1380	1515	1645	80
33. Elbow span	805	890	975	52	720	800	880	48
34. Vertical grip reach (standing)	1770	1915	2060	88	1640	1770	1900	80
35. Vertical grip reach (sitting)	1065	1175	1280	66	985	1085	1180	60
36. Forward grip reach	675	735	795	38	605	660	720	35

Notes: See notes in Sections 10.1 to 10.3.

Dimensions marked with an asterisk (*) are quoted from the original source. The remainder have been estimated.

TABLE 10.7
Anthropometric Estimates for Swedish Adults (all dimensions in millimetres, except for body weight, given in kilograms)

	Dimension	Men				Women			
		5th %ile	50th %ile	95th %ile	SD	5th %ile	50th %ile	95th %ile	SD
1.	Stature	1630	1740	1850	68	1540	1640	1740	62*
2.	Eye height	1520	1630	1740	68	1435	1535	1635	62*
3.	Shoulder height	1345	1445	1545	62	1255	1355	1455	60*
4.	Elbow height	1020	1100	1180	49	905	1025	1145	73*
5.	Hip height	815	890	965	45	745	830	915	52*
6.	Knuckle height	720	760	800	25	675	735	795	36
7.	Fingertip height	595	655	715	37	570	635	700	38
8.	Sitting height	830	900	970	43	805	860	915	33*
9.	Sitting eye height	715	785	855	42	705	755	805	30*
10.	Sitting shoulder height	545	600	655	34	525	575	625	30*
11.	Sitting elbow height	175	225	275	31	165	215	265	31*
12.	Thigh thickness	120	152	180	18	130	155	180	16*
13.	Buttock–knee length	545	595	645	30	525	585	645	35*
14.	Buttock–popliteal length	430	480	530	30	430	485	540	33*
15.	Knee height	480	530	580	30	455	500	545	28*
16.	Popliteal height	385	430	475	27	350	400	450	29
17.	Shoulder breadth (bideltoid)	420	465	510	27	355	390	425	20
18.	Shoulder breadth (biacromial)	365	400	435	20	325	350	375	15*(W)
19.	Hip breadth	310	360	410	29	315	365	415	31
20.	Chest (bust) depth	185	220	255	21	185	241	300	35
21.	Abdominal depth	190	240	290	31	180	245	310	40*
22.	Shoulder–elbow length	330	365	400	20	305	335	365	17
23.	Elbow–fingertip length	440	475	510	20	413	442	471	19*(W)
24.	Upper limb length	720	780	840	35	660	705	750	28*(W)
25.	Shoulder–grip length	615	665	715	31	555	595	635	24
26.	Head length	185	195	205	7	170	180	190	7
27.	Head breadth	145	155	165	6	135	145	155	6
28.	Hand length	175	190	205	10	165	180	195	10*(W)
29.	Hand breadth	75	85	95	5	70	75	80	4*(W)
30.	Foot length	240	265	290	14	225	245	265	11*(W)
31.	Foot breadth	85	95	105	6	85	95	105	7*(W)
32.	Span	1660	1790	1920	80	1525	1640	1755	71
33.	Elbow span	870	945	1020	45	795	865	935	43
34.	Vertical grip reach (standing)	1930	2060	2190	78	1825	1940	2055	70
35.	Vertical grip reach (sitting)	1150	1245	1340	58	1090	1175	1260	53
36.	Forward grip reach	725	780	835	33	665	715	765	31
	Body weight					48	59	70	7*

Notes: See notes in Sections 10.1, 10.2 and 10.4.

Dimensions marked with an asterisk (*) are quoted from the original source (W for women only). The remainder have been estimated.

TABLE 10.8
Anthropometric Estimates for Dutch Adults Aged 20 to 60 Years (all dimensions in millimetres, except for body weight, given in kilograms)

	Dimension	Men				Women			
		5th %ile	50th %ile	95th %ile	SD	5th %ile	50th %ile	95th %ile	SD
1.	Stature	1690	1795	1900	65	1545	1650	1755	65*
2.	Eye height	1575	1670	1765	59	1435	1530	1625	59*
3.	Shoulder height	1400	1495	1590	58	1265	1365	1465	61*
4.	Elbow height	1055	1135	1215	48	980	1050	1120	43*
5.	Hip height	885	960	1035	46	780	845	910	40*
6.	Knuckle height	745	795	845	30	705	775	845	42*
7.	Fingertip height	645	690	735	28	605	675	745	42
8.	Sitting height	885	940	995	34	820	875	930	33*
9.	Sitting eye height	770	820	875	32	695	750	805	32*
10.	Sitting shoulder height	570	620	670	31	515	565	615	30
11.	Sitting elbow height	195	240	280	26	200	240	280	26*
12.	Thigh thickness	120	140	160	12	125	150	175	17*
13.	Buttock–knee length	575	620	665	28	550	600	650	31*
14.	Buttock–popliteal length	470	520	570	30	440	495	550	32*
15.	Knee height	520	565	610	28	450	505	560	32
16.	Popliteal height	415	455	495	25	370	405	445	25*
17.	Shoulder breadth (bideltoid)	430	475	520	28	355	400	445	27
18.	Shoulder breadth (biacromial)	385	410	445	20	330	360	390	20*
19.	Hip breadth	340	375	410	20	340	395	450	34
20.	Chest (bust) depth	240	285	330	26	230	290	350	36*
21.	Abdominal depth	245	310	375	38	230	295	360	40
22.	Shoulder–elbow length	340	375	405	19	305	335	365	18
23.	Elbow–fingertip length	455	490	525	20	405	440	475	20
24.	Upper limb length	750	805	860	33	665	720	775	34
25.	Shoulder–grip length	635	685	735	30	560	610	660	30
26.	Head length	190	200	210	7	175	185	195	7
27.	Head breadth	150	160	170	6	140	150	160	6
28.	Hand length	180	195	210	9	160	175	190	9*
29.	Hand breadth	80	90	100	5	70	80	90	4
30.	Foot length	255	275	295	13	220	240	260	13
31.	Foot breadth	90	100	110	6	80	90	100	6
32.	Span	1720	1845	1970	77	1510	1635	1760	75
33.	Elbow span	905	975	1045	44	790	865	920	45
34.	Vertical grip reach (standing)	2000	2125	2250	76	1780	1905	2030	76*
35.	Vertical grip reach (sitting)	1190	1280	1890	55	1085	1175	1265	56
36.	Forward grip reach	680	745	810	38	635	705	780	44*
	Body weight	60	76	92	10	49	65	81	10*

Notes: See notes in Sections 10.1, 10.2 and 10.4.

Dimensions marked with an asterisk (*) are quoted from the original source. The remainder have been estimated.

TABLE 10.9
Anthropometric Estimates for French Drivers (all dimensions in millimetres, except for body weight, given in kilograms)

	Men				Women			
Dimension	5th %ile	50th %ile	95th %ile	SD	5th %ile	50th %ile	95th %ile	SD
1. Stature	1600	1715	1830	69	1500	1600	1700	61*
2. Eye height	1450	1560	1670	68	1400	1500	1600	60
3. Shoulder height	1300	1405	1510	65	1210	1305	1400	57
4. Elbow height	995	1080	1165	51	925	1000	1075	45
5. Hip height	815	895	975	49	750	820	890	43
6. Knuckle height	680	745	810	40	655	715	775	35
7. Fingertip height	580	645	710	38	560	620	680	37
8. Sitting height	850	910	970	35	810	860	910	31*
9. Sitting eye height	735	795	855	35	700	750	800	30*
10. Sitting shoulder height	570	620	670	31	535	580	625	27*
11. Sitting elbow height	190	240	290	30	185	230	275	28
12. Thigh thickness	150	180	210	17	135	165	195	17*
13. Buttock–knee length	550	595	640	28	520	565	610	28*
14. Buttock–popliteal length	435	480	525	26	415	460	505	26*
15. Knee height	485	530	575	26	455	495	535	24*
16. Popliteal height	385	425	465	25	350	390	430	23*
17. Shoulder breadth (bideltoid)	425	470	515	26	380	425	470	27*
18. Shoulder breadth (biacromial)	360	395	430	20	325	355	385	18
19. Hip breadth	330	370	410	24	330	380	430	30*
20. Chest (bust) depth	210	245	280	22	205	250	295	26
21. Abdominal depth	220	270	320	31	205	255	305	30
22. Shoulder–elbow length	325	360	395	20	300	330	360	17*
23. Elbow–fingertip length	435	470	505	21	395	425	455	19*
24. Upper limb length	710	770	830	35	650	705	760	32*
25. Shoulder–grip length	600	655	710	32	550	600	650	29
26. Head length	175	190	205	8	170	180	190	7
27. Head breadth	140	150	160	6	130	140	150	5
28. Hand length	170	185	200	10	160	175	190	9
29. Hand breadth	75	85	95	5	70	75	80	4
30. Foot length	235	260	285	14	215	235	255	12
31. Foot breadth	85	95	105	6	80	90	100	5
32. Span	1630	1765	1900	81	1485	1600	1715	70
33. Elbow span	855	930	1005	46	775	845	915	42
34. Vertical grip reach (standing)	1900	2030	2160	79	1780	1895	2010	70
35. Vertical grip reach (sitting)	1130	1225	1320	59	1060	1145	1230	52
36. Forward grip reach	715	770	825	34	645	700	755	32
Body weight	58	73	95	11	47	58	78	10*

Notes: See notes in Sections 10.1, 10.2 and 10.4.

Dimensions marked with an asterisk (*) are quoted from the original source. The remainder have been estimated.

TABLE 10.10
Anthropometric Estimates for Polish Industrial Workers (all dimensions in millimetres)

	Dimension	Men 5th %ile	Men 50th %ile	Men 95th %ile	Men SD	Women 5th %ile	Women 50th %ile	Women 95th %ile	Women SD
1.	Stature	1595	1695	1795	61	1480	1575	1670	58*
2.	Eye height	1505	1600	1695	58	1390	1485	1580	57*
3.	Shoulder height	1275	1365	1455	54	1170	1280	1390	68*
4.	Elbow height	990	1065	1140	45	915	985	1055	43
5.	Hip height	810	880	950	43	745	810	875	41
6.	Knuckle height	545	595	640	29	535	570	605	22*
7.	Fingertip height	615	675	735	36	590	645	700	32*
8.	Sitting height	830	885	940	34	770	825	880	33*
9.	Sitting eye height	720	780	840	36	665	725	785	35*
10.	Sitting shoulder height	555	605	655	31	515	565	615	31*
11.	Sitting elbow height	195	240	285	27	185	230	275	27
12.	Thigh thickness	110	140	170	19	115	140	165	16*
13.	Buttock–knee length	540	585	630	26	515	565	615	29*
14.	Buttock–popliteal length	405	455	505	29	360	450	540	54*
15.	Knee height	485	530	575	27	445	485	525	24*
16.	Popliteal height	410	445	480	21	390	420	450	19*
17.	Shoulder breadth (bideltoid)	405	440	475	21	350	380	410	18*
18.	Shoulder breadth (biacromial)	360	390	420	18	320	350	380	17*
19.	Hip breadth	305	345	385	25	315	360	405	26*
20.	Chest (bust) depth	215	245	275	19	205	245	285	25
21.	Abdominal depth	220	265	310	27	205	250	295	28
22.	Shoulder–elbow length	310	330	350	13	280	300	320	12*
23.	Elbow–fingertip length	430	460	490	18	390	420	450	18
24.	Upper limb length	705	755	805	30	655	700	745	28*
25.	Shoulder–grip length	640	675	710	22	595	625	655	18*
26.	Head length	175	185	195	7	170	180	190	5*
27.	Head breadth	145	155	165	6	140	150	160	5*
28.	Hand length	175	190	205	8	160	175	190	8*
29.	Hand breadth	80	90	100	5	75	80	85	4*
30.	Foot length	240	260	280	12	210	230	250	11
31.	Foot breadth	85	95	105	5	75	85	95	5
32.	Span	1640	1755	1870	70	1505	1610	1715	65*
33.	Elbow span	795	860	925	38	720	785	850	41*
34.	Vertical grip reach (standing)	2065	2205	2345	84	1875	2005	2135	79*
35.	Vertical grip reach (sitting)	1210	1290	1370	50	1115	1185	1255	42*
36.	Forward grip reach	730	795	860	38	680	735	790	34*

Notes: See notes in Sections 10.1, 10.2 and 10.4.

Dimensions marked with an asterisk (*) are quoted from the original source. The remainder have been estimated.

TABLE 10.11
Anthropometric Estimates for U.S. Adults Aged 19 to 65 Years (all dimensions in millimetres, except for body weight, given in kilograms)

	Men				Women			
Dimension	5th %ile	50th %ile	95th %ile	SD	5th %ile	50th %ile	95th %ile	SD
1. Stature	1640	1755	1870	71	1520	1625	1730	64*
2. Eye height	1529	1644	1759	70	1416	1519	1622	63
3. Shoulder height	1330	1440	1550	67	1225	1325	1425	60
4. Elbow height	1020	1105	1190	53	945	1020	1095	47
5. Hip height	835	915	995	50	760	835	910	45
6. Knuckle height	700	765	830	41	670	730	790	37
7. Fingertip height	595	660	725	39	565	630	695	40
8. Sitting height	855	915	975	36	800	860	920	36
9. Sitting eye height	740	800	860	35	690	750	810	35
10. Sitting shoulder height	545	600	655	32	510	565	620	32
11. Sitting elbow height	195	245	295	31	185	235	285	29
12. Thigh thickness	135	160	185	16	125	155	185	17
13. Buttock–knee length	550	600	650	31	525	575	625	31
14. Buttock–popliteal length	445	500	555	33	440	490	540	31
15. Knee height	495	550	605	32	460	505	550	28
16. Popliteal height	395	445	495	29	360	405	450	28
17. Shoulder breadth (bideltoid)	425	470	515	28	360	400	440	25
18. Shoulder breadth (biacromial)	365	400	435	21	330	360	390	19
19. Hip breadth	310	360	410	30	310	375	440	39
20. Chest (bust) depth	220	255	290	22	210	255	300	28
21. Abdominal depth	220	275	330	32	210	260	310	31
22. Shoulder–elbow length	330	365	400	21	305	335	365	18
23. Elbow–fingertip length	445	480	515	21	400	435	470	20
24. Upper limb length	730	790	850	36	655	715	775	35
25. Shoulder–grip length	615	670	725	33	560	610	660	30
26. Head length	180	195	210	8	165	180	195	8
27. Head breadth	145	155	165	6	135	145	155	6
28. Hand length	175	191	205	10	160	175	190	10
29. Hand breadth	80	90	100	5	65	75	85	5
30. Foot length	240	265	290	14	220	240	260	13
31. Foot breadth	90	100	110	6	80	90	100	6
32. Span	1670	1810	1950	84	1505	1625	1745	73
33. Elbow span	875	955	1035	48	790	860	930	44
34. Vertical grip reach (standing)	1950	2080	2210	80	1805	1925	2045	73
35. Vertical grip reach (sitting)	1155	1255	1355	61	1070	1160	1250	55
36. Forward grip reach	725	785	845	35	655	710	765	32
Body weight	55	78	102	14	41	65	89	15*

Notes: See notes in Sections 10.1, 10.2 and 10.4.

Dimensions marked with an asterisk (*) are quoted from the original source. The remainder have been estimated.

TABLE 10.12
Anthropometric Estimates for Brazilian Industrial Workers (all dimensions in millimetres, except for body weight, given in kilograms)

	Dimension	Men			
		5th %ile	50th %ile	95th %ile	SD
1.	Stature	1595	1700	1810	66*
2.	Eye height	1490	1595	1700	66*
3.	Shoulder height	1315	1410	1510	60*
4.	Elbow height	965	1045	1120	49*
5.	Hip height	800	880	960	47
6.	Knuckle height	655	720	785	40
7.	Fingertip height	565	625	690	37*
8.	Sitting height	825	880	940	35*
9.	Sitting eye height	720	775	830	34*
10.	Sitting shoulder height	550	595	645	29*
11.	Sitting elbow height	185	230	275	28*
12.	Thigh thickness	120	150	180	16*
13.	Buttock–knee length	550	595	650	30*
14.	Buttock–popliteal length	435	480	530	29*
15.	Knee height	490	530	575	27*
16.	Popliteal height	390	425	465	24*
17.	Shoulder breadth (bideltoid)	400	445	490	27*
18.	Shoulder breadth (biacromial)	355	385	415	19
19.	Hip breadth	305	340	385	25*
20.	Chest depth	205	235	275	22*
21.	Abdominal depth	220	245	305	33*
22.	Shoulder–elbow length	335	365	405	21*
23.	Elbow–fingertip length	440	475	510	22
24.	Upper limb length	725	785	850	38*
25.	Shoulder–grip length	615	670	725	34
26.	Head length	175	190	205	8
27.	Head breadth	140	150	160	6
28.	Hand length	170	185	200	9
29.	Hand breadth	75	85	95	5
30.	Foot length	240	260	280	12*
31.	Foot breadth	95	100	110	5*
32.	Span	1625	1755	1885	78
33.	Elbow span	855	925	995	44
34.	Vertical grip reach (standing)	1895	2020	2145	75
35.	Vertical grip reach (sitting)	1130	1220	1310	56
36.	Forward grip reach	710	765	820	32
	Body weight	52	66	86	11*

Notes: See notes in Sections 10.1, 10.2 and 10.4.

Dimensions marked with an asterisk (*) are quoted from the original source. The remainder have been estimated.

TABLE 10.13
Anthropometric Estimates for Sri Lankan Workers (all dimensions in millimetres, except for body weight, given in kilograms)

		Men				Women			
	Dimension	5th %ile	50th %ile	95th %ile	SD	5th %ile	50th %ile	95th %ile	SD
1.	Stature	1535	1640	1745	64	1425	1525	1620	59*
2.	Eye height	1430	1505	1640	63	1325	1420	1520	59*
3.	Shoulder height	1280	1375	1475	60	1180	1270	1360	60*
4.	Elbow height	930	1015	1100	70	875	940	1015	62*
5.	Hip height	885	970	1060	59	840	920	985	77*
6.	Knuckle height	630	700	770	42	580	655	730	45
7.	Fingertip height	555	605	665	39	505	570	675	45*
8.	Sitting height	780	835	890	33	725	775	830	32*
9.	Sitting eye height	680	730	785	32	625	675	725	30*
10.	Sitting shoulder height	520	570	620	30	475	525	575	29*
11.	Sitting elbow height	160	200	240	25	150	185	220	20*
12.	Thigh thickness	100	120	140	11	70	85	100	9*
13.	Buttock–knee length	510	555	620	32	485	535	570	29*
14.	Buttock–popliteal length	415	460	510	31	400	445	495	35*
15.	Knee height	410	455	500	29	375	420	465	27*
16.	Popliteal height	325	370	410	26	290	335	380	27
17.	Shoulder breadth (bideltoid)	330	370	400	23	300	330	360	18*
18.	Shoulder breadth (biacromial)	295	320	345	16	270	295	320	14
19.	Hip breadth	225	250	280	18	210	245	280	22*
20.	Chest (bust) depth	145	170	205	20	130	170	210	25*(M)
21.	Abdominal depth	140	185	235	29	130	175	220	28
22.	Shoulder–elbow length	290	345	400	33	270	315	355	26
23.	Elbow–fingertip length	415	450	490	34	380	410	450	29*
24.	Upper limb length	680	735	790	33	615	665	715	30
25.	Shoulder–grip length	615	665	710	29	560	605	650	28
26.	Head length	155	180	195	12	150	170	190	20*
27.	Head breadth	130	145	155	12	125	135	145	17*
28.	Hand length	165	180	195	12	150	165	180	14*
29.	Hand breadth	90	100	110	6	80	90	90	5*
30.	Foot length	230	250	280	17	210	235	255	13*
31.	Foot breadth	90	105	120	11	80	95	110	12*
32.	Span	1505	1690	1815	89	1405	1545	1680	99*
33.	Elbow span	780	880	940	51	720	795	870	46*
34.	Vertical grip reach (standing)	1895	1940	2060	73	1695	1805	1915	68
35.	Vertical grip reach (sitting)	1085	1175	1265	55	1010	1090	1170	50
36.	Forward grip reach	685	735	785	31	620	670	720	30
	Body weight	39	51	67	8	34	43	56	7

Notes: See notes in Sections 10.1, 10.2 and 10.4.

Dimensions marked with an asterisk (*) are quoted from the original source (or for men only). The remainder have been estimated.

TABLE 10.14
Anthropometric Estimates for Indian Agricultural Workers (all dimensions in millimetres, except for body weight, given in kilograms)

		Men			
	Dimension	**5th %ile**	**50th %ile**	**95th %ile**	**SD**
1.	Stature	1540	1620	1700	50*
2.	Eye height	1425	1510	1595	52*
3.	Shoulder height	1265	1345	1425	49*
4.	Elbow height	940	1025	1105	40*
5.	Hip height	800	865	930	38
6.	Knuckle height	635	685	730	29*
7.	Fingertip height	540	585	630	28*
8.	Sitting height	795	840	880	25*
9.	Sitting eye height	695	740	780	26*
10.	Sitting shoulder height	520	555	590	21*
11.	Sitting elbow height	170	205	235	20*
12.	Thigh thickness	110	135	160	13*
13.	Buttock–knee length	520	555	590	21*
14.	Buttock–popliteal length	435	465	495	18*
15.	Knee height	460	510	560	30*
16.	Popliteal height	380	415	450	21*
17.	Shoulder breadth (bideltoid)	375	410	440	19*
18.	Shoulder breadth (biacromial)	320	355	395	24*
19.	Hip breadth	280	310	335	16*
20.	Chest depth	145	170	205	20
21.	Abdominal depth	140	185	235	33
22.	Shoulder–elbow length	325	355	385	19
23.	Elbow–fingertip length	425	460	490	20*
24.	Upper limb length	700	755	810	34
25.	Shoulder–grip length	655	710	760	32*
26.	Head length	170	180	190	6
27.	Head breadth	140	145	150	4
28.	Hand length	170	185	195	8*
29.	Hand breadth	75	85	90	4*
30.	Foot length	235	250	265	10*
31.	Foot breadth	90	95	105	4*
32.	Span	1595	1705	1810	66*
33.	Elbow span	825	880	935	33*
34.	Vertical grip reach (standing)	1875	1995	2110	72
35.	Vertical grip reach (sitting)	1120	1190	1265	44*
36.	Forward grip reach	685	725	765	24
	Body weight	40	49	59	6*

Notes: See notes in Sections 10.1, 10.2 and 10.4.

Dimensions marked with an asterisk (*) are quoted from the original source. The remainder have been estimated.

TABLE 10.15
Anthropometric Estimates for Hong Kong Chinese Industrial Workers (all dimensions in millimetres, except for body weight, given in kilograms)

	Men				Women			
Dimension	5th %ile	50th %ile	95th %ile	SD	5th %ile	50th %ile	95th %ile	SD
1. Stature	1585	1680	1775	58	1455	1555	1655	60*
2. Eye height	1470	1555	1640	52	1330	1425	1520	57*
3. Shoulder height	1300	1380	1460	50	1180	1265	1350	51*
4. Elbow height	950	1015	1080	39	870	935	1000	41*
5. Hip height	790	855	920	41	715	785	855	42
6. Knuckle height	685	750	815	40	650	715	780	41
7. Fingertip height	575	640	705	38	540	610	680	44
8. Sitting height	845	900	955	34	780	840	900	37*
9. Sitting eye height	720	780	840	35	660	720	780	35*
10. Sitting shoulder height	555	605	655	31	510	560	610	29*
11. Sitting elbow height	190	240	290	31	165	230	295	38*
12. Thigh thickness	110	135	160	14	105	130	155	14
13. Buttock–knee length	505	550	595	26	470	520	570	30*
14. Buttock–popliteal length	405	450	495	26	385	435	485	29*
15. Knee height	450	495	540	26	410	455	500	27*
16. Popliteal height	365	405	445	25	325	375	425	29*
17. Shoulder breadth (bideltoid)	380	425	470	26	335	385	435	29*
18. Shoulder breadth (biacromial)	335	365	395	19	315	350	385	22
19. Hip breadth	300	335	370	22	295	330	365	21*(M)
20. Chest (bust) depth	155	195	235	25	160	215	270	34
21. Abdominal depth	150	210	270	36	150	215	280	39
22. Shoulder–elbow length	310	340	370	19	290	315	340	16*
23. Elbow–fingertip length	410	445	480	22	360	400	440	24*
24. Upper limb length	680	730	780	30	615	660	705	26
25. Shoulder–grip length	580	620	660	25	525	560	595	22
26. Head length	175	190	205	8	160	175	190	9
27. Head breadth	150	160	170	7	135	150	165	8
28. Hand length	165	180	195	9	150	165	180	9*
29. Hand breadth	70	80	90	5	60	70	80	5*
30. Foot length	235	250	265	10	205	225	245	11*
31. Foot breadth	85	95	105	5	80	85	90	4*
32. Span	1480	1635	1790	95	1350	1480	1610	80*
33. Elbow span	805	885	965	48	690	775	860	51
34. Vertical grip reach (standing)	1835	1970	2105	83	1685	1825	1965	86
35. Vertical grip reach (sitting)	1110	1205	1300	58	855	940	1025	51
36. Forward grip reach	640	705	770	38	580	635	690	32
Body weight	47	60	75	9	39	47	62	7

Notes: See notes in Sections 10.1, 10.2 and 10.4.

Dimensions marked with an asterisk (*) are quoted from the original source (or for men only). The remainder have been estimated.

TABLE 10.16
Anthropometric Estimates for Japanese Adults (all dimensions in millimetres, except for body weight, given in kilograms)

	Dimension	Men				Women			
		5th %ile	50th %ile	95th %ile	SD	5th %ile	50th %ile	95th %ile	SD
1.	Stature	1560	1655	11750	58	1450	1530	1610	48*
2.	Eye height	1445	1540	1635	57	1350	1425	1500	47
3.	Shoulder height	1250	1340	1430	54	1075	1145	1215	44
4.	Elbow height	965	1035	1105	43	895	955	1015	36
5.	Hip height	765	830	895	41	700	755	810	33
6.	Knuckle height	675	740	805	40	650	705	760	33
7.	Fingertip height	565	630	695	38	540	600	660	35
8.	Sitting height	850	900	950	31	800	845	890	28
9.	Sitting eye height	735	785	835	31	690	735	780	28
10.	Sitting shoulder height	545	590	635	28	510	555	600	26
11.	Sitting elbow height	220	260	300	23	215	250	285	20
12.	Thigh thickness	110	135	160	14	105	130	155	14
13.	Buttock–knee length	500	550	600	29	485	530	575	26
14.	Buttock–popliteal length	410	470	510	31	405	450	495	26
15.	Knee height	450	490	530	23	420	450	480	18
16.	Popliteal height	360	400	440	24	325	360	395	21
17.	Shoulder breadth (bideltoid)	405	440	475	22	365	395	425	18
18.	Shoulder breadth (biacromial)	350	380	410	18	315	340	365	15
19.	Hip breadth	280	305	330	14	270	305	340	20
20.	Chest (bust) depth	180	205	230	16	175	205	235	18
21.	Abdominal depth	185	220	255	22	170	205	240	20
22.	Shoulder–elbow length	295	330	365	21	270	300	330	17
23.	Elbow–fingertip length	405	440	475	20	370	400	430	17
24.	Upper limb length	665	715	765	29	605	645	685	25
25.	Shoulder–grip length	565	610	655	26	515	550	585	22
26.	Head length	170	185	200	8	160	170	180	7
27.	Head breadth	145	155	165	7	140	150	160	6
28.	Hand length	165	180	195	10	150	165	180	9
29.	Hand breadth	75	85	95	6	65	75	85	5
30.	Foot length	230	245	260	10	210	225	240	9
31.	Foot breadth	95	105	115	5	90	95	100	4
32.	Span	1540	1655	1770	70	1395	1485	1575	56
33.	Elbow span	790	870	950	48	715	780	845	41
34.	Vertical grip reach (standing)	1805	1940	2075	83	1680	1795	1910	69
35.	Vertical grip reach (sitting)	1105	1185	1265	49	1030	1095	1160	41
36.	Forward grip reach	630	690	750	37	570	620	670	31
	Body weight	41	60	74	9	40	51	63	7

Notes: See notes in Sections 10.1 to 10.2 and 10.4.

Dimensions marked with an asterisk (*) are quoted from the original source. The remainder have been estimated.

TABLE 10.17
Anthropometric Estimates for Newborn Infants (all dimensions in millimetres)

	Dimension	5th %ile	50th %ile	95th %ile	SD
1.	Crown–heel length (1)[a]	465	500	535	20
2.	Crown—rump length (8)	330	350	370	11
3.	Rump–knee length (13)	105	125	140	10
4.	Knee–sole length (15)	120	135	145	7
5.	Shoulder breadth (bideltoid) (17)	135	150	160	8
6.	Hip breadth (19)	105	120	130	8
7.	Shoulder–elbow length (22)	85	95	105	6
8.	Elbow–fingertip length (23)	120	135	145	7
9.	Head length (26)	115	120	125	4
10.	Head breadth (27)	90	95	100	3
11.	Hand length (28)	55	60	65	3
12.	Hand breadth (29)	30	35	35	2
13.	Foot length (30)	65	75	80	4
14.	Foot breadth (31)	30	30	35	2

Note: See notes in Sections 10.1, 10.2 and 10.5.

[a] Numbers in parentheses are the equivalent dimensions in adult tables.

TABLE 10.18
Anthropometric Estimates for Infants Less than 6 Months of Age (all dimensions in millimetres)

	Dimension	5th %ile	50th %ile	95th %ile	SD
1.	Crown–heel length (1)[a]	510	600	690	54
2.	Crown–rump length (8)	360	410	460	31
3.	Rump–knee length (13)	105	150	195	28
4.	Knee–sole length (15)	125	160	190	19
5.	Shoulder breadth (bideltoid) (17)	140	180	215	21
6.	Hip breadth (19)	100	140	175	22
7.	Shoulder–elbow length (22)	90	115	145	16
8.	Elbow–fingertip length (23)	125	160	190	19
9.	Head length (26)	120	140	160	11
10.	Head breadth (27)	100	110	120	7
11.	Hand length (28)	55	70	85	9
12.	Hand breadth (29)	30	40	45	4
13.	Foot length (30)	70	85	105	12
14.	Foot breadth (31)	30	40	50	6

Note: See notes in Sections 10.1, 10.2 and 10.5.

[a] Numbers in parentheses are the equivalent dimensions in adult tables.

TABLE 10.19
Anthropometric Estimates for Infants from 6 Months to 1 Year (all dimensions in millimetres)

	Dimension	5th %ile	50th %ile	95th %ile	SD
1.	Crown–heel length (1)[a]	655	715	775	37
2.	Crown–rump length (8)	435	470	505	21
3.	Rump–knee length (13)	155	185	215	19
4.	Knee–sole length (15)	170	190	210	13
5.	Shoulder breadth (bideltoid) (17)	185	210	230	14
6.	Hip breadth (19)	140	165	190	15
7.	Shoulder–elbow length (22)	120	140	155	11
8.	Elbow–fingertip length (23)	170	190	210	13
9.	Head length (26)	145	160	170	8
10.	Head breadth (27)	115	120	130	5
11.	Hand length (28)	75	85	95	6
12.	Hand breadth (29)	40	45	50	3
13.	Foot length (30)	90	105	120	8
14.	Foot breadth (31)	40	45	50	4

Note: See notes in Sections 10.1, 10.2 and 10.5.

[a] Numbers in parentheses are the equivalent dimensions in adult tables.

TABLE 10.20
Anthropometric Estimates for Infants from 1 Year to 18 Months (all dimensions in millimetres)

	Dimension	5th %ile	50th %ile	95th %ile	SD
1.	Crown–heel length (1)[a]	690	745	800	35
2.	Crown–rump length (8)	440	475	505	20
3.	Rump–knee length (13)	170	195	225	18
4.	Knee–sole length (15)	175	195	215	12
5.	Shoulder breadth (bideltoid) (17)	185	205	230	14
6.	Hip breadth (19)	140	165	185	14
7.	Shoulder–elbow length (22)	130	145	160	10
8.	Elbow–fingertip length (23)	175	195	215	12
9.	Head length (26)	150	160	170	7
10.	Head breadth (27)	115	120	130	5
11.	Hand length (28)	80	90	100	6
12.	Hand breadth (29)	40	45	50	3
13.	Foot length (30)	100	115	125	8
14.	Foot breadth (31)	40	45	55	4

Note: See notes in Sections 10.1, 10.2 and 10.5.

[a] Numbers in parentheses are the equivalent dimensions in adult tables.

TABLE 10.21
Anthropometric Estimates for Infants from 18 Months to 2 Years (all dimensions in millimetres)

	Dimension	5th %ile	50th %ile	95th %ile	SD
1.	Crown–heel length (1)[a]	780	840	900	36
2.	Crown–rump length (8)	490	525	555	20
3.	Rump–knee length (13)	200	230	260	18
4.	Knee–sole length (15)	210	230	255	13
5.	Shoulder breadth (bideltoid) (17)	205	230	250	14
6.	Hip breadth (19)	150	175	200	15
7.	Shoulder–elbow length (22)	145	165	180	11
8.	Elbow–fingertip length (23)	200	220	240	12
9.	Head length (26)	160	175	185	7
10.	Head breadth (27)	125	130	140	5
11.	Hand length (28)	85	95	105	6
12.	Hand breadth (29)	45	50	55	3
13.	Foot length (30)	115	125	140	8
14.	Foot breadth (31)	45	55	60	4

Notes: See notes in Sections 10.1, 10.2 and 10.5.

[a] Numbers in parentheses are the equivalent dimensions in adult tables.

TABLE 10.22
Anthropometric Estimates for British 2-Year-Olds (all dimensions in millimetres)

	Boys				Girls			
Dimension	5th %ile	50th %ile	95th %ile	SD	5th %ile	50th %ile	95th %ile	SD
1. Stature	850	930	1010	49	825	890	955	40*
2. Eye height	760	840	920	49	725	805	885	48
3. Shoulder height	675	735	795	37	630	695	760	38
4. Elbow height	495	555	615	36	480	530	580	30
5. Hip height	360	420	480	37	365	415	465	30
6. Knuckle height	340	385	430	26	335	375	415	25
7. Fingertip height	275	315	360	26	270	310	350	25
8. Sitting height	505	545	585	24	485	520	555	21
9. Sitting eye height	410	445	480	20	370	410	450	24
10. Sitting shoulder height	305	340	375	22	275	310	345	20
11. Sitting elbow height	105	140	175	20	105	130	155	15
12. Thigh thickness	65	80	95	10	60	75	90	10
13. Buttock–knee length	245	275	305	19	250	280	310	17
14. Buttock–popliteal length	210	235	260	16	185	245	305	36
15. Knee height	235	270	305	20	230	260	290	17
16. Popliteal height	155	205	255	29	170	205	240	20
17. Shoulder breadth (bideltoid)	215	245	275	17	210	235	260	14
18. Shoulder breadth (biacromial)	190	215	240	15	190	210	230	12
19. Hip breadth	170	190	210	13	165	185	205	11
20. Chest (bust) depth	100	120	140	12	100	115	130	10
21. Abdominal depth	130	145	160	10	135	145	155	7
22. Shoulder–elbow length	160	185	205	13	160	175	190	10
23. Elbow–fingertip length	215	245	275	17	210	235	260	14
24. Upper limb length	365	410	455	28	335	380	425	27
25. Shoulder–grip length	295	340	390	28	270	315	360	27
26. Head length	170	180	190	7	160	165	170	4
27. Head breadth	130	140	150	6	125	130	135	4
28. Hand length	90	105	120	8	90	100	110	6
29. Hand breadth	50	55	60	4	40	45	50	4
30. Foot length	130	145	160	10	130	145	160	9
31. Foot breadth	60	65	70	4	50	55	60	4
32. Span	835	925	1015	54	785	865	945	49
33. Elbow span	435	490	540	31	410	455	505	30
34. Vertical grip reach (standing)	920	1045	1170	77	965	1045	1125	49
35. Vertical grip reach (sitting)	605	675	745	42	550	620	690	42
36. Forward grip reach	340	400	460	35	345	385	425	25

Notes: See notes in Sections 10.1, 10.2 and 10.6.

Dimensions marked with an asterisk (*) are quoted from the original source. The remainder have been estimated.

TABLE 10.23
Anthropometric Estimates for British 3-Year-Olds (all dimensions in millimetres)

	Dimension	Boys				Girls			
		5th %ile	50th %ile	95th %ile	SD	5th %ile	50th %ile	95th %ile	SD
1.	Stature	910	990	1070	48	895	970	1045	46*
2.	Eye height	810	890	970	48	785	875	965	55
3.	Shoulder height	720	780	840	36	690	760	830	43
4.	Elbow height	535	595	655	35	520	580	640	35
5.	Hip height	400	460	520	35	405	460	515	33
6.	Knuckle height	365	410	455	26	360	410	460	29
7.	Fingertip height	295	340	380	26	290	340	385	29
8.	Sitting height	530	570	610	25	515	555	595	25
9.	Sitting eye height	425	465	505	24	400	445	490	28
10.	Sitting shoulder height	310	350	390	23	295	335	375	23
11.	Sitting elbow height	115	150	185	20	110	140	170	17
12.	Thigh thickness	65	85	105	11	60	80	100	12
13.	Buttock–knee length	270	300	330	19	270	305	340	20
14.	Buttock–popliteal length	225	250	275	16	215	260	305	26
15.	Knee height	255	290	325	20	250	285	320	20
16.	Popliteal height	195	230	265	21	200	230	260	17
17.	Shoulder breadth (bideltoid)	230	255	280	16	225	250	275	15
18.	Shoulder breadth (biacromial)	200	225	250	14	205	225	245	13
19.	Hip breadth	175	195	215	13	175	195	215	13
20.	Chest (bust) depth	105	125	145	12	105	120	140	12
21.	Abdominal depth	135	150	165	10	135	150	165	10
22.	Shoulder–elbow length	175	195	220	13	175	195	215	12
23.	Elbow–fingertip length	235	260	286	16	230	255	280	16
24.	Upper limb length	390	435	480	27	365	415	465	31
25.	Shoulder–grip length	320	365	410	27	295	345	395	31
26.	Head length	170	180	190	7	155	165	175	6
27.	Head breadth	130	140	150	6	120	130	140	5
28.	Hand length	95	110	125	8	100	110	120	7
29.	Hand breadth	50	55	60	4	45	50	55	4
30.	Foot length	140	155	170	10	140	155	170	10
31.	Foot breadth	60	65	70	4	55	60	65	4
32.	Span	890	980	1070	56	850	940	1030	56
33.	Elbow span	465	515	570	32	440	495	555	34
34.	Vertical grip reach (standing)	1005	1130	1255	75	1025	1125	1225	61
35.	Vertical grip reach (sitting)	640	705	775	41	605	675	740	41
36.	Forward grip reach	360	420	480	35	360	415	470	33

Notes: See notes in Sections 10.1, 10.2 and 10.6.

Dimensions marked with an asterisk (*) are quoted from the original source. The remainder have been estimated.

TABLE 10.24
Anthropometric Estimates for British 4-Year-Olds (all dimensions in millimetres)

	Dimension	Boys				Girls			
		5th %ile	50th %ile	95th %ile	SD	5th %ile	50th %ile	95th %ile	SD
1.	Stature	975	1050	1125	47	965	1050	1135	52*
2.	Eye height	865	940	1015	47	845	945	1045	62
3.	Shoulder height	765	825	885	35	745	825	905	48
4.	Elbow height	580	635	690	34	565	630	695	40
5.	Hip height	445	500	555	33	445	505	565	36
6.	Knuckle height	390	435	480	26	390	445	500	33
7.	Fingertip height	315	360	405	26	315	365	420	33
8.	Sitting height	550	595	640	26	540	590	640	29
9.	Sitting eye height	440	485	530	28	425	480	535	32
10.	Sitting shoulder height	320	360	400	24	315	360	405	26
11.	Sitting elbow height	125	160	195	20	120	150	180	19
12.	Thigh thickness	70	90	110	12	60	85	110	14
13.	Buttock–knee length	295	325	355	19	290	330	370	23
14.	Buttock–popliteal length	240	265	290	16	250	275	300	16
15.	Knee height	275	310	345	20	270	310	350	23
16.	Popliteal height	235	255	275	13	230	255	280	14
17.	Shoulder breadth (bideltoid)	240	265	290	15	240	265	290	16
18.	Shoulder breadth (biacromial)	215	235	255	13	215	240	265	14
19.	Hip breadth	180	200	220	13	180	205	230	15
20.	Chest (bust) depth	110	130	150	12	110	130	150	13
21.	Abdominal depth	140	155	170	10	135	155	175	13
22.	Shoulder–elbow length	190	210	230	13	185	210	235	14
23.	Elbow–fingertip length	250	275	300	15	245	275	305	18
24.	Upper limb length	415	460	505	26	390	450	510	35
25.	Shoulder–grip length	340	385	430	26	315	370	430	35
26.	Head length	170	180	190	7	150	165	180	8
27.	Head breadth	130	140	150	6	125	135	145	6
28.	Hand length	100	115	130	8	105	120	135	8
29.	Hand breadth	50	55	60	4	50	55	60	4
30.	Foot length	150	165	180	10	145	165	185	11
31.	Foot breadth	60	65	70	4	60	65	70	4
32.	Span	940	1035	1130	58	910	1015	1120	63
33.	Elbow span	490	545	600	33	475	535	600	38
34.	Vertical grip reach (standing)	1095	1215	1335	73	1085	1205	1325	73
35.	Vertical grip reach (sitting)	670	735	805	40	660	725	795	40
36.	Forward grip reach	380	440	500	35	380	445	510	41

Notes: See notes in Sections 10.1, 10.2 and 10.6.

Dimensions marked with an asterisk (*) are quoted from the original source. The remainder have been estimated.

TABLE 10.25
Anthropometric Estimates for British 5-Year-Olds (all dimensions in millimetres)

	Dimension	Boys				Girls			
		5th %ile	50th %ile	95th %ile	SD	5th %ile	50th %ile	95th %ile	SD
1.	Stature	1025	1110	1195	52	1015	1100	1185	53*
2.	Eye height	910	995	1080	53	885	990	1095	64
3.	Shoulder height	810	875	940	39	785	865	945	48
4.	Elbow height	605	670	735	39	595	660	725	41
5.	Hip height	490	550	610	36	490	540	590	31
6.	Knuckle height	405	455	505	29	410	465	520	34
7.	Fingertip height	325	375	420	29	330	385	445	34
8.	Sitting height	575	620	665	28	560	610	660	29
9.	Sitting eye height	455	505	555	29	450	500	550	31
10.	Sitting shoulder height	340	380	420	25	325	370	415	26
11.	Sitting elbow height	130	165	200	22	125	155	185	19
12.	Thigh thickness	75	90	105	10	70	90	110	12
13.	Buttock–knee length	310	345	380	21	310	350	390	23
14.	Buttock–popliteal length	250	280	310	17	265	295	325	19
15.	Knee height	300	335	370	22	295	330	365	21
16.	Popliteal height	240	270	300	18	245	270	295	16
17.	Shoulder breadth (bideltoid)	245	275	305	17	245	270	295	16
18.	Shoulder breadth (biacromial)	225	250	275	15	230	250	270	12
19.	Hip breadth	185	210	235	15	185	210	235	16
20.	Chest (bust) depth	110	135	160	14	110	135	155	14
21.	Abdominal depth	135	155	175	12	135	160	185	14
22.	Shoulder–elbow length	205	225	250	14	200	220	245	13
23.	Elbow–fingertip length	265	295	325	17	260	290	320	17
24.	Upper limb length	435	485	535	29	410	470	530	35
25.	Shoulder–grip length	355	405	450	29	335	390	450	35
26.	Head length	165	180	195	8	150	165	180	8
27.	Head breadth	130	140	150	5	120	130	140	5
28.	Hand length	110	125	140	9	105	120	135	8
29.	Hand breadth	55	60	65	4	50	55	60	4
30.	Foot length	155	175	195	11	155	170	185	10
31.	Foot breadth	60	70	80	5	60	65	70	4
32.	Span	995	1095	1195	60	955	1060	1165	64
33.	Elbow span	520	575	635	34	495	560	625	39
34.	Vertical grip reach (standing)	1180	1305	1430	77	1170	1290	1410	72
35.	Vertical grip reach (sitting)	700	775	850	45	685	755	830	45
36.	Forward grip reach	420	470	520	30	400	460	520	36

Notes: See notes in Sections 10.1, 10.2 and 10.6.

Dimensions marked with an asterisk (*) are quoted from the original source. The remainder have been estimated.

TABLE 10.26
Anthropometric Estimates for British 6-Year-Olds (all dimensions in millimetres)

		Boys				Girls			
	Dimension	5th %ile	50th %ile	95th %ile	SD	5th %ile	50th %ile	95th %ile	SD
1.	Stature	1070	1170	1270	60	1070	1160	1250	53*
2.	Eye height	950	1050	1150	60	935	1045	1155	67
3.	Shoulder height	845	920	995	45	825	910	995	52
4.	Elbow height	635	705	775	44	625	695	765	43
5.	Hip height	520	595	670	45	420	475	530	32
6.	Knuckle height	425	480	535	33	430	490	550	36
7.	Fingertip height	340	395	450	33	350	410	470	36
8.	Sitting height	585	640	695	32	585	635	685	31
9.	Sitting eye height	475	525	575	31	470	525	580	32
10.	Sitting shoulder height	340	390	440	29	335	380	425	28
11.	Sitting elbow height	130	170	210	25	125	160	195	21
12.	Thigh thickness	75	95	115	13	75	95	115	11
13.	Buttock–knee length	330	370	410	25	330	370	410	25
14.	Buttock–popliteal length	270	305	340	21	275	310	345	20
15.	Knee height	320	360	400	25	320	355	390	21
16.	Popliteal height	260	295	330	22	265	290	315	16
17.	Shoulder breadth (bideltoid)	245	285	325	23	250	285	320	20
18.	Shoulder breadth (biacromial)	235	265	295	18	240	260	280	13
19.	Hip breadth	180	215	250	21	190	220	250	19
20.	Chest (bust) depth	110	140	170	19	110	140	170	18
21.	Abdominal depth	135	160	185	16	135	165	195	18
22.	Shoulder–elbow length	215	240	265	16	215	235	255	13
23.	Elbow–fingertip length	275	310	345	21	275	305	335	18
24.	Upper limb length	455	510	565	34	430	495	560	38
25.	Shoulder–grip length	370	425	480	34	350	415	475	38
26.	Head length	165	180	195	9	160	170	180	7
27.	Head breadth	130	140	150	6	125	135	145	6
28.	Hand length	115	130	145	10	110	125	140	8
29.	Hand breadth	50	60	70	5	55	60	65	4
30.	Foot length	165	185	205	13	160	180	200	11
31.	Foot breadth	65	75	85	6	60	70	80	5
32.	Span	1045	1160	1275	70	1010	1120	1230	68
33.	Elbow span	545	610	675	40	525	590	660	41
34.	Vertical grip reach (standing)	1235	1390	1545	93	1255	1380	1505	76
35.	Vertical grip reach (sitting)	720	805	890	52	705	790	875	52
36.	Forward grip reach	435	495	555	35	435	485	535	31

Notes: See notes in Sections 10.1, 10.2 and 10.6.

Dimensions marked with an asterisk (*) are quoted from the original source. The remainder have been estimated.

TABLE 10.27
Anthropometric Estimates for British 7-Year-Olds (all dimensions in millimetres)

	Dimension	Boys				Girls			
		5th %ile	50th %ile	95th %ile	SD	5th %ile	50th %ile	95th %ile	SD
1.	Stature	1140	1230	1320	56	1125	1220	1315	59*
2.	Eye height	1020	1115	1210	57	995	1105	1215	66
3.	Shoulder height	885	975	1065	54	870	960	1050	54
4.	Elbow height	680	745	810	40	665	735	805	42
5.	Hip height	570	635	700	39	555	615	675	35
6.	Knuckle height	460	510	560	31	465	525	585	37
7.	Fingertip height	370	420	475	31	375	435	500	37
8.	Sitting height	615	665	715	30	610	660	710	31
9.	Sitting eye height	505	550	595	28	500	555	610	32
10.	Sitting shoulder height	360	405	450	27	350	395	440	26
11.	Sitting elbow height	140	175	210	20	140	170	200	19
12.	Thigh thickness	85	105	125	13	85	105	125	13
13.	Buttock–knee length	355	395	435	24	355	400	445	26
14.	Buttock–popliteal length	280	325	370	27	290	335	380	27
15.	Knee height	340	380	420	25	335	375	415	23
16.	Popliteal height	285	315	345	19	275	310	345	21
17.	Shoulder breadth (bideltoid)	265	300	335	22	255	295	335	24
18.	Shoulder breadth (biacromial)	250	275	300	15	245	270	295	15
19.	Hip breadth	190	225	260	21	195	235	275	23
20.	Chest (bust) depth	110	145	180	20	110	145	180	21
21.	Abdominal depth	135	165	195	19	130	170	210	23
22.	Shoulder–elbow length	230	255	280	15	225	250	275	15
23.	Elbow–fingertip length	295	325	355	19	290	320	350	18
24.	Upper limb length	485	540	595	32	470	525	580	34
25.	Shoulder–grip length	400	450	505	32	380	435	495	34
26.	Head length	170	185	200	8	160	170	180	6
27.	Head breadth	130	140	150	5	125	135	145	5
28.	Hand length	120	135	150	9	120	135	150	8
29.	Hand breadth	60	65	70	4	55	60	65	4
30.	Foot length	175	195	215	11	170	190	210	12
31.	Foot breadth	65	75	85	5	65	75	85	5
32.	Span	1125	1230	1335	64	1095	1195	1295	62
33.	Elbow span	590	650	710	36	570	630	695	38
34.	Vertical grip reach (standing)	1350	1475	1600	76	1325	1455	1585	79
35.	Vertical grip reach (sitting)	770	850	925	48	745	825	905	48
36.	Forward grip reach	470	520	570	31	455	505	555	29

Notes: See notes in Sections 10.1, 10.2 and 10.6.

Dimensions marked with an asterisk (*) are quoted from the original source. The remainder have been estimated.

TABLE 10.28
Anthropometric Estimates for British 8-Year-Olds (all dimensions in millimetres)

	Boys				Girls			
Dimension	5th %ile	50th %ile	95th %ile	SD	5th %ile	50th %ile	95th %ile	SD
1. Stature	1180	1280	1380	60	1185	1280	1375	59*
2. Eye height	1070	1165	1260	59	1070	1165	1260	58
3. Shoulder height	930	1020	1110	54	930	1015	1100	53
4. Elbow height	705	780	855	45	705	775	845	42
5. Hip height	605	665	725	35	585	650	715	38
6. Knuckle height	480	535	590	32	495	555	615	37
7. Fingertip height	390	445	495	32	405	465	525	37
8. Sitting height	630	680	730	31	640	685	730	28
9. Sitting eye height	520	570	620	31	525	580	635	32
10. Sitting shoulder height	380	425	470	27	370	410	450	25
11. Sitting elbow height	145	180	215	21	145	175	205	19
12. Thigh thickness	85	110	135	14	90	110	130	13
13. Buttock–knee length	355	415	455	25	375	420	465	26
14. Buttock–popliteal length	305	340	375	22	310	355	400	27
15. Knee height	360	400	440	25	355	395	435	24
16. Popliteal height	295	325	355	18	295	330	365	20
17. Shoulder breadth (bideltoid)	275	310	345	21	270	310	350	24
18. Shoulder breadth (biacromial)	265	285	305	13	255	280	305	16
19. Hip breadth	200	235	270	20	205	245	285	23
20. Chest (bust) depth	115	150	185	20	120	150	180	20
21. Abdominal depth	135	170	205	20	140	180	220	24
22. Shoulder–elbow length	240	265	290	15	240	260	285	14
23. Elbow–fingertip length	310	340	370	19	305	335	365	19
24. Upper limb length	515	565	615	30	495	555	615	35
25. Shoulder–grip length	425	475	525	30	405	465	520	35
26. Head length	170	185	200	8	165	175	185	5
27. Head breadth	130	140	150	5	125	135	145	5
28. Hand length	125	140	155	9	125	140	155	8
29. Hand breadth	60	65	70	4	60	65	70	4
30. Foot length	180	200	220	12	180	200	220	12
31. Foot breadth	70	80	90	5	65	75	85	5
32. Span	1165	1280	1395	69	1150	1250	1350	60
33. Elbow span	610	675	740	39	600	660	720	36
34. Vertical grip reach (standing)	1425	1550	1675	75	1405	1535	1665	78
35. Vertical grip reach (sitting)	805	890	975	52	785	870	955	52
36. Forward grip reach	475	535	595	35	475	530	585	34

Notes: See notes in Sections 10.1, 10.2 and 10.6.

Dimensions marked with an asterisk (*) are quoted from the original source. The remainder have been estimated.

TABLE 10.29
Anthropometric Estimates for British 9-Year-Olds (all dimensions in millimetres)

	Dimension	Boys				Girls			
		5th %ile	50th %ile	95th %ile	SD	5th %ile	50th %ile	95th %ile	SD
1.	Stature	1225	1330	1435	63	1220	1330	1440	68*
2.	Eye height	1005	1110	1215	64	1105	1215	1325	67
3.	Shoulder height	965	1065	1165	60	955	1060	1165	63
4.	Elbow height	740	820	900	50	720	815	910	57
5.	Hip height	635	700	765	40	610	690	770	48
6.	Knuckle height	505	565	625	36	530	590	650	37
7.	Fingertip height	410	470	530	36	435	495	555	37
8.	Sitting height	650	700	750	31	645	700	755	33
9.	Sitting eye height	530	585	640	33	540	595	650	33
10.	Sitting shoulder height	390	440	490	29	385	430	475	28
11.	Sitting elbow height	150	190	230	24	140	180	220	25
12.	Thigh thickness	90	115	140	15	90	115	140	15
13.	Buttock–knee length	395	440	485	26	395	445	495	30
14.	Buttock–popliteal length	325	365	405	25	330	380	430	31
15.	Knee height	375	420	465	27	375	420	465	27
16.	Popliteal height	300	340	380	23	300	340	380	24
17.	Shoulder breadth (bideltoid)	280	320	360	23	285	320	355	20
18.	Shoulder breadth (biacromial)	270	295	320	15	265	295	325	19
19.	Hip breadth	205	245	285	24	210	255	300	27
20.	Chest (bust) depth	120	155	190	22	115	155	195	24
21.	Abdominal depth	145	180	215	21	140	185	230	26
22.	Shoulder–elbow length	250	275	305	16	245	275	300	17
23.	Elbow–fingertip length	320	355	390	21	310	350	390	23
24.	Upper limb length	530	585	640	33	500	575	650	45
25.	Shoulder–grip length	435	490	545	33	405	480	555	45
26.	Head length	170	185	200	8	165	175	185	7
27.	Head breadth	135	145	155	5	125	135	145	6
28.	Hand length	130	145	160	9	130	145	160	10
29.	Hand breadth	60	65	70	4	60	65	70	4
30.	Foot length	185	210	235	14	185	210	235	14
31.	Foot breadth	70	80	90	5	70	80	90	6
32.	Span	1200	1330	1460	78	1180	1300	1420	74
33.	Elbow span	630	700	775	44	615	685	760	45
34.	Vertical grip reach (standing)	1475	1610	1745	83	1460	1615	1770	94
35.	Vertical grip reach (sitting)	830	920	1010	54	815	905	995	54
36.	Forward grip reach	495	555	615	36	485	555	625	42

Notes: See notes in Sections 10.1, 10.2 and 10.6.

Dimensions marked with an asterisk (*) are quoted from the original source. The remainder have been estimated.

TABLE 10.30
Anthropometric Estimates for British 10-Year-Olds (all dimensions in millimetres)

	Boys				Girls			
Dimension	5th %ile	50th %ile	95th %ile	SD	5th %ile	50th %ile	95th %ile	SD
1. Stature	1290	1390	1490	63	1270	1390	1510	72*
2. Eye height	1180	1275	1370	58	1155	1275	1395	72
3. Shoulder height	1025	1120	1215	57	1015	1120	1225	65
4. Elbow height	770	860	950	55	765	860	955	57
5. Hip height	660	735	810	46	650	730	810	50
6. Knuckle height	540	595	650	33	555	615	675	36
7. Fingertip height	445	500	550	33	460	520	575	36
8. Sitting height	670	725	780	32	665	725	785	36
9. Sitting eye height	550	600	650	29	555	615	675	35
10. Sitting shoulder height	410	455	500	28	400	450	500	30
11. Sitting elbow height	160	195	230	21	150	190	230	25
12. Thigh thickness	100	120	140	13	95	120	145	16
13. Buttock–knee length	415	460	505	27	415	470	525	32
14. Buttock–popliteal length	340	380	420	25	350	400	450	29
15. Knee height	395	440	485	26	395	440	485	28
16. Popliteal height	330	360	390	19	325	365	405	25
17. Shoulder breadth (bideltoid)	290	335	380	27	280	330	380	31
18. Shoulder breadth (biacromial)	275	305	335	18	275	305	335	19
19. Hip breadth	215	260	305	28	215	265	315	30
20. Chest (bust) depth	120	165	210	26	115	165	215	31
21. Abdominal depth	145	185	225	25	145	190	235	27
22. Shoulder–elbow length	265	290	315	16	260	290	320	18
23. Elbow–fingertip length	335	370	405	22	330	370	410	25
24. Upper limb length	540	610	680	42	520	590	660	44
25. Shoulder–grip length	445	515	580	42	420	495	565	44
26. Head length	170	185	200	8	160	170	180	7
27. Head breadth	135	145	155	5	125	135	145	5
28. Hand length	135	150	165	9	135	150	165	10
29. Hand breadth	65	70	75	4	60	70	80	5
30. Foot length	195	220	245	14	190	215	240	14
31. Foot breadth	70	85	95	5	70	80	90	7
32. Span	1275	1395	1515	73	1240	1365	1490	77
33. Elbow span	665	735	805	41	645	720	800	47
34. Vertical grip reach (standing)	1540	1680	1820	86	1540	1705	1870	101
35. Vertical grip reach (sitting)	870	955	1045	52	850	935	1020	52
36. Forward grip reach	525	580	635	33	520	585	650	40

Notes: See notes in Sections 10.1, 10.2 and 10.6.

Dimensions marked with an asterisk (*) are quoted from the original source. The remainder have been estimated.

TABLE 10.31
Anthropometric Estimates for British 11-Year-Olds
(all dimensions in millimetres)

	Boys				Girls			
Dimension	5th %ile	50th %ile	95th %ile	SD	5th %ile	50th %ile	95th %ile	SD
1. Stature	1325	1430	1535	65	1310	1440	1570	79*
2. Eye height	1215	1315	1415	62	1195	1325	1455	78
3. Shoulder height	1060	1160	1260	60	1050	1165	1280	69
4. Elbow height	795	890	985	57	800	890	980	56
5. Hip height	685	765	845	50	670	750	830	48
6. Knuckle height	560	620	680	35	575	645	715	42
7. Fingertip height	460	520	575	35	475	545	615	42
8. Sitting height	685	740	795	34	680	745	810	41
9. Sitting eye height	575	620	665	28	570	635	700	39
10. Sitting shoulder height	425	470	515	26	415	470	525	33
11. Sitting elbow height	160	200	240	24	155	200	245	26
12. Thigh thickness	100	120	140	11	100	125	150	16
13. Buttock–knee length	435	480	525	28	430	490	550	37
14. Buttock–popliteal length	345	395	445	30	365	410	455	26
15. Knee height	420	460	500	25	405	455	505	30
16. Popliteal height	330	375	420	26	335	375	415	24
17. Shoulder breadth (bideltoid)	300	345	390	26	285	340	395	34
18. Shoulder breadth (biacromial)	280	315	350	21	280	315	350	21
19. Hip breadth	220	265	310	27	225	280	335	34
20. Chest (bust) depth	130	170	210	24	115	175	240	38
21. Abdominal depth	150	190	230	23	145	195	245	29
22. Shoulder–elbow length	270	300	325	16	265	300	330	20
23. Elbow–fingertip length	350	385	420	22	340	385	430	28
24. Upper limb length	560	630	700	43	555	630	705	46
25. Shoulder–grip length	460	530	600	43	455	530	605	46
26. Head length	170	185	200	8	155	170	185	8
27. Head breadth	135	145	155	5	125	135	145	5
28. Hand length	140	155	170	10	135	155	175	11
29. Hand breadth	60	70	80	5	60	70	80	5
30. Foot length	205	225	245	13	195	220	245	14
31. Foot breadth	75	85	95	7	75	85	95	7
32. Span	1310	1440	1570	78	1270	1415	1560	87
33. Elbow span	685	760	830	44	660	750	835	53
34. Vertical grip reach (standing)	1575	1740	1905	100	1575	1760	1945	111
35. Vertical grip reach (sitting)	895	990	1080	56	900	990	1085	56
36. Forward grip reach	535	595	655	37	530	600	670	42

Notes: See notes in Sections 10.1, 10.2 and 10.6.

Dimensions marked with an asterisk (*) are quoted from the original source. The remainder have been estimated.

TABLE 10.32
Anthropometric Estimates for British 12-Year-Olds (all dimensions in millimetres)

	Dimension	Boys				Girls			
		5th %ile	50th %ile	95th %ile	SD	5th %ile	50th %ile	95th %ile	SD
1.	Stature	1360	1490	1620	78	1370	1500	1630	79*
2.	Eye height	1245	1375	1505	78	1255	1385	1515	80
3.	Shoulder height	1095	1215	1335	72	1100	1215	1330	69
4.	Elbow height	840	930	1020	55	840	940	1040	60
5.	Hip height	720	805	890	53	705	780	855	47
6.	Knuckle height	580	645	710	40	590	665	740	46
7.	Fingertip height	470	540	605	40	480	560	635	46
8.	Sitting height	700	765	830	39	700	775	850	45
9.	Sitting eye height	590	650	710	37	600	665	730	40
10.	Sitting shoulder height	440	490	540	30	435	490	545	32
11.	Sitting elbow height	160	205	250	27	155	205	255	31
12.	Thigh thickness	105	125	145	13	100	130	160	17
13.	Buttock–knee length	445	500	555	32	450	510	570	36
14.	Buttock–popliteal length	375	415	455	23	380	435	490	33
15.	Knee height	430	480	530	30	420	470	520	29
16.	Popliteal height	350	390	430	23	345	385	425	24
17.	Shoulder breadth (bideltoid)	315	355	395	25	305	355	405	29
18.	Shoulder breadth (biacromial)	290	325	360	21	290	325	360	21
19.	Hip breadth	230	275	320	26	235	295	355	35
20.	Chest (bust) depth	135	175	215	24	135	190	240	33
21.	Abdominal depth	165	200	235	22	155	200	245	27
22.	Shoulder–elbow length	280	310	340	18	280	315	345	20
23.	Elbow–fingertip length	360	400	440	25	355	400	445	27
24.	Upper limb length	600	665	730	41	575	660	745	52
25.	Shoulder–grip length	490	560	625	41	465	555	640	52
26.	Head length	170	185	200	8	165	175	185	7
27.	Head breadth	135	145	155	5	130	140	150	6
28.	Hand length	150	165	180	10	145	165	185	11
29.	Hand breadth	65	75	85	5	60	70	80	5
30.	Foot length	215	235	255	13	205	230	255	14
31.	Foot breadth	80	90	100	7	75	85	95	7
32.	Span	1355	1510	1665	93	1320	1480	1640	96
33.	Elbow span	710	795	885	53	685	780	880	58
34.	Vertical grip reach (standing)	1655	1835	2015	110	1650	1835	2020	112
35.	Vertical grip reach (sitting)	925	1035	1145	67	925	1035	1145	67
36.	Forward grip reach	550	620	690	42	550	625	700	45

Notes: See notes in Sections 10.1, 10.2 and 10.6.

Dimensions marked with an asterisk (*) are quoted from the original source. The remainder have been estimated.

TABLE 10.33
Anthropometric Estimates for British 13-Year-Olds (all dimensions in millimetres)

Dimension	Boys				Girls			
	5th %ile	50th %ile	95th %ile	SD	5th %ile	50th %ile	95th %ile	SD
1. Stature	1400	1550	1700	91	1430	1550	1670	73*
2. Eye height	1285	1435	1585	90	1315	1435	1555	74
3. Shoulder height	1130	1265	1400	81	1145	1255	1365	68
4. Elbow height	870	970	1070	61	875	970	1065	57
5. Hip height	740	835	930	57	725	805	885	50
6. Knuckle height	600	670	740	43	605	675	745	43
7. Fingertip height	490	560	630	43	495	565	635	43
8. Sitting height	710	790	870	49	740	805	870	41
9. Sitting eye height	605	680	755	47	630	695	760	39
10. Sitting shoulder height	450	510	570	37	455	510	565	34
11. Sitting elbow height	165	210	255	28	155	210	265	34
12. Thigh thickness	105	130	155	15	110	135	160	15
13. Buttock–knee length	465	525	585	35	480	530	580	31
14. Buttock–popliteal length	375	435	495	35	400	445	490	27
15. Knee height	440	500	560	35	440	485	530	27
16. Popliteal height	355	405	455	30	350	390	430	25
17. Shoulder breadth (bideltoid)	325	375	425	29	325	370	415	26
18. Shoulder breadth (biacromial)	295	335	375	24	300	335	370	21
19. Hip breadth	245	290	335	28	265	315	365	30
20. Chest (bust) depth	135	185	235	29	150	200	245	29
21. Abdominal depth	165	205	245	24	170	210	250	24
22. Shoulder–elbow length	290	325	360	22	295	325	355	18
23. Elbow–fingertip length	370	420	470	29	375	410	445	22
24. Upper limb length	620	695	770	47	605	680	755	47
25. Shoulder–grip length	505	585	660	47	490	570	645	47
26. Head length	175	190	205	8	165	175	185	6
27. Head breadth	140	150	160	5	130	140	150	5
28. Hand length	150	170	190	12	155	170	185	10
29. Hand breadth	70	80	90	6	70	75	80	4
30. Foot length	220	245	270	16	210	230	250	13
31. Foot breadth	80	90	100	7	80	90	100	6
32. Span	1400	1580	1760	110	1385	1540	1695	93
33. Elbow span	730	835	935	62	720	815	905	56
34. Vertical grip reach (standing)	1720	1905	2090	112	1700	1890	2080	114
35. Vertical grip reach (sitting)	955	1080	1210	78	945	1070	1200	78
36. Forward grip reach	575	655	735	48	575	640	705	41

Notes: See notes in Sections 10.1, 10.2 and 10.6

Dimensions marked with an asterisk (*) are quoted from the original source. The remainder have been estimated.

TABLE 10.34
Anthropometric Estimates for British 14-Year-Olds (all dimensions in millimetres)

	Boys				Girls			
Dimension	5th %ile	50th %ile	95th %ile	SD	5th %ile	50th %ile	95th %ile	SD
1. Stature	1480	1630	1780	90	1480	1590	1700	66*
2. Eye height	1360	1510	1660	91	1365	1475	1585	66
3. Shoulder height	1205	1335	1465	80	1190	1295	1400	64
4. Elbow height	915	1015	1115	60	900	985	1070	53
5. Hip height	795	870	945	46	735	810	885	45
6. Knuckle height	630	700	770	44	640	705	770	40
7. Fingertip height	510	585	655	44	530	595	660	40
8. Sitting height	750	835	920	52	770	830	890	36
9. Sitting eye height	640	720	800	48	660	720	780	37
10. Sitting shoulder height	470	535	600	39	470	525	580	32
11. Sitting elbow height	165	215	265	30	165	220	275	33
12. Thigh thickness	115	140	165	16	115	140	165	14
13. Buttock–knee length	495	550	605	34	495	545	595	29
14. Buttock–popliteal length	405	460	515	33	415	455	495	25
15. Knee height	465	520	575	33	450	495	540	27
16. Popliteal height	380	425	470	28	355	395	435	25
17. Shoulder breadth (bideltoid)	345	395	445	29	345	385	425	25
18. Shoulder breadth (biacromial)	320	355	390	22	315	345	375	19
19. Hip breadth	260	305	350	28	285	330	375	26
20. Chest (bust) depth	145	195	245	29	165	210	255	27
21. Abdominal depth	175	215	255	23	175	215	255	23
22. Shoulder–elbow length	310	345	380	21	305	335	360	17
23. Elbow–fingertip length	400	445	490	28	385	420	455	21
24. Upper limb length	660	735	810	47	640	700	760	36
25. Shoulder–grip length	540	620	695	47	530	590	650	36
26. Head length	180	190	200	7	165	175	185	6
27. Head breadth	140	150	160	6	130	140	150	5
28. Hand length	160	180	200	11	155	170	185	9
29. Hand breadth	75	85	95	6	70	75	80	4
30. Foot length	230	255	280	15	215	235	255	12
31. Foot breadth	85	95	105	7	80	90	100	6
32. Span	1480	1670	1860	114	1450	1580	1710	79
33. Elbow span	775	880	985	65	755	835	915	48
34. Vertical grip reach (standing)	1825	1990	2155	101	1765	1930	2095	101
35. Vertical grip reach (sitting)	1015	1140	1270	77	980	1105	1235	77
36. Forward grip reach	615	680	745	41	595	655	715	36

Notes: See notes in Sections 10.1, 10.2 and 10.6.

Dimensions marked with an asterisk (*) are quoted from the original source. The remainder have been estimated.

TABLE 10.35
Anthropometric Estimates for British 15-Year-Olds (all dimensions in millimetres)

	Boys				Girls			
Dimension	5th %ile	50th %ile	95th %ile	SD	5th %ile	50th %ile	95th %ile	SD
1. Stature	1555	1690	1825	83	1510	1610	1710	62*
2. Eye height	1430	1570	1710	84	1395	1495	1595	62
3. Shoulder height	1265	1385	1505	73	1215	1310	1405	58
4. Elbow height	965	1055	1145	56	915	995	1075	48
5. Hip height	825	895	965	44	745	815	885	42
6. Knuckle height	650	725	800	45	650	715	780	39
7. Fingertip height	530	605	680	45	540	605	670	39
8. Sitting height	785	870	955	51	790	845	900	33
9. Sitting eye height	680	755	830	47	680	735	790	33
10. Sitting shoulder height	495	555	615	36	490	535	580	27
11. Sitting elbow height	170	225	280	34	180	225	270	28
12. Thigh thickness	115	140	165	15	115	140	165	14
13. Buttock–knee length	515	570	625	32	505	550	595	27
14. Buttock–popliteal length	425	480	535	32	435	470	505	22
15. Knee height	485	535	585	31	450	495	540	26
16. Popliteal height	385	430	475	28	360	400	440	25
17. Shoulder breadth (bideltoid)	354	415	465	30	350	390	430	23
18. Shoulder breadth (biacromial)	330	370	410	23	320	350	380	18
19. Hip breadth	275	320	365	26	295	335	375	25
20. Chest (bust) depth	155	205	255	30	175	215	260	26
21. Abdominal depth	180	220	260	24	185	220	255	22
22. Shoulder–elbow length	325	355	385	19	305	335	365	17
23. Elbow–fingertip length	420	460	500	25	395	425	455	17
24. Upper limb length	695	770	845	47	650	705	760	33
25. Shoulder–grip length	570	650	725	47	540	595	650	33
26. Head length	185	195	205	7	170	180	190	7
27. Head breadth	145	155	165	6	130	140	150	5
28. Hand length	170	185	200	10	155	170	185	9
29. Hand breadth	75	85	95	5	70	75	80	4
30. Foot length	240	260	280	13	215	235	255	13
31. Foot breadth	85	95	105	7	80	90	100	5
32. Span	1560	1740	1920	109	1490	1600	1710	67
33. Elbow span	815	915	1020	62	780	845	910	41
34. Vertical grip reach (standing)	1900	2060	2220	97	1810	1960	2110	91
35. Vertical grip reach (sitting)	1075	1190	1310	71	1005	1120	1240	71
36. Forward grip reach	635	700	765	40	600	665	730	39

Note: See notes in Sections 10.1, 10.2 and 10.6.

Dimensions marked with an asterisk (*) are quoted from the original source. The remainder have been estimated.

TABLE 10.36
Anthropometric Estimates for British 16-Year-Olds (all dimensions in millimetres)

		Boys				Girls			
	Dimension	5th %ile	50th %ile	95th %ile	SD	5th %ile	50th %ile	95th %ile	SD
1.	Stature	1620	1730	1840	68	1520	1620	1720	61*
2.	Eye height	1500	1610	1720	67	1410	1510	1610	60
3.	Shoulder height	1315	1415	1515	62	1225	1315	1405	55
4.	Elbow height	995	1075	1155	49	930	1005	1080	45
5.	Hip height	830	910	990	49	755	820	885	40
6.	Knuckle height	675	740	805	40	660	720	780	36
7.	Fingertip height	555	620	685	40	545	605	665	36
8.	Sitting height	830	895	960	39	800	855	910	33
9.	Sitting eye height	725	785	845	35	685	740	795	32
10.	Sitting shoulder height	520	570	620	29	500	545	590	27
11.	Sitting elbow height	190	235	280	28	185	230	275	26
12.	Thigh thickness	125	150	175	15	120	145	170	14
13.	Buttock–knee length	530	580	630	29	510	555	600	27
14.	Buttock–popliteal length	435	490	545	32	435	480	525	26
15.	Knee height	500	545	590	27	450	595	540	26
16.	Popliteal height	395	440	485	28	365	405	445	25
17.	Shoulder breadth (bideltoid)	380	430	480	29	360	395	430	21
18.	Shoulder breadth (biacromial)	340	380	420	23	330	355	380	16
19.	Hip breadth	290	330	370	23	305	345	385	25
20.	Chest (bust) depth	165	215	265	29	180	225	265	25
21.	Abdominal depth	185	225	265	24	185	220	255	21
22.	Shoulder–elbow length	335	365	395	18	310	335	365	17
23.	Elbow–fingertip length	435	470	505	22	395	425	455	17
24.	Upper limb length	725	790	855	40	660	710	760	29
25.	Shoulder–grip length	605	670	735	40	550	595	645	29
26.	Head length	185	195	205	7	165	180	195	8
27.	Head breadth	145	155	165	5	135	145	155	5
28.	Hand length	170	185	200	9	160	175	190	9
29.	Hand breadth	80	85	90	4	70	75	80	4
30.	Foot length	240	260	280	12	220	240	260	12
31.	Foot breadth	90	100	110	6	80	90	100	5
32.	Span	1640	1785	1930	88	1500	1610	1720	67
33.	Elbow span	860	940	1025	50	785	850	920	41
34.	Vertical grip reach (standing)	1945	2100	2255	93	1820	1965	2110	88
35.	Vertical grip reach (sitting)	1130	1225	1320	58	1035	1135	1230	58
36.	Forward grip reach	650	720	790	42	605	670	735	41

Notes: See notes in Sections 10.1, 10.2 and 10.6.

Dimensions marked with an asterisk (*) are quoted from the original source. The remainder have been estimated.

TABLE 10.37
Anthropometric Estimates for British 17-Year-Olds (all dimensions in millimetres)

	Dimension	Boys				Girls			
		5th %ile	50th %ile	95th %ile	SD	5th %ile	50th %ile	95th %ile	SD
1.	Stature	1640	1750	1860	66	1520	1620	1720	61*
2.	Eye height	1530	1635	1740	65	1420	1515	1610	58
3.	Shoulder height	1335	1435	1535	62	1235	1320	1405	52
4.	Elbow height	1010	1090	1170	50	935	1005	1075	43
5.	Hip height	845	925	1005	50	755	820	885	39
6.	Knuckle height	690	755	820	41	670	725	780	33
7.	Fingertip height	565	630	700	41	555	610	665	33
8.	Sitting height	850	910	970	35	800	855	910	33
9.	Sitting eye height	745	795	845	31	690	740	790	30
10.	Sitting shoulder height	535	585	635	29	515	555	565	25
11.	Sitting elbow height	195	240	285	28	190	230	270	25
12.	Thigh thickness	125	155	185	17	120	145	170	16
13.	Buttock–knee length	535	585	635	30	515	560	605	28
14.	Buttock–popliteal length	445	495	545	30	435	480	525	27
15.	Knee height	505	550	595	27	455	500	545	26
16.	Popliteal height	405	445	485	25	365	405	445	25
17.	Shoulder breadth (bideltoid)	400	445	490	28	360	395	430	21
18.	Shoulder breadth (biacromial)	350	385	420	21	335	360	385	16
19.	Hip breadth	295	335	375	24	300	345	390	28
20.	Chest (bust) depth	180	225	270	27	190	230	270	24
21.	Abdominal depth	195	235	275	25	185	220	255	21
22.	Shoulder–elbow length	335	365	395	19	305	335	365	18
23.	Elbow–fingertip length	440	475	510	21	395	425	455	17
24.	Upper limb length	730	790	850	36	660	710	760	29
25.	Shoulder–grip length	605	665	725	36	550	595	645	29
26.	Head length	185	200	215	8	165	180	195	8
27.	Head breadth	145	155	165	6	135	145	155	5
28.	Hand length	175	190	205	9	160	175	190	9
29.	Hand breadth	80	90	100	5	70	75	80	4
30.	Foot length	240	265	290	15	220	240	260	12
31.	Foot breadth	90	100	110	5	80	90	100	5
32.	Span	1660	1795	1930	81	1510	1615	1720	64
33.	Elbow span	870	945	1020	46	790	855	915	39
34.	Vertical grip reach (standing)	1980	2125	2270	87	1830	1970	2110	85
35.	Vertical grip reach (sitting)	1145	1240	1330	57	1050	1145	1235	57
36.	Forward grip reach	655	730	805	46	610	670	730	37

Notes: See notes in Sections 10.1, 10.2 and 10.6.

Dimensions marked with an asterisk (*) are quoted from the original source. The remainder have been estimated.

TABLE 10.38
Anthropometric Estimates for British 18-Year-Olds (all dimensions in millimetres)

	Dimension	Men				Women			
		5th %ile	50th %ile	95th %ile	SD	5th %ile	50th %ile	95th %ile	SD
1.	Stature	1660	1760	1860	60	1530	1620	1710	56*
2.	Eye height	1555	1650	1745	59	1430	1520	1610	55
3.	Shoulder height	1355	1445	1535	54	1235	1320	1405	52
4.	Elbow height	1010	1105	1175	44	940	1010	1080	42
5.	Hip height	865	935	1005	43	755	820	885	40
6.	Knuckle height	705	765	825	35	670	725	780	32
7.	Fingertip height	585	640	700	35	560	610	665	32
8.	Sitting height	860	915	970	32	800	855	910	32
9.	Sitting eye height	745	800	855	32	695	745	795	30
10.	Sitting shoulder height	550	600	650	30	515	560	605	28
11.	Sitting elbow height	200	245	290	26	185	230	275	26
12.	Thigh thickness	135	160	185	15	120	145	170	14
13.	Buttock–knee length	545	590	635	26	515	560	605	28
14.	Buttock–popliteal length	450	500	550	29	435	480	525	27
15.	Knee height	505	550	595	26	455	500	545	26
16.	Popliteal height	405	445	485	25	365	405	445	25
17.	Shoulder breadth (bideltoid)	415	455	495	23	360	395	430	21
18.	Shoulder breadth (biacromial)	365	395	425	17	335	360	385	16
19.	Hip breadth	300	340	380	25	300	345	390	27
20.	Chest (bust) depth	190	225	260	21	195	235	275	24
21.	Abdominal depth	205	240	275	21	185	220	255	20
22.	Shoulder–elbow length	340	370	395	17	310	335	360	16
23.	Elbow–fingertip length	450	480	510	18	395	425	455	17
24.	Upper limb length	740	790	840	31	660	710	760	29
25.	Shoulder–grip length	615	665	715	31	550	595	645	29
26.	Head length	185	200	215	8	170	180	190	7
27.	Head breadth	145	155	165	5	135	145	155	5
28.	Hand length	175	190	205	8	160	175	190	8
29.	Hand breadth	85	90	95	4	70	75	80	4
30.	Foot length	250	270	290	12	220	240	260	11
31.	Foot breadth	90	100	110	5	80	90	100	5
32.	Span	1695	1810	1925	71	1520	1620	1720	62
33.	Elbow span	890	955	1020	40	795	855	920	38
34.	Vertical grip reach (standing)	2045	2150	2255	65	1830	1970	2110	85
35.	Vertical grip reach (sitting)	1170	1250	1335	52	1065	1150	1235	52
36.	Forward grip reach	675	740	805	41	610	670	730	37

Notes: See notes in Sections 10.1, 10.2 and 10.6.

Dimensions marked with an asterisk (*) are quoted from the original source. The remainder have been estimated.

APPENDIX

A Mathematical Synopsis of Anthropometrics

The following sections present the terminology and mathematical basis of the main parameters used in anthropometrics, together with practical methods for calculating them and for making approximations where data is sparse. As throughout the rest of the book, the convention used to define the distribution of a dimension is mean [standard deviation] or μ [σ], making the assumption that it can be treated as a normal (Gaussian) distribution. Further guidance on some of the statistical issues can be found in Mascie-Taylor (1994) as well as in statistics textbooks.

A.1 THE NORMAL DISTRIBUTION

Consider a variable x which is normally distributed in a population, shown in Figure A1.

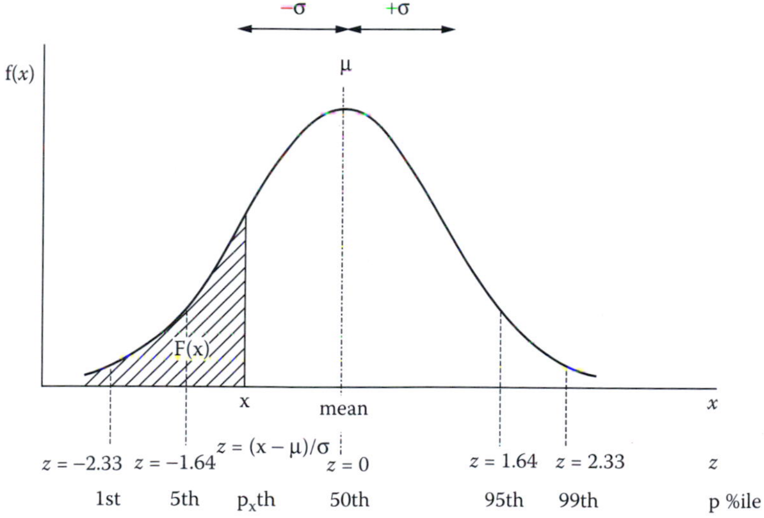

FIGURE A1 Some of the parameters used to describe a normal distribution.

Its probability density function is

$$f(x) = (1/\sigma \sqrt{2\pi}\ ) \exp[-(x-\mu)^2/2\sigma^2] \tag{A1}$$

where μ is the mean, and σ is the standard deviation of x for the population. $f(x)$ is a measure of the relative probability or relative frequency of the variable having a given value of x; it will be found in books of statistical tables as the 'ordinate of the normal curve'.

A standardised form of this distribution can be defined if the variable x is replaced by the standard normal deviate (z), such that

$$z = (x - \mu)/\sigma \tag{A2}$$

The standard normal deviate is simply the distance of a value of x from the mean value μ, normalized by dividing by the standard deviation σ. Equation A1 then becomes

$$f(z) = 1/\sqrt{2\pi}\ \exp(-z^2/2) \tag{A3}$$

which is known as the standardised form of the normal distribution (having zero mean and unit standard deviation).

The probability that x is less than or equal to a certain value is given by

$$F(x) = \int_{-\infty}^{x} f(x)\ dx \tag{A4}$$

That is, $F(x)$ corresponds to the area between the abscissa and the curve from $-\infty$ to x. This is the cumulative normal curve or normal ogive, and in its standardized form it would be expressed as $F(z)$. The probabilities of given values of z are tabulated in Table A1, in which $F(z)$ is given as a percentage (p).

The percentage probabilities $p\%$ indicate what are known as percentiles (which are commonly abbreviated as %ile). Thus, for example, $p = 5\%$ is the 5th percentile of the distribution, with 5% of the population having values less than or equal to $z = -1.64$ and 95% of the population having values greater than this, as is illustrated in Figure A1.

A.2 SAMPLES, POPULATIONS AND ERRORS

In reality we cannot know μ and σ, the parameters of a population (except in very special circumstances). We can only infer or estimate them from m and s — the mean and standard deviation of a sample of individuals deemed to be representative of the population, such that

TABLE A1
p (%) and z Values of the Normal Distribution

p	z	p	z	p	z	p	z
1	−2.33	26	−0.64	51	0.03	76	0.71
2	−2.05	27	−0.61	52	0.05	77	0.74
3	−1.88	28	−0.58	53	0.08	78	0.77
4	−1.75	29	−0.55	54	0.10	79	0.81
5	−1.64	30	−0.52	55	0.13	80	0.84
6	−1.55	31	−0.50	56	0.15	81	0.88
7	−1.48	32	−0.47	57	0.18	82	0.92
8	−1.41	33	−0.44	58	0.20	83	0.95
9	−1.34	34	−0.41	59	0.23	84	0.99
10	−1.28	35	−0.39	60	0.25	85	1.04
11	−1.23	36	−0.36	61	0.28	86	1.08
12	−1.18	37	−0.33	62	0.31	87	1.13
13	−1.13	38	−0.31	63	0.33	88	1.18
14	−1.08	39	−0.28	64	0.36	89	1.23
15	−1.04	40	−0.25	65	0.39	90	1.28
16	−0.99	41	−0.23	66	0.41	91	1.34
17	−0.95	42	−0.20	67	0.44	92	1.41
18	−0.92	43	−0.18	68	0.47	93	1.48
19	−0.88	44	−0.15	69	0.50	94	1.55
20	−0.84	45	−0.13	70	0.52	95	1.64
21	−0.81	46	−0.10	71	0.55	96	1.75
22	−0.77	47	−0.08	72	0.58	97	1.88
23	−0.74	48	−0.05	73	0.61	98	2.05
24	−0.71	49	−0.03	74	0.64	99	2.33
25	−0.67	50	0	75	0.67		

p	z	p	z
2.5	−1.96	97.5	1.96
0.5	−2.58	99.5	2.58
0.1	−3.09	99.9	3.09
0.01	−3.72	99.99	3.72
0.001	−4.26	99.999	4.26

$$m = \sum \frac{x}{n} \tag{A5}$$

and

$$s = \sqrt{\frac{\sum (x-m)^2}{n}} \tag{A6}$$

where n is the number of subjects in the sample.
In many cases it is more convenient to calculate the standard deviation by substituting the following identity (Equation A7) within Equation A6.

$$\sum (x-m)^2 = \sum x^2 - \frac{(\sum x)^2}{n} \tag{A7}$$

When the sample is small ($n \leq 30$), it is conventional to make an arbitrary correction (Bessel's correction) to the sample standard deviation to obtain the best estimate for the population standard deviation (σ'). Then

$$\sigma' = \sqrt{\frac{\sum (x-m)^2}{n-1}} \tag{A8}$$

As the sample size n increases, m and s become more reliable estimates of μ and σ; that is, the likely magnitude of random sampling errors diminishes. (Note that we are not talking about errors of bias due to nonrepresentative sampling — this is a more complex matter.) Sampling errors in estimating population parameters may be shown to be normally distributed with a mean of zero and a standard deviation known as the standard error (SE) of the parameter concerned, such that

$$\text{SE of mean} = \frac{s}{\sqrt{n}} \tag{A9}$$

$$\text{SE of standard deviation} = \frac{s}{\sqrt{2n}} = 0.71 \text{ SE of mean} \tag{A10}$$

$$\text{SE of } p\text{th \%ile} = \frac{p(100-p)s}{100 f_p n} \tag{A11}$$

where f_p is the ordinate of the normal curve at the pth %ile.

Probable magnitudes of sampling errors are commonly expressed in terms of the 95% confidence limits of the parameter concerned, which are calculated as ±1.96 SE; that is, the true values of a population parameter will lie within ±1.96 standard errors of the sample statistic 95 times out of every 100 times the sample is drawn. (Alternatively, if we are concerned with errors in one direction only, we use 1.645 SE.)

To simplify matters we may summarize this by saying that, in any anthropometric survey, the 95% confidence limits of a statistic ($\pm U_{95}$) are given by

TABLE A2
Values of the Parameter *k*, as Used in Equations A12 and A13

Statistic	k
Mean	1.96
Standard deviation	1.39
Percentiles	
40th and 60th	2.49
30th and 70th	2.58
20th and 80th	2.80
10th and 90th	3.35
5th and 95th	4.14
1st and 99th	7.33

$$U_{95} = \frac{ks}{\sqrt{n}} \tag{A12}$$

where k is a constant for the statistic concerned, as given in Table A2. Alternatively, the equation

$$n = \left(\frac{ks}{U_{95}} \right)^2 \tag{A13}$$

gives an indication of the number of subjects we need to measure in order for a particular statistic to have a certain desired degree of accuracy. However, for this, as can be seen from Equation A13, we need to have at least an estimate of the standard deviation s expected in our sample of subjects.

A.3 THE COEFFICIENT OF VARIATION

The coefficient of variation (CV) is a useful index of the inherent variability of a dimension. This is given by

$$\text{CV} = \frac{s}{m} \times 100\% \tag{A14}$$

It is independent both of absolute magnitude and of units of measurement.

In most populations stature has a lower CV than any other dimension. (Does this reflect a biological phenomenon or is it an artefact of measurement?) Characteristic ranges of CV of various types of anthropometric data are shown in Table A3. The

TABLE A3
Characteristic Coefficients of Variation of Anthropometric Data

Dimension	CV (%)
Stature	3–4
Body heights (sitting height, elbow height, etc.)	3–5
Parts of limbs	4–5
Body breadths (hips, shoulder, etc.)	5–9
Body depths (abdominal, chest, etc.)	6–9
Dynamic reach	4–11
Weight	10–21
Joint ranges	7–28
Muscular strength (static)	13–85

figures were gathered from a number of sources (Damon et al., 1966; Roebuck et al., 1975; Grieve and Pheasant, 1982) and do not reflect any specific population. They should rather be seen as a general guide to the approximate levels that we might anticipate. The high CVs of the lower part of the table are indicative of a skewed distribution, which is characteristic of anthropometric dimensions including soft tissue (fat) and of functional measures such as strength.

Roebuck et al. (1975) have demonstrated that for body length and breadth dimensions, in general, the relationship between standard deviation and mean will tend to be curvilinear (i.e., CV declines with increasing mean value). The reasons behind this observation are obscure but may be concerned with measurement error. Figure A2 shows this relationship plotted out for the 36 dimensions of Table 10.1.

A.4 SOME INDICES USED IN ANTHROPOMETRICS

A few indices appear in the literature as representing aspects of anthropometric characteristics. Two of these are the ponderal index and the body mass index (BMI), defined in Equations A15 and A16. The ponderal index is an index of body size, although rarely used now.

$$\text{Ponderal index (PI)} = \frac{Stature}{\sqrt[3]{Mass}} \tag{A15}$$

The BMI (sometimes known as the Quetelet index) is a measure expressing the relation of weight-to-height, which is correlated with body fat (adipose tissue) in relation to lean body mass and is commonly used as an indicator of whether a person is overweight or obese.

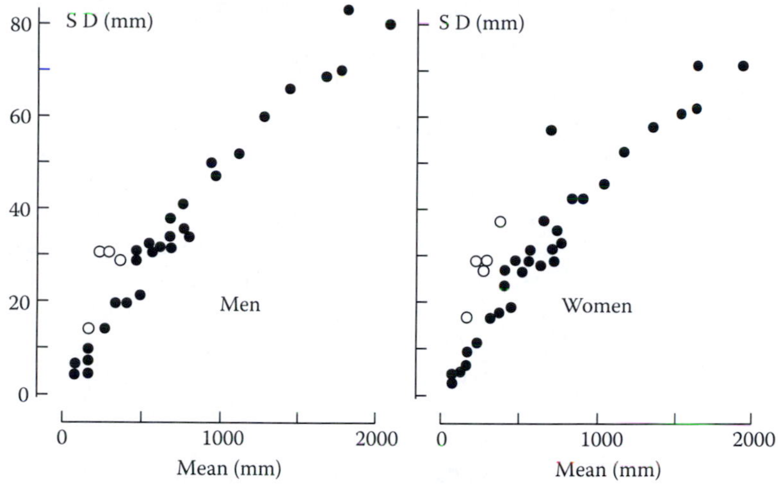

FIGURE A2 Relationship between standard deviation (SD) and mean in the anthropometric data of Table 10.1. ○, body depths, thigh thickness, sitting elbow height and hip breadth; ●, all other dimensions.

$$\text{Body mass index (BMI)} = \frac{Mass}{Stature^2} \quad \text{(measured in kg/m}^2\text{)} \qquad \text{(A16)}$$

BMI is typically in the range of 20 to 25. Individuals with a BMI between 25 and 30 have generally been considered to be overweight and those with a BMI of 30 or more as obese (NRC, 1989). The World Health Organization has introduced a new graded classification system to take account of more recent knowledge of health risks (WHO, 1998b). The classification is as follows:

Underweight: <18.5
Normal range: 18.5 to 24.9
Overweight: ≥25
  Pre-obese: 25.0 to 29.9
  Obese class I: 30.0 to 34.9
  Obese class II: 35.0 to 39.9
  Obese class III: ≥40

However, there is evidence that there may be ethnic variations in the relationship between BMI and body fat content, so these ranges may not apply exactly to all populations (Antipatis and Gill, 2001).

Individual three-dimensional (3-D) or body circumference parameters have sometimes been normalized (on an ad hoc basis) by dividing by $^3\sqrt{}$Stature or $^2\sqrt{}$Stature, respectively, in an attempt to provide population characteristics which are less affected by the variation in the dominant factor of overall body size. However, the research findings on allometric effects discussed in Chapter 3 indicate that such approaches are likely to be oversimplified.

A.5 COMBINING DISTRIBUTIONS FROM TWO OR MORE SAMPLES

There are numerous situations in which the parameters of two or more normally distributed samples or populations must be combined to give a single lumped distribution. For example, Al-Haboubi (1992, 1997, 1999) argued for such a community-based approach to design for nonhomogenous populations, whether for an educational campus, a city or a country. If anthropometric data are available for the subgroups within the population (perhaps by age or for ethnic groups), the data for the combined population can be estimated as an alternative to carrying out a full-scale anthropometric survey.

Strictly speaking, the new lumped distribution cannot be normal (Gaussian). To describe it, we should calculate percentiles iteratively. Consider two sample distributions m_1 [s_1] and m_2 [s_2] of n_1 and n_2 subjects, respectively. For any value of x, calculate standard normal deviates z_1 and z_2 in the two distributions (using Equation A2) and convert to percentiles p_1 and p_2 using Table A1. The percentile p' in the lumped distribution is given by

$$p' = \frac{p_1 n_1 + p_2 n_2}{n_1 + n_2} \tag{A17}$$

To describe the complete lumped distribution, this process is repeated for as many values of x as is required. It may then be convenient to plot this as a cumulative graph of the distribution (see Figure 2.3 for an example of a cumulative graph) from which any required percentile can be estimated by interpolation.

However, in many cases, the lumped distribution may be approximated by a new normal distribution m' [s']. In order to do this we must recalculate the sum ($\sum x$) and the sum of squares ($\sum x^2$) of the original raw data. For each of the distributions of the original samples (or populations)

$$\sum x = nm \tag{A18}$$

and from Equations A6 and A7

$$\sum x^2 = ns^2 + nm^2 \tag{A19}$$

These parameters for the lumped distribution, combining the k samples (described by m_i, s_i, n_i, etc.) are then found as follows:

$$\sum x = \sum_{i=1}^{k} n_i m_i \tag{A20}$$

$$\sum x^2 = \sum_{i=1}^{k} (n_i s_i^2 + n_i m_i^2) \tag{A21}$$

and

$$n' = \sum_{i=1}^{k} n_i \tag{A22}$$

Therefore,

$$m' = \frac{\sum (n_i m_i)}{\sum n_i} \tag{A23}$$

$$(s')^2 = \frac{\left(\sum (n_i s_i^2 + n_i m_i^2) - \frac{\left[\sum n_i m_i \right]^2}{\sum n_i} \right)}{n_i} \tag{A24}$$

The validity of the latter approach of approximation is greatest when the constituent sample standard deviations are large and the differences between the means are small.

In the case when n is the same for all k samples, Equations A21 and A22 can be simplified to

$$m' = \frac{\sum m_i}{k} \tag{A25}$$

and

$$(s')^2 = \frac{\sum \left(m_i^2 + s_i^2 \right) \frac{\left(\sum m_i \right)^2}{k}}{k} \tag{A26}$$

A.6 THE BIVARIATE DISTRIBUTION COMBINING DATA FOR TWO DIMENSIONS

In most design situations, two or more dimensions have to be considered simultaneously. Taking the simplest case of a situation where the interaction between two dimensions is being considered, the bivariate distribution can be described and the probabilities of combinations of values of the two dimensions can be calculated.

Consider two normally distributed variables x and y. Their joint probability density function is given by

$$f_{(xy)} = \frac{1}{2\pi\sigma_x\sigma_y\sqrt{1-\rho^2}}\ \mathrm{e}^w \qquad \text{(A27)}$$

where

$$w = -\frac{(x-\mu_x)^2}{\sigma_x} - \frac{2\rho(x-\mu_x)(y-\mu_y)}{\sigma_x\sigma_y} + \frac{(y-\mu_y)^2\sigma_y{}^2}{2(1-\rho^2)} \qquad \text{(A28)}$$

This distribution has five parameters of which μ_x, μ_y, σ_x, σ_y are self-explanatory, and ρ is the correlation coefficient between the two variables for the population, which is best estimated by the sample correlation coefficient (r), where

$$r = \frac{\sum(x-m_x)(y-m_y)}{\sqrt{\sum(x-m_x)^2\sum(y-m_y)^2}} \qquad \text{(A29)}$$

or

$$r = \frac{s_{xy}}{s_x s_y} \qquad \text{(A30)}$$

where

$$s_{xy} = \frac{\sum(x-m_x)(y-m_y)}{n} \qquad \text{(A31)}$$

s_{xy} is known as the covariance of x and y, and s_x and s_y are the sample standard deviations of x and y, respectively.

The bivariate probability function can be plotted as a 3-D graph. If the surface of the bivariate probability function is cut by a plane parallel to the y axis, the intersection will describe a normal probability curve. This curve defines the distribution of values of y as a population of subjects who all have the same value of x; the mean of this distribution is the most probable value of y for a given value of x. The means of all such distributions fall on a straight line known as the regression line of y on x given by the equation

$$y = a + bx \qquad \text{(A32)}$$

where

$$b = \frac{\sum (x - m_x)(y - m_y)}{\sum (x - m_x)^2} = \frac{s_{xy}}{s_x^{\ 2}} = \frac{rs_y}{s_x} \tag{A33}$$

and

$$a = m_y - bm_x \tag{A34}$$

Equations A32 to A34 may also be written as

$$y - m_y = r(s_x/s_y)(x - m_x) \tag{A35}$$

At any given value of x, y is normally distributed with a mean defined by the regression line. The standard error of this estimate of the mean is given by

$$\text{SE of mean} = s_y\sqrt{1 - r^2} \tag{A36}$$

In a similar way, the means of normal distributions given by sections cutting the bivariate distribution parallel to the x axis define the regression line of x on y. The two regression lines (y on x, x on y) would be coincident if r were 1 (perfect correlation between the two variables) and perpendicular if r were 0 (and the two variables uncorrelated). These two extreme situations, of course, are never encountered in practice, but they do show that regressions of both y on x and x on y should be calculated when making estimates from bivariate distributions, particularly when the correlation between the two variables is low (McConville and Churchill [1976] found that the vast majority of correlations between anthropometric variables have coefficients of less than 0.4).

The general concept of percentile values can be used to define percentage of accommodation for particular combinations of values for variables x and y, but their definition and interpretation is more complex than for a single variable. The 3-D graph of the probability density function is transected by a plane with a given percentage of the population to one side of that plane, but the choice of plane depends upon the use for which the percentile information is required. This approach has been used, for example, to set requirements for forward vision in cars by defining limiting percentiles of drivers' eye locations, using bivariate distributions of fore/aft and vertical (or fore/aft and lateral) eye positions to produce models called 'eyellipses' (described in Haslegrave, 1993). The mathematical basis of this approach is complex and will not be discussed further here.

A.7 MULTIVARIATE ANALYSIS

The approach used in bivariate analysis can obviously be extended to analysing accommodation for design problems involving multiple dimensions. Various statistical and modeling techniques have been developed for multivariate analysis.

Multiple regression equations dealing with three, four or more dimensions have been developed to extend the application of survey data and can be useful in many contexts, such as clothing design (McConville et al., 1979), fit of head and face personal protective equipment (Xiao et al., 1998), confined workspace design (Meindl et al., 1993; Zehner et al., 1993) and creating sets of human models or manikins (Robinette and McConville, 1981). Gordon (2002) has shown the improvement in accommodation that can be gained by using multivariate analysis to define the design parameters for nine dimensions of an office workstation, pointing out that it has become increasingly important as adjustment mechanisims have been added to office chairs, desks and VDUs (some of these adjustments possibly being interrelated). Among other multivariate techniques used to evaluate percentage population accommodation are principal components analysis (used to define boundary limits) and Monte Carlo generation (used to generate a set of random user models that have different percentiles for each dimension, within constraints from the correlation coefficients between all pairs of dimensions), as in the CAPE model developed by Bittner [1975, 1978]). Factor analysis has been used to identify the relationships between dimensions that might be used for characterising body proportions (Haslegrave, 1980) and discriminant function analysis for determining clothing sizes (Meunier, 2000).

Multivariate analysis techniques rely on complex computer processing and analysis of the data, which is beyond the scope of this book. However, the multivariate analysis of traditional anthropometric dimensions is being superseded by the developments in collection of 3-D anthropometric data (discussed in Section 2.6.4), which will allow the complete 3-D records for individual subjects (collected in an electronic database) to be analysed to determine degree of accommodation for large-scale samples representing real people.

A.8 ESTIMATING UNKNOWN DISTRIBUTIONS FROM DATA AVAILABLE FOR SIMILAR POPULATIONS OR FROM DATA AVAILABLE FOR RELATED DIMENSIONS

The practical anthropometrist is frequently required to estimate the distribution of a dimension that, for reasons of practical expediency, may not be measured directly in a particular population. Some useful techniques will be described.

A.8.1 Estimating the Parameters of the Unknown Distribution by Correlation and Regression Parameters of Data from a Similar Population

If the parameters m_x, m_y, s_x, s_y and r are known in sample 1, and the parameters m_x and s_x are known in sample 2, then m_y and s_y may be estimated for sample 2 from Equations A35 and A36 on the assumption that r is the same in both samples. (These samples may, of course, be deemed to be representative of populations.)

A.8.2 Sum and Difference Dimensions

When an unknown dimension is anatomically equivalent to the sum of two known dimensions (x and y), then the mean ($m_{(x+y)}$) and standard deviation ($s_{(x+y)}$) for the combined segments are

$$m_{(x+y)} = m_x + m_y \tag{A37}$$

$$s^2_{(x+y)} = s_x^2 + s_y^2 + 2rs_xs_y \tag{A38}$$

When an unknown dimension is anatomically equivalent to the difference between two known dimensions, then

$$m_{(x-y)} = m_x - m_y \tag{A39}$$

$$s^2_{(x-y)} = s_x^2 + s_y^2 - 2rs_xs_y \tag{A40}$$

A.8.3 Empirical Estimation of the Parameters of the Unknown Distribution by the Method of Ratio Scaling from Data for a Similar Population

If the parameters of variables x and y are known in a reference population A (or more precisely in a sample drawn from it) but only the parameters of x are known in population B (which we shall call the 'target population'), then

$$m_y/m_x \text{ (in reference population A)} \approx m_y/m_x \text{ (in target population B)} \tag{A41}$$

and

$$s_y/s_x \text{ (in reference population A)} \approx s_y/s_x \text{ (in target population B)} \tag{A42}$$

provided that populations A and B are similar in terms of age range, gender and ethnicity.

Although these equations cannot be justified mathematically, they have been widely employed, both in the present text and elsewhere, on the grounds of practical expediency (e.g., Barkla, 1961). We may call the dimension x, which is known in both populations, the 'scaling dimension'. Stature is most commonly used for this purpose since it is commonly available for populations in which other data are sparse. However, it is best to use the most closely correlated dimension for which data is available.

The simplest technique is to collect, from a variety of reference populations, the coefficients

$$E_1 = \frac{\text{mean of required dimension}}{\text{mean stature}} \tag{A43}$$

$$E_2 = \frac{\text{standard deviation of required dimension}}{\text{standard deviation of stature}} \quad \text{(A44)}$$

and simply multiply the average of each of these by the relevant parameter for stature in the target population to obtain the estimates of the mean [standard deviation] for the unkown dimension. Pheasant (1982a) conducted a validation study of this technique and found that its errors are acceptable for most purposes. Moreover, no consistent pattern was found in the errors, and the size of error was not associated with either size of dimension or with whether or not the dimension was closely correlated with stature.

One possibility that has been considered is whether self-reported stature from a sample might be used in estimating the parameters of a population. Caution is advised in this case. Buckle (1985) investigated the relationship between actual and self-reported stature and weight and found that his two groups of male subjects overestimated their stature by about 10 mm and 15 mm, respectively, and underestimated their weight by 1.1 kg and 1.3 kg. Further analysis suggested that the tendency to underestimate weight increased with body weight. An earlier study by Schlichting et al. (1981) showed similar results for weight and also that the responses of men and women were similar in relation to both height and weight. Their results for height, however, differed from those of Buckle (1985) in that there was a tendency among their subjects for tall people to underestimate their height and for small people to overestimate theirs. Under- or overestimating dimensions may be influenced by many factors, probably predominantly psychosocial but also due to experiential and environmental influences — hence the need for caution in extrapolating results on the basis of self-reported anthropometric data.

A.8.4 Empirical Estimation of Standard Deviation When Only the Mean is Known

If only the mean value (m) for a dimension is known, the standard deviation (s) may be estimated by one of two methods.

The data plotted in Figure A2 may be empirically fitted with the following regression equations. For body heights, lengths and breadths:

$$\text{men: } s = 0.05703m - 0.000008347m^2 \quad \text{(A45)}$$

$$\text{women: } s = 0.05783m - 0.000010647m^2 \quad \text{(A46)}$$

For body depths, thigh thickness, sitting elbow height and hip breadth:

$$\text{men: } s = 7.864 + 0.06977m \quad \text{(A47)}$$

$$\text{women: } s = 4.249 + 0.09467m \quad \text{(A48)}$$

These equations may then be used as a first estimate of the standard deviation of a dimension for which the mean is known or can be calculated.

Alternatively, if the coefficient of variation of a similar or related dimension is known, it may be assumed that the CV of the required dimension is the same and the standard deviation may again be calculated from the mean by using Equation A14. (If Equations A45 to A48 hold, then this latter assumption will tend to overestimate the standard deviation of large dimensions and underestimate that of small ones.)

A.9 ESTIMATING DIMENSIONS FOR A COMBINATION OF PEOPLE OR VARIABLES

Consider distributions m_a [s_a] and m_b [s_b], which might be for the same variable in different samples of individuals or for different variables in the same sample.)

If members of the two distributions meet at random (i.e., chance encounters occur) the distribution of differences is given by

$$m_{(a-b)} = m_a - m_b \tag{A49}$$

$$s^2_{(a-b)} = s_a^{\,2} + s_b^{\,2} \tag{A50}$$

and the distribution of sums is given by

$$m_{(a+b)} = m_a + m_b \tag{A51}$$

$$s^2_{(a+b)} = s_a^{\,2} + s_b^{\,2} \tag{A52}$$

In certain design applications it is necessary to know the breadth of two or more people placed side by side, for example upon a bench seat. If the body breadth concerned has the distribution m [s] and there are n people in the group, then the parameters of the group distribution m_g [s_g] are given by

$$m_g = nm \tag{A53}$$

$$s_g = \mathrm{s}\sqrt{n} \tag{A54}$$

References

Aarås, A., Dainoff, M., Ro, O., and Thoresen, M. (2002). Can a more neutral position of the forearm when operating a computer mouse reduce the pain level for VDU operators? *International Journal of Industrial Ergonomics,* 30(4-5), 307–324.

Aarås, A., Fostervold, K. I., Ro, O., Thoresen, M., and Larsen, S. (1997). Postural load during VDU work: a comparison between various work postures. *Ergonomics,* 40, 1255–1268.

Aarås, A., Horgen, G., Bjørset, H.-H., Ro, O., and Thoresen, M. (1998). Musculoskeletal, visual and psychosocial stress in VDU operators before and after multidisciplinary ergonomic interventions. *Applied Ergonomics,* 29(5), 335–354.

Abeysekera, J. D. A. and Shanavaz, H. (1987). Body size of Sri Lankan workers and their variability with other populations in the world: its impact on the use of imported goods. *Journal of Human Ergology,* 16, 193–208.

Abraham, S. (1979). Weight and height of adults 18-74 years of age. United States, 1971-1974. *Vital and Health Statistics,* Series 11, No. 211, U.S. Department of Health Education and Welfare, Hyattsville, MD.

Adams, G. A. (1961). A comparative anthropometric study of hard labor during youth as a stimulator of physical growth of young colored women. *Research Quarterly AAHPER,* 9, 102–8.

Al-Haboubi, M. H. (1992). Anthropometry for a mix of different populations. *Applied Ergonomics,* 23(3), 191–196.

Al-Haboubi, M. H. (1997). Statistics for a composite distribution in anthropometric studies. *Ergonomics,* 40(2), 189–198.

Al-Haboubi, M. H. (1999). Statistics for a composite distribution in anthropometric studies: the general case. *Ergonomics,* 42(4), 565–572.

Al-Haboubi, M. H. and Selim, S. Z. (1997). Design to minimize congestion around the Ka'aba. *Computers Industrial Engineering,* 32(2), 419–428.

American Academy of Orthopaedic Surgeons (1965). *Joint Motion: Method of Measuring and Recording.* Singapore: Churchill Livingstone.

Andersson, G. B. J., Örtengren, R., Nachemson, A., and Elfström, G. (1974). Lumbar disc pressure and myoelectric back muscle activity during sitting. I. Studies on an experimental chair. *Scandinavian Journal of Rehabilitation Medicine,* 6, 3, 104–114.

Andersson, M., Berns, T., and Klusell, L. (2000). The TCO "Office Checker": a tool for ergonomic workplace evaluation. In *Proceedings of the XIVth Triennial Congress of the International Ergonomics Association and the 44th Meeting of the Human Factors and Ergonomics Society, 27 July–6 August 2000, San Diego.* Santa Monica: Human Factors and Ergonomics Society, Volume 1, 659–662.

Andersson, M., Hwang, S. G., and Green, W. T. (1965). Growth of the normal trunk in boys and girls during the second decade of life. *Journal of Bone and Joint Surgery,* 47A, 1554–1564.

Ankrum, D. R. (1997). A challenge to eye-level, perpendicular-to-gaze, monitor placement. In *Proceedings of the 13th Triennial Congress of the International Ergonomics Association, 29 June–4 July 1997.* Tampere, Helsinki: Finnish Institute of Occupational Health, Volume 5, 35–37.

Annis, J. F. (1996). Aging effects on anthropometric dimensions important to workplace design. *International Journal of Industrial Ergonomics,* 18(5-6), 381–388.

Annis, J. F. and McConville, J. T. (1990). Applications of anthropometric data in sizing and design. In B. Das (Ed.), *Advances in Industrial Ergonomics and Safety II.* London: Taylor & Francis, 309–314.

ANSI/HFS (1988). *ANSI/HFS 100-1988 American National Standard for Human Factors Engineering of Visual Display Terminal Workstations.* Santa Monica, CA: Human Factors Society.

Antipatis, V. J. and Gill, T. P. (2001). In P. Bjorntorp (Ed.), *International Textbook of Obesity.* New York: John Wiley, 3–22.

Archea, J. C. (1985). Environmental factors associated with stair accidents by the elderly. *Clinics in Geriatric Medicine,* 1(3), 555–569.

Armstrong, T. J., Buckle, P., Fine, L. J., Hagberg, M., Jonsson, B., Kilbom, A., Kuorinka, I. A. A., Silverstein, B. A., Sjøgaard, G., and Viikari-Juntura, E. R. A. (1993). A conceptual model for work-related neck and upper-limb musculoskeletal disorders. *Scandinavian Journal of Work, Environment and Health,* 19, 73–84.

Armstrong, T. J., Foulke, J. A., Joseph, B. S., and Goldstein, S. A. (1982). Investigation of cumulative trauma disorders in a poultry processing plant. *American Industrial Hygiene Association Journal,* 43(2), 103–116.

Asmussen, E. and Heebøll-Nielsen, K. (1962). Isometric muscle strength in relation to age in men and women. *Ergonomics,* 5, 167–176.

Atherton, J., Clarke, A. K., and Harrison, E. (1982). *Office Seating for the Arthritic and Low Back Pain Patients.* Bath, U.K.: Royal National Hospital for Rheumatic Diseases.

Ayoub, M. M. and McDaniel, J. W. (1973). Effects of operator stance on pushing and pulling tasks. *AIIE Transactions,* 6, 185–195.

Backwin, H. and McLaughlin, S. D. (1964). Increase in stature: is the end in sight? *Lancet,* ii, 1195–1197.

Barkla, D. (1961). The estimation of body measurements of the British population in relation to seat design. *Ergonomics,* 4, 123–132.

Barnes, R. M. (1958). *Motion and Time Study.* New York: Wiley.

Barter, T., Emmanuel, I., and Truett, B. (1957). A Statistical Evaluation of Joint Range Data. WADC Technical Note 53-311, Wright Patterson Air Force Base, OH.

Batogowska, A. and Slowikowski, J. (1974). *Anthropometric Atlas of the Polish Adult Population for Designer Use.* Warsaw: Instytut Wzornictwa Przemystowego (in Polish).

Bauer, D. and Cavonius, C. R. (1980). Improving the legibility of visual display units through contrast reversal. In E. Grandjean and E. Vigliani (Eds.), *Ergonomic Aspects of Visual Display Terminals,* London: Taylor & Francis, 137–42.

Beasley, R., Raymond, N., Hill, S., Nowitz, M., and Hughes, R. (2003). eThrombosis: the 21st century variant of venous thromboembolism associated with immobility. *European Respiratory Journal,* 21(2), 374–376.

Bendix, A. F., Jensen, C. V., and Bendix, T. (1988). Posture, acceptability and energy consumption on a tiltable and a knee-support chair. *Clinical Biomechanics,* 3, 66–73.

Bendix, T. and Biering-Sørensen, F. (1983). Posture of the trunk when sitting on forward inclining seats. *Scandinavian Journal of Rehabilitation Medicine,* 15, 197–203.

Bendix, T., Jessen, F., and Krohn, L. (1988). Biomechanics of forward-reaching movements while sitting on fixed forward- or backward-inclining or tiltable seats. *Spine,* 13, 193–196.

Bendix, T., Winkel, J., and Jessen, F. (1985). Comparison of office chairs with fixed forwards or backwards inclining, or tiltable seats. *European Journal of Applied Physiology,* 54, 378–385.

Bermùdez de Castro, J. M., Bromage, T. G., and Fernández Jalvo, Y. (1988). Buccal striations on fossil human anterior teeth: evidence of handedness in the Middle and early Upper Pleistocene. *Journal of Human Evolution,* 17, 403–412.

Bernard, B. P. (Ed.). (1997). Musculoskeletal Disorders and Workplace Factors. A Critical Review of the Epidemiologic Evidence for WMSDs of the Neck, Upper Extremity and Low Back. NIOSH Publication No. 97-141, Cincinnati, OH: National Institute for Occupational Safety and Health.

Berns, T. and Klusell, L. (2000). Computer workplaces for primary school children: what about ergonomics? In *Proceedings of the XIVth Triennial Congress of the International Ergonomics Association and 44th Meeting of the Human Factors and Ergonomics Society, 27 July–6 August 2000, San Diego.* Santa Monica, CA: Human Factors and Ergonomics Society, Volume 5, 415–418.

Bittner, A. C. (1975). Computerized Accommodated Percentage Evaluation (CAPE) Model for Cockpit Analysis and Exclusion Studies. Technical Publication TP-75-49/TIP-03. Point Mugu, CA: Pacific Missile Test Center.

Bittner, A. C. (1978). Toward a Computerized Accommodated Percentage Evaluation (Cape) Model for Automotive Vehicle Interiors. SAE Paper 780281. Warrendale, PA: Society of Automotive Engineers.

Blackwell, S., Robinette, K., Daanen, H., Boehmer, M., Fleming, S., Kelly, S., Brill, T., Hoeferlin, D., and Burnsides, D. (2002). *Civilian American and European Surface Anthropometry Resource (CAESAR), Final Report, Volume II: Descriptions,* AFRL-HE-WP-TR-2002-0173, United States Air Force Research Laboratory, Human Effectiveness Directorate, Crew System Interface Division, Wright-Patterson AFB, OH.

BLS (Bureau of Labor Statistics). (2001). Statistics on Nonfatal Occupational Injuries and Illnesses with Days away from Work 2001. Bureau of Labor Statistics, U.S. Department of Labor. http://www.bls.org/. Accessed 12 September 2004.

Boas, F. (1912). *Changes in Bodily Form of Descendants of Immigrants.* New York: Columbia University Press.

Bobjer, O. (1989). Ergonomic knives. In A. Mital (Ed.), *Advances in Industrial Ergonomics and Safety I.* London: Taylor & Francis, 291–298.

Bongers, P. M., De Winter, C. R., Kompier, M. A. J., and Hildebrandt, V. H. (1993). Psychosocial factors at work and musculoskeletal disease. *Scandinavian Journal of Work, Environment and Health,* 19, 297–312.

Borg, G., Holmgren, A., and Lindblad, I. (1981). Quantitative evaluation of chest pain. *Acta Scandinavica,* Supplementum 644, 43–45.

Borkan, G. A., Hults, D. E., and Glynn, R. J. (1983). Role of longitudinal change and secular trend in age differences in male body dimensions. *Human Biology,* 55, 629–641.

Borkan, G. A. and Norris, A. H. (1977). Fat redistribution and the changing body dimensions of the adult male. *Human Biology,* 49, 495–514.

Bottoms, D. J. and Butterworth, D. J. (1990). Foot reach under guard rails on agricultural machinery. *Applied Ergonomics,* 21(3), 179–186.

Bradtmiller, B., Gordon, C., Kouchi, M., Jürgens, H., and Lee, Y.-S. (2004). Traditional anthropometry. In N. J. Delleman, C. M. Haslegrave, and D. B. Chaffin (Eds.), *Working Postures and Movements: Tools for Evaluation and Engineering.* Boca Raton, FL: CRC Press, 18–29, 45–49.

Branton, P. (1969). Behaviour, body mechanics and discomfort. *Ergonomics,* 12, 316–327.

Bridger, B. and Bendix, T. (2004). Pelvis and neighbouring segments. In N. J. Delleman, C. M. Haslegrave, and D. B. Chaffin (Eds.), *Working Postures and Movements: Tools for Evaluation and Engineering.* Boca Raton, FL: CRC Press, 168–188.

Briggs, A., Straker, L., and Greig, A. (2004). Upper quadrant postural changes of school children in response to interaction with different information technologies. *Ergonomics,* 47(7), 790–819.

Brown, C. H. and Wilmore, J. H. (1974). The effects of maximal resistance training on the strength and body composition of women athletes. *Medicine and Science in Sports,* 6, 174–177.

Brown, C. R. and Schaum, D. L. (1980). User-adjusted VDU parameters. In E. Grandjean and E. Vigliani (Eds.), *Ergonomic Aspects of Visual Display Terminals.* London: Taylor & Francis, 195–100.

BRS/HFES. (2004). Draft Standard BRS/HFES 100 Human Factors Engineering of Computer Workstations. Santa Monica, CA: Human Factors and Ergonomics Society.

Brumfield, R. H. and Champoux, J. A. (1984). A biomechanical study of normal functional wrist motion. *Clinical Orthopaedics,* 187, 23–25.

Brunswic, M. (1981). How Seat Design and Task Affect the Posture of the Spine in Unsupported Sitting, MSc. dissertation. Ergonomics Unit, University College, London.

BSI. (1972). BS 3705: Recommendations for Provision of Space for Domestic Kitchen Equipment. London: British Standards Institution. (Standard now withdrawn)

BSI. (1982). BS 6222: Part I Domestic Kitchen Equipment: Specification for Co-ordinating Dimensions. London: British Standards Institution. (Standard now withdrawn)

BSI. (1990a). BS 3044:1990 Guide to Ergonomics Principles in the Design and Selection of Office Furniture. London: British Standards Institution.

BSI. (1990b). BS 7231-1:1990 Body Measurements of Boys and Girls from Birth up to 16.9 Years. Information in the Form of Tables. London: British Standards Institution.

BSI. (1990c). BS 7231-2:1990 Body Measurements of Boys and Girls from Birth up to 16.9 Years. Recommendations for Body Dimensions for Children. London: British Standards Institution.

BSI. (1991). BS 4467:1991 Dimensions in Designing for Elderly People. London: British Standards Institution.

BSI. (2000). BS 5395:2000 Stairs, Ladders and Walkways. Part 1 Code of Practice for the Design, Construction and Maintenance of Straight Stairs and Winders. London: British Standards Institution.

BSI. (2001). BS 8300: 2001 Access and Facilities for Disabled People. London: British Standards Institution.

Buckle, P. and Devereux, J. (1999). Research on Work-Related Neck and Upper Limb Musculoskeletal Disorders. Report for the European Agency for Safety and Health at Work. Luxembourg: Statistical Office of the European Communities.

Buckle, P. W. (1985). Self-reported anthropometry. *Ergonomics,* 28(11), 1575–1577.

Buckle, P. W. and Devereux, J. J. (2002). The nature of work-related neck and upper limb musculoskeletal disorders. *Applied Ergonomics,* 33(3), 207–217.

Burgess-Limerick, R., Plooy, A., and Mon-Williams, M. (1998). The effect of vertical visual target location on head and neck posture. In M. A. Hanson (Ed.), *Contemporary Ergonomics 1998,* London: Taylor & Francis, 123–127.

Cai, D. and You, M. (1998). An ergonomic approach to public squatting-type toilet design. *Applied Ergonomics,* 29(2), 147–153.

Çakir, A. (1995). Acceptance of the adjustable keyboard. *Ergonomics,* 38(9), 1728–1744.

Çakir, A., Hart, D. J., and Stewart, T. F. M. (1980). *Visual Display Terminals.* Chichester: Wiley.

Cameron, N. (1979). The growth of London schoolchildren 1904-1966: an analysis of secular trend and intra-county variation. *Annals of Human Biology,* 6, 505–525.

Cameron, N., Tanner, J. M., and Whitehouse, R. A. (1982). A longitudinal analysis of the growth of limb segments in adolescence, *Annals of Human Biology,* 9, 211–220.

Cannon, L. J., Bernacki, E. J., and Walter, S. D. (1981). Personal and occupational risk factors associated with the carpal tunnel syndrome. *Journal of Occupational Medicine,* 23, 255–258.

CEN. (1991a). EN 292-1:1991 Safety of Machinery: Basic Concepts, General Principles for Design. Brussels: European Committee for Standardization (CEN).

CEN. (1991b). EN 292-2:1991 Safety of Machinery: Technical Principles and Specifications. Brussels: European Committee for Standardization (CEN).

CEN. (1992). EN 294: 1992 Safety of Machinery: Safety Distances to Prevent Danger Zones Being Reached by the Upper Limbs. Brussels: European Committee for Standardization (CEN).

CEN. (1996). EN 547-1: 1996 Safety of Machinery: Human Body Measurements. Part 1: Principles for Determining the Dimensions Required for Openings for Whole Body Access into Machinery. Brussels: European Committee for Standardization (CEN).

CEN. (1997a). EN 547-2: 1997 Safety of Machinery: Human Body Measurements. Part 2: Principles for Determining the Dimensions Required for Access Openings. Brussels: European Committee for Standardization (CEN).

CEN. (1997b). EN 547-3: 1997 Safety of Machinery: Human Body Measurements. Part 3: Anthropometric Data. Brussels: European Committee for Standardization (CEN).

CEN. (2000). EN 1335-1: 2000 Office Furniture: Office Work Chair. Part 1: Dimensions: Determination of Dimensions. Brussels: European Committee for Standardization (CEN).

CEN. (2002). EN 1005-3:2002 Safety of Machinery: Human Physical Performance. Part 3: Recommended Force Limits for Machinery Operation. Brussels: European Committee for Standardization (CEN).

CEN. (2003a). EN 1005-2: 2003 Safety of Machinery: Human Physical Performance. Part 2: Manual Handling of Machinery and Component Parts of Machinery. Brussels: European Committee for Standardization (CEN).

CEN. (2003b). prEN 547-4: 1997 Safety of Machinery: Human Physical Performance. Part 4: Evaluation of Working Postures and Movements in Relation to Machinery. Brussels: European Committee for Standardization (CEN).

Cerney, M. M. and Adams, D. C. (2004). Sequestering size: the role of allometry and gender in digital human modeling. In *Proceedings of the Digital Human Modeling for Design and Engineering Symposium, 15–17 June 2004, Rochester, MI.* Warrendale, PA: Society of Automotive Engineers. (on CD ROM)

Chaffin, D. B. (1973). Localized muscle fatigue: definition and measurement. *Journal of Occupational Medicine,* 15(4), 346–354.

Chaffin, D. B. (2001). *Digital Human Modeling for Vehicle and Workplace Design.* Warrendale, PA: Society of Automotive Engineers.

Chaffin, D. B. (2002). Simulation of human reach motions for ergonomics analyses. In *Proceedings of SAE Digital Human Modeling Conference, Munich , 18–22 June 2002.* Dusseldorf: VDI Verlag, 9–23.

Chaffin, D. B. (2004). Digital human models for ergonomic design and engineering. In N. J. Delleman, C. M. Haslegrave, and D. B. Chaffin (Eds.), *Working Postures and Movements.* Boca Raton, FL: CRC Press, 425–431, 466–470.

Chaffin, D. B., Andersson, G. B. J., and Martin, B. J. (1999). *Occupational Biomechanics,* 3rd ed. New York: John Wiley & Sons.

Chaffin, D. B., Faraway, J. J., Zhang, X., and Woolley, C. (2000). Stature, age, and gender effects on reach motion postures. *Human Factors,* 42(3), 408–420.

Che Doi, M. A. and Haslegrave, C. M. (2004). Modelling the effects on head and trunk postures of an obstruction in the line of sight. In *Proceedings of the Digital Human Modeling for Design and Engineering Symposium, 15–17 June 2004, Rochester, Michigan.* Warrendale, PA: Society of Automotive Engineers. (on CD-ROM)

Chinn, S. and Rona, R. J. (1994). Trends in weight-for-height and triceps skinfold thickness for English and Scottish children, 1972-1982 and 1982-1990. *Paediatric and Perinatal Epidemiology,* 8, 90–106.

Chinn, S., Rona, R. J., and Price, C. E. (1989). The secular trend in height of primary school children in England and Scotland, 1972-1979 and 1979-1986. *Annals of Human Biology,* 16, 387–395.

Cohen, H. H. (2000). A field study of stair descent. *Ergonomics in Design,* 8(2), 11–15.

Cohen, M. L., Arroyo, J. F., Champion, G. D., and Browne, C. D. (1992). In search of the pathogenesis of refractory cervicobrachial pain syndrome: a deconstruction of the RSI phenomenon, *Medical Journal of Australia,* 156, 432–436.

Coleman, N., Hull, B. P., and Ellitt, G. (1998). An empirical study of preferred settings for lumbar support on adjustable office chairs. *Ergonomics,* 41(4), 401–419.

Colombini, D. and Occhipinti, E. (2004). Multiple factor models. In N. J. Delleman, C. M. Haslegrave, and D. B. Chaffin (Eds.), *Working Postures and Movements: Tools for Evaluation and Engineering,* Boca Raton, FL: CRC Press, 312–330, 355–366.

Cook, C. J. and Kothiyal, K. (1998). Influence of mouse position on muscular activity in the neck, shoulder and arm in computer users. *Applied Ergonomics,* 29(6), 439–443.

Coole, C. and Haslegrave, C. M. (2000). Reducing risks for work-related musculoskeletal disorders in school nurseries. In P. T. McCabe, M. A. Hanson, and S. A. Robertson (Eds.), *Contemporary Ergonomics 2000,* London: Taylor & Francis, 317–321.

Cooper, A. and Straker, L. (1998). Mouse versus keyboard use: a comparison of shoulder muscle load. *International Journal of Industrial Ergonomics,* 22(4-5), 351–357.

Cooper, C., McAlindon, T., Coggon, D., Egger, P., and Dieppe, P. (1994). Occupational activity and osteoarthritis of the knee. *Annals of the Rheumatic Diseases,* 53, 90–93.

Corlett, E. N. (1983). Analysis and evaluation of working posture. In T. O. Kvålseth (Ed.), *Ergonomics of Workstation Design,* London: Butterworths, 1–18.

Corlett, E. N. (1999). Are you sitting comfortably? *International Journal of Industrial Ergonomics,* 24(1), 7–12.

Corlett, E. N. (2005). The evaluation of industrial seating. In J. R. Wilson and E. N. Corlett (Eds.), *Evaluation of Human Work: A Practical Ergonomics Methodology,* 3rd ed. London: Taylor & Francis, 729–742.

Corlett, E. N. and Bishop, R. P. (1976). A technique for assessing postural discomfort. *Ergonomics,* 19, 175–182.

Corlett, E. N. and Clark, T. S. (1995). *The Ergonomics of Workspaces and Machines: A Design Manual,* 2nd ed. London: Taylor & Francis.

Corlett, E. N. and Gregg, H. (1994). Seating and access to work. In R. Lueder and K. Noro (Eds.), *Hard Facts about Soft Machines: The Ergonomics of Seating.* London: Taylor & Francis, 335–345.

Culver, C. C. and Viano, D. C. (1990). Anthropometry of seated women during pregnancy: defining a fetal region for crash protection research. *Human Factors,* 32(6), 625–636.

Dainoff, M. J. (1994). Three myths of ergonomic seating. In R. Lueder and K. Noro (Eds.), *Hard Facts about Soft Machines: The Ergonomics of Seating.* London: Taylor & Francis, 37–46.

Dainoff, M. J. and Balliet, J. (1991). Seated posture and workstation configuration. In M. Kumashiro and E. D. Megaw (Eds.), *Towards Human Work.* London: Taylor & Francis, 156–163.

Dalassio, D. J. (Ed.). (1980). *Wolfs Headache and Other Head Pain.* Oxford: Oxford University Press.

Damon, A. (1973). Ongoing human evolution. In I. H. Porter and R. E. Skalko (Eds), *Heredity and Society.* New York: Academic Press, 45–74.

Damon, A., Seltzer, C. C., Stoudt, H. W., and Bell, B. (1972). Age and physique in healthy white veterans at Boston. *Journal of Gerontology,* 27, 202–208.

Damon, A., Stoudt, H. W., and McFarland, R. A. (1966). *The Human Body in Equipment Design.* Cambridge, MA: Harvard University Press.

Daniels, G. S. (1952). The "Average Man"? Technical Note WCRD 53-7. Wright Air Development Center, Wright-Patterson Air Force Base, OH.

Das, B. and Grady, R. M. (1983). The normal working area in the horizontal plane: a comparative analysis between Farley's and Squires' concepts, *Ergonomics,* 26(5), 449–459.

David, G. C. (1987). Intra-abdominal pressure measurements and load capacities for females. *Ergonomics,* 28, 345–348.

Davies, B. T., Abada, A., Benson, K., Courtney, A., and Minto, I. (1980). Female hand dimensions and guarding of machines. *Ergonomics,* 23, 79–84.

Davis, P. R. and Stubbs, D. A. (1977a). Safe levels of manual forces for young males (1). *Applied Ergonomics,* 8(3), 141–150.

Davis, P. R. and Stubbs, D. A. (1977b). Safe levels of manual forces for young males (2). *Applied Ergonomics,* 8(4), 219–228.

Davis, P. R. and Stubbs, D. A. (1978). Safe levels of manual forces for young males (3). *Applied Ergonomics,* 9(1), 33–37.

De Croon, E. M., Sluiter, J. K., Kuijer, P. P. F. M., and Frings-Dresen, M. H. W. (2005). The effect of office concepts on worker health and performance: a systematic review of the literature. *Ergonomics,* 48, 2, 119–134.

Deeb, J. M., Drury, C. G., and Begbie, K. L. (1985). Handle positions in a holding task as a function of task height. *Ergonomics,* 28(5), 747–763.

Delisle, A., Imbeau, D., Santos, B., Plamondon, A., and Montpetit, Y. (2004). Left-handed versus right-handed computer mouse use: effect on upper-extremity posture. *Applied Ergonomics,* 35(1), 21–28.

Delleman, N. J. (1999). Working Postures: Prediction and Evaluation. Doctoral thesis, Vrije Universiteit, Amsterdam, The Netherlands.

Delleman, N. J. (2004). Head and neck. In N. J. Delleman, C. M. Haslegrave, and D. B. Chaffin (Eds.), *Working Postures and Movements.* Boca Raton, FL: CRC Press, 87–107.

Dempsey, P. G. and Leamon, T. B. (1995). Implementing bent-handled tools in the workplace. *Ergonomics in Design,* October, 15–21.

Dempsey, P. G., McGorry, R. W., Leamon, T. B., and O'Brien, N. (2002). Bending the tool and the effect on human performance: further investigation of a simulated wire-twisting task. *American Industrial Hygiene Association Journal,* 63, 586–593.

Dempsey, P. G., McGorry, R. W., and O'Brien, N. V. (2004). The effects of work height, workpiece orientation, gender, and screwdriver type on productivity and wrist deviation. *International Journal of Industrial Ergonomics,* 33(4), 339–346.

Dempster, W. T. (1955). Space Requirements of the Seated Operator: Geometrical, Kinematic and Mechanical Aspects of the Body with Special Reference to the Limbs. WADC Tech. Note 55-159. Wright Patterson Air Force Base, OH.

Department of Defense. (1981). Human Engineering Design Criteria Standard for Military Systems, Equipment and Facilities, MIL-STD-1472C. Washington, DC: Department of Defense.

Department of Defense. (1999). Human Engineering Design Criteria Standard for Military Systems, Equipment and Facilities, MIL-STD-1472F. Washington, DC: Department of Defense.

Department of Education and Science. (1972). British School Population Dimensional Survey. Building Bulletin 46. London: HMSO.

Department of Education and Science. (1985). Body Dimensions of the School Population. Building Bulletin 62. London: HMSO.

Department of the Environment. (1972). Space in the Home. Design Bulletin 6. London: HMSO.

Department of Health. (1999). Health Survey for England: Health of Young People '95-'97. London: The Stationery Office.

Department of Trade and Industry. (2000). Home Accident Surveillance System Including Leisure Activities. 22nd Annual Report 1998 Data. London: Department of Trade and Industry.

Devereux, J., Buckle, P., and Vlachonokilis, I. (1999). Interactions between physical and psychosocial risk factors at work increase the risk of back disorders: an epidemiological approach. *Occupational and Environmental Health,* 56, 343–353.

De Wall, M., Van Riel, M. P. J. M., Snijders, C. J., and Van Wingerden, J. P. (1991). The effect on sitting posture of a desk with a 10° inclination for reading and writing. *Ergonomics,* 34(5), 575–584.

Drury, C. G. and Coury, B. G. (1982). A methodology for chair evaluation. *Applied Ergonomics,* 13(30), 195–202.

Drury, C. G., Begbie, K., Ulate, C., and Deeb, J. M. (1985). Experiments on wrist deviation in manual materials handling. *Ergonomics*, 28, 3, 577–589.

Drury, C. G. and Francher, M. (1985). Evaluation of a forward-sloping chair. *Applied Ergonomics,* 16, 41–47.

Duke, K., Mirka, G. A., and Sommerich, C. M. (2004). Productivity and ergonomic investigation of bent-handle pliers. *Human Factors,* 46(2), 234–243.

Dukic, T., Rönnäng, M., Örtengren, R., Christmannson, M., and Johansson Davidsson, A. (2002). Virtual evaluation of human factors for assembly line work: a case study in an automotive industry. In *Proceedings of the Digital Human Modeling for Design and Engineering Symposium, 18–20 June 2002, Munich.* Dusseldorf: VDI Verlag, 129–150.

Duncan, J. and Ferguson, D. (1974). Keyboard operating posture and symptoms in operating. *Ergonomics,* 17, 651–662.

Durnin, J. V. G. A. and Rahaman, M. M. (1967). The assessment of the amount of fat in the human body from measurements of skinfold thickness. *British Journal of Nutrition,* 21, 681–689.

Durnin, J. V. G. A. and Womersley, J. (1974). Body fat assessed from total body density and its estimation from skinfold thickness: measurements on 481 men and women aged from 16 to 72 years. *British Journal of Nutrition,* 32, 77–97.

Edgren, C. S. and Radwin, R. G. (2000). Power grip force magnitude and direction for varying cylindrical handle diameters and hand size. In *Proceedings of the XIVth Triennial Congress of the International Ergonomics Association and 44th Annual Meeting of the Human Factors and Ergonomics Society, San Diego, 20 July – 4 August 2000, Volume 5.* Santa Monica, CA: Human Factors and Ergonomics Society, 25–28.

Eklund, J. A. E. (1986). Industrial Seating and Spinal Loading. PhD thesis, University of Nottingham, Nottingham, U.K.

Ellis, R. and Parsons, C. (2000). Post Office counter customer interface: a design challenge. In P. T. McCabe, M.A. Hanson, and S. A. Robertson (Eds.), *Contemporary Ergonomics 2000.* London: Taylor and Francis, 380–384.

Era, P., Schroll, M., Heikkinen, E., and Steen, B. (1992). Functional abilities in old age and occupational background: experiences of a comparative study. In M. Kumashiro (Ed.), *The Paths to Productive Aging.* London: Taylor and Francis, 294–299.

Erdelyi, A., Sihvonen, T., Helin, P., and Hanninen, O. (1988). Shoulder strain in keyboard workers and its alleviation by arm supports. *International Archives of Occupational and Environmental Health,* 60, 119–124.

Eveleth, P. B. (1975). Differences between ethnic groups in sex dimorphism of adult height. *Annals of Human Biology.* 2, 35–39.

Eveleth, P. B. and Tanner, J. M. (1976). *Worldwide Variation in Human Growth,* Cambridge: Cambridge University Press.

Fernandez, J. E. and Poonawala, M. F. (1998). How long should it take to evaluate seats subjectively? *International Journal of Industrial Ergonomics,* 22(6), 483–487.

Fisk, J. (1993). Design for the elderly: a biological perspective. *Applied Ergonomics,* 24(1), 47–50.

Fitch, J. M., Templar, J., and Corcoran, P. (1974). The dimensions of stairs. *Scientific American,* 231(4), 82–90.

Freivalds, A. (1987). The ergonomics of tools. *International Reviews of Ergonomics,* 1, 43–75.

Friedlander, J. S., Costa, P. T., Bosse, R., Ellis, E., Rhoads, J. G., and Stoudt, H. W. (1977). Longitudinal physique changes among healthy white veterans at Boston. *Human Biology,* 49, 541–558.

Friess, M. and Corner, B. D. (2004). From XS to XL: statistical modeling of humam body shape using 3D surface scans. In *Proceedings of the Digital Human Modeling for Design and Engineering Symposium, 15–17 June 2004, Rochester, MI.* Warrendale, PA: Society of Automotive Engineers. (on CD ROM)

Fruin, J. (1971). *Pedestrian Planning and Design,* New York: Metropolitan Association of Urban Designers and Environmental Planners.

Garmer, K., Sperling, L., and Forsberg, A. (2002). A hand-ergonomics training kit: development and evaluation of a package to support improved awareness and critical thinking. *Applied Ergonomics,* 33(1), 39–49.

Garner, D. M., Garfinkle, P. E., Schwartz, D., and Thompson, M. (1980). Cultural expectations of thinness in women. *Psychological Reports,* 47, 483–491.

Garonzik, R (1989). Hand dominance and the implications for left-handed operation of controls. *Ergonomics,* 32(10), 1185–1192.

Garrett, J. W. (1971). The adult human hand: some anthropometric and biomechanical considerations. *Human Factors,* 13(2), 117–131.

Gerard, M. J., Jones, S. K., Smith, L. A., Thomas, R. E., and Wang, T. (1994). An ergonomic evaluation of the Kinesis ergonomic computer keyboard. *Ergonomics,* 37(10), 1661–1668.

Gibson, S. J., Le Vasseur, S. A., and Helme, R. D. (1991). Cerebral event-related responses induced by $C0_2$ laser stimulation in subjects suffering from cervico-brachial syndrome. *Pain,* 47, 173–182.

Gilbert, B. G., Hahn, H. A., Gilmore, W. E., and Schurman, D. L. (1988). Thumbs up: anthropometry of the first digit. *Human Factors,* 30, 747–750.

Gite, L. P. and Yadav, B. G. (1989). Anthropometric survey for agricultural machine design. *Applied Ergonomics,* 20, 191–196.

Goldsmith, S. (2000). *Universal Design: A Manual of Practical Guidance for Architects.* Oxford: Architectural Press.

Goldstein, H. (1971). Factors influencing the height of 7 year old children. *Human Biology,* 43, 92–111.

Gooderson, C. Y. and Beebee, M. (1977). A Comparison of the Anthropometry of 100 Guardsmen with that of 500 Infantrymen, 500 RAC Servicemen and 2000 RAF Aircrew. Report APRE 37/76, Army Personnel Research Establishment, Farnborough, Hants, U.K.

Gooderson, C. Y., Knowles, D. J., and Gooderson, P. M. E. (1982). The Hand Anthropometry of Male and Female Military Personnel. APRE Memorandum 82M510, Army Personnel Research Establishment, Farnborough, Hants, U.K.

Goossens, R. H. M., Snijders, C. J., Roelofs, G. Y., and Van Buchem, F. (2003). Free shoulder space requirements in the design of high backrests. *Ergonomics,* 46(5), 518–530.

Gordon, C. C. (2002). Multivariate anthropometric models for seated workstation design. In: *Contemporary Ergonomics* 2002 (Ed. P.T. McCabe), London: Taylor & Francis, 582–589.

Gould, S. J. (1984). *The Mismeasure of Man.* Harmondsworth, U.K.: Penguin.

Government Statistical Service/Central Statistical Office. (1990). *Social Trends 20.* London: HMSO.

Grahame, R. and Jenkins, J. M. (1972). Joint hypermobility: asset or liability? A study of joint mobility in ballet dancers. *Annals of the Rheumatic Diseases,* 31, 109–111.

Grandjean, E. (1973). *Ergonomics of the Home.* London: Taylor & Francis.

Grandjean, E. (1987). *The Ergonomics of Computerized Offices.* London: Taylor & Francis.

Grandjean, E. (1988). *Fitting the Task to the Man: A Textbook of Occupational Ergonomics,* 4th ed. London: Taylor & Francis.

Grandjean, E., Hünting, W., and Piderman, M. (1983). VDT workstation design: preferred settings and their effects. *Human Factors,* 25(2), 161–175.

Grandjean, E., Nishiyama, K., Hünting, W., and Piderman, M. (1984). A laboratory study on preferred and imposed settings of a VDT workstation. *Behaviour and Information Technology,* 3, 289–304.

Grant, C. and Goldberg, N. (1994). Towards systematic descriptions of chair performance. In R. Lueder and K. Noro (Eds.), *Hard Facts about Soft Machines: The Ergonomics of Seating.* London: Taylor & Francis, 405–411.

Gray, H, G. and Macmillan, F. (2003). Can ergonomic microscopes reduce the risks of work-related musculo-skeletal disorders? In P. T. McCabe (Ed.), *Contemporary Ergonomics 2003.* London: Taylor and Francis, 35–40.

Green, R. A. and Briggs, C. A. (1989). Anthropometric dimensions and overuse injury among Australian keyboard operators. *Journal of Occupational Medicine,* 31, 747–750.

Greenberg, L. and Chaffin, D. B. (1976). *Workers and Their Tools.* Midland, MI: Pendell.

Greening, J. and Lynn, B. (1998). Vibration sense in the upper limb in patients with repetitive strain injury and a group of at-risk office workers. *International Archives of Occupational and Environmental Health,* 71(1), 29–34.

Greening, J., Lynn, B., and Leary, R. (2003). Sensory and autonomic function in the hands of patients with non-specific arm pain (NSAP) and asymptomatic office workers. *Pain,* 104, 275–281.

Greening, J., Smart, S., Leary, R., Hall-Craggs, M., O'Higgins, P., and Lynn, B. (1999). Reduced movement of median nerve in carpal tunnel during wrist flexion in patients with non-specific arm pain. *Lancet,* 354, 217–218.

Greulich, H. (1957). A comparison of the physical growth and development of American born and native Japanese children. *American Journal of Physical Anthropology,* 15, 489–516.

Grieve, D. W. and Pheasant, S. T. (1982). Biomechanics. In W. T. Singleton (Ed.), *The Body at Work*. Cambridge: Cambridge University Press, 71–200.

Grimshaw v. Ford Motor Co. (1981). 119 Cal App 3d 757. Cited in Jones, M. A. (1986). *Textbook on Torts*. London: Blackstone.

Guven, M., Elalmis, D. D., Binokay, S., and Tan, U. (2003). Population-level right-paw preference in rats assessed by a new computerized food-reaching test. *International Journal of Neuroscience,* 113(12), 1675–1689.

Habes, D. J. and Grant, K. A. (1997). An electromyographic study of maximum torques and upper extremity muscle activity in simulated screwdriving tasks. *International Journal of Industrial Ergonomics,* 20(4), 339–346.

Hagberg, M., Silverstein, B., Wells, R., Smith, M. J., Hendrick, H. W., Carayon, P., and Pérusse, M. (1995). I. Kuorinka and L. Forcier (Eds.), *Work Related Musculoskeletal Disorders (WMSDs): A Reference Book for Prevention*. London: Taylor & Francis.

Haigh, R. (1993). The ageing process: a challenge for design. *Applied Ergonomics,* 24(1), 9–14.

Haigh, R. and Haslegrave, C. M. (1992). The older worker in industry. In E. J. Lovesey (Ed.), *Contemporary Ergonomics 1992*. London: Taylor and Francis, 181–185.

Haines, H. and McAtamney, L. (1993). Applying ergonomics to improve microscopy work. *Microscopy and Analysis,* July, 15–17.

Hall, E. T. (1969). *The Hidden Dimension*. New York: Doubleday.

Hallbeck, M. S. and Kadefors, R. (2004). Hand/handle coupling. In N. J. Delleman, C. M. Haslegrave, and D. B. Chaffin (Eds.), *Working Postures and Movements*. Boca Raton, FL: CRC Press, 284–297, 306–310.

Han, J.-A. (2003). Study of the Effects of Handle Configuration on Wrist Posture and Muscle Activity. PhD thesis, University of Nottingham, Nottingham, U.K.

Hänel, S-E., Dartman, T., and Shishoo, R. (1997). Measuring methods for comfort rating of seats and beds. *International Journal of Industrial Ergonomics,* 20(2), 163–172.

Hardyk, C., Goldman, R., and Petrinovich, L. (1975). Handedness and sex, race and age. *Human Biology,* 47(3), 369–375.

Harrington, J. M., Carter, J. T., Birrell, L., and Gompertz, D. (1998). Surveillance case definitions for work related upper limb pain syndromes. *Occupational and Environmental Medicine,* 55, 264–271.

Harris, C. and Straker, L. (2000). Survey of physical ergonomics issues associated with school children's use of laptop computers. *International Journal of Industrial Ergonomics,* 26, 337–346.

Haslam, R. A., Hill, L. D., Howarth, P. A., Brooke-Wavell, K., and Sloane, J. E. (2001). Actions of older people affect their risk of falling on stairs. In M.A. Hanson (Ed.), *Contemporary Ergonomics 2001*. London: Taylor and Francis, 159–163.

Haslegrave, C. M. (1979). An anthropometric survey of British drivers. *Ergonomics,* 22, 145–154.

Haslegrave, C. M. (1980). Anthropometric profile of the British car driver. *Ergonomics,* 23(5), 437–467.

Haslegrave, C. M. (1986). Characterising the anthropometric extremes of the British population. *Ergonomics,* 29(2), 281–301.

Haslegrave, C. M. (1993). Visual aspects in vehicle design. In B. Peacock and W. Karwowski (Eds.), *Automotive Ergonomics*. London: Taylor & Francis, 79–98.

Haslegrave, C. M. (2004). Force exertion. In N. J. Delleman, C. M. Haslegrave, and D. B. Chaffin (Eds.), *Working Postures and Movements*. Boca Raton, FL: CRC Press, 367–402.

Haslegrave, C. M., Tracy, M. F., and Corlett, E. N. (1997a). Force exertion in awkward working postures: strength capability while twisting or working overhead. *Ergonomics,* 40(12), 1335–1362.

Haslegrave, C. M., Tracy, M. F., and Corlett, E. N. (1997b). Strength capability while kneeling. *Ergonomics,* 40(12), 1363–1379.

Hauspie, R. C., Vercauteren, M., and Susanne, C. (1996). Secular changes in growth. *Hormone Research,* 45(Suppl. 2), 8–17.

Healy, M. J. R. (1962). The effect of age-grouping on the distribution of a measurement effected by growth. *American Journal of Physical Anthropology,* 20, 49–50.

Heasman, T., Brooks, A., and Stewart, T. (2000). Health and Safety of Portable Display Screen Equipment. Contract Research Report 304/2000. Sudbury, U.K.: HSE Books.

Hedge, A., Morimoto, S., and McCrobie, D. (1999). Effects of keyboard tray geometry on upper body posture and comfort. *Ergonomics,* 42(10), 1333–1349.

Helander, M. G. and Furtado, D. (1992). Product design for manual assembly. In M. Helander and M. Nagamachi (Eds.), *Design for Manufacturability: A Systems Approach to Concurrent Engineering and Ergonomics.* London: Taylor & Francis, 171–188.

Helander, M., Little, S. E., and Drury, C. G. (2000). Adaptation and sensitivity to postural change in sitting. *Human Factors,* 42(4), 617–629.

Helander, M. G. and Zhang, L. (1997). Field studies of comfort and discomfort in sitting. *Ergonomics,* 40(9), 895–915.

Helme, R. D., Levasseur, S. A., and Gibson, S. J. (1992). RSI revisited: evidence for psychological and physiological differences from an age, sex and occupation matched control group. *Australian and New Zealand Journal of Medicine,* 22, 23–29.

HES (Health Examination Survey). (1962). User Data Set No. 3: Physical Measurements. Washington, DC: Health Examination Survey, U.S. Department of Health, Education and Welfare.

Hettinger, T. (1961). *Physiology of Strength,* Springfield, IL: Charles C. Thomas.

Hinojosa, T., Sheu, C. F., and Michel, G. F. (2003). Infant hand-use preferences for grasping objects contributes to the development of a hand-use preference for manipulating objects. *Developmental Psychobiology,* 43(4), 328–334.

Hirao, N. and Kajiyama, M. (1994). Seating for pregnant workers based on subjective symptoms and motion analysis. In R. Lueder and K. Noro (Eds.), *Hard Facts about Soft Machines: The Ergonomics of Seating,* London: Taylor & Francis, 317–331.

Hirsch, J. and O'Donnell, M. (2001). Venous thromboembolism after long flights: are airlines to blame? *Lancet,* 357(May 12), 1461–1462.

Holden, J. M., Fernie, G., and Lunau, K. (1988). Chairs for the elderly: design considerations. *Applied Ergonomics,* 19(4), 281–288.

Hopkins, A. (1990). Stress, the quality of work and repetition strain injury in Australia. *Work and Stress,* 4, 129–138.

Horgen, G., Aarås, A., Fagerthun, H. E., and Larsen, S. E. (1989). The work posture and the postural load of the neck/shoulder muscles when correcting presbyopia with different types of multifocal lenses on VDU-workers. In M. J. Smith and G. Salvendy (Eds.), *Work with Computers: Organizational, Management, Stress and Health Aspects.* Amsterdam: Elsevier Science Publishers, 338–347.

Hornibrook, F. A. (1934). *The Culture of the Abdomen.* New York: Doubleday.

Hsaio, H. and Keyserling, W. M. (1991). Evaluating posture behaviour during seated tasks. *International Journal of Industrial Ergonomics,* 8, 313–334.

Hsaio, H., Long, D., and Snyder, K. (2002). Anthropometric differences among occupational groups. *Ergonomics,* 45(2), 136–152.

HSC (Health and Safety Commission). (2000). Management of health and safety at work. In *Management of Health and Safety at Work Regulations 1992. Approved code of practice and guidance L21,* 2nd ed. Sudbury, U.K.: HSE Books.

HSC (Health and Safety Commission). (2003). *Health and Safety Statistics Highlights 2002/2003.* Sudbury, U.K.: HSE Books.

HSE. (1982). *The Lifting of Patients in the Health Service.* London: HMSO.

HSE. (1995). Economic Impact: Revised Data from the Self-Reported Work-Related Survey in 1995. Information sheet SW 195. Sudbury, U.K.: HSE Books.

HSE. (1996). *Managing Crowds Safely.* London: HMSO.

HSE. (1997). *Seating at Work*, 3rd ed., HSG57. Sudbury, U.K.: HSE Books.

HSE. (2002). *Upper Limb Disorders in the Workplace*, 2nd ed., HSG60(rev). Sudbury, U.K.: HSE Books.

HSE. (2003a). Work with Display Screen Equipment: Health and Safety (Display Screen Equipment) Regulations 1992 as Amended by the Health and Safety (Miscellaneous Amendments) Regulations 2002: Guidance on Regulations. HSE Publication L26. Sudbury, U.K.: HSE Books.

HSE. (2003b). Manual Handling Assessment Charts. HSE Publication INDG383. Sudbury, U.K.: HSE Books.

HSE. (2004). Manual Handling. Manual Handling Operations Regulations 1992 (as amended): Guidance on Regulations, 3rd ed. HSE Publication L23. Sudbury, U.K.: HSE Books.

Hsu, S.-H. and Chen, Y.-H. (1999). Evaluation of bent-handled files. *International Journal of Industrial Ergonomics,* 25(1), 1–10.

Hult, L. (1954). Cervical, dorsal and lumbar spinal syndromes. *Acta Orthopaedica Scandinavia,* Supplement 17.

Hunter, D. (1955). *Diseases of the Occupations.* London: English Universities Press.

Ikai, M. and Fukunaga, T. (1968). Calculation of muscle strength per unit cross-sectioned area of human muscle by means of ultrasonic measurement. *Internationale Zeitschrift für Angewandte Physiologie Einschliesslich Arbeits-physiologie,* 26, 26–32.

Ilmarinen, J. (1997). Aging and work — coping with strengths and weaknesses. *Scandinavian Journal of Work, Environment and Health,* 23, 3–5.

Ilmarinen, J., Tuomi, K., and Klockars, M. (1997). Changes in the work ability of active employees over an 11-year period. *Scandinavian Journal of Work, Environment and Health,* 23,(Suppl. 1), 49–57.

ILO (1990). Maximum Weights in Load Lifting and Carrying. Occupational Safety and Health Series No. 59. Geneva: International Labour Office.

Imrhan, S. N. (1994). Muscular strength in the elderly — implications for ergonomic design. *International Journal of Industrial Ergonomics,* 13(2), 125–138.

Imrhan, S. N. and Sundararajan, K. (1992). An investigation of finger pull strengths. *Ergonomics,* 35(3), 289–299.

Inglemark, B. E. and Lewin, T. (1968). Anthropometrical studies on Swedish women. *Acta Morphologica Neerlando-Scandinavica,* III (2), 145–166.

Institute for Consumer Ergonomics. (1983). Seating for Elderly and Disabled People. Report No. 2, Anthropometric Survey. Loughborough: University of Technology.

Instituto Nacional de Tecnologia. (1989). Pesquisa Antropometrica e Biomecacica dos Operarios da Industria Transformmacao - RJ. Rio de Janeiro: Institute Nacional De Tecnologia.

Irvine, C. H., Snook, S. H., and Sparshatt, J. H. (1990). Stairway risers and treads: acceptable and preferred dimensions. *Applied Ergonomics,* 21(3), 215–225.

ISO. (1985). ISO 3055: 1985 Kitchen Equipment: Coordinating Sizes, Geneva: International Standards Organisation.

ISO. (1992a). ISO 2860: 1992 Earth-Moving Machinery: Minimum Access Dimensions. Geneva: International Standards Organisation.

ISO. (1992b). ISO 9241 Part 2:1992 Ergonomic Requirements for Office Work with Visual Display Terminals (VDTs): Guidance on Task Requirements. Geneva: International Standards Organisation.

ISO. (1994). ISO 5725: 1994 Accuracy (Trueness and Precision) of Measurement Methods and Results. Geneva: International Standards Organisation.

ISO. (1996). ISO 7250: 1996 Basic Human Body Measurements for Technological Design. Geneva: International Standards Organisation.

ISO. (1998a). ISO 9241 Part 4:1998 Ergonomic Requirements for Office Work with Visual Display Terminals (VDTs): Keyboard Requirements. Geneva: International Standards Organisation.

ISO. (1998b). ISO 9241 Part 5:1998 Ergonomic Requirements for Office Work with Visual Display Terminals (VDTs): Workstation Layout and Postural Requirements. Geneva: International Standards Organisation.

ISO. (2000a). ISO 9241 Part 3:1992/Amd 1:2000 Ergonomic Requirements for Office Work with Visual Display Terminals (VDTs): Visual Display Requirements. Geneva: International Standards Organisation.

ISO. (2000b). ISO 9241 Part 9:2000 Ergonomic Requirements for Office Work with Visual Display Terminals (VDTs). Requirements for Non-Keyboard Input Devices. Geneva: International Standards Organisation.

ISO. (2000c). ISO 11226 Ergonomics: Evaluation of Static Working Postures. Geneva: International Standards Organisation.

ISO. (2001). ISO 9241 Part 1:1997/Amd 1:2001 Ergonomic Requirements for Office Work with Visual Display Terminals (VDTs): General Introduction. Geneva: International Standards Organisation.

ISO. (2002a). ISO/FDIS 11228-1:2002 Ergonomics: Manual Handling Part 1: Lifting and Carrying. Geneva: International Standards Organisation.

ISO. (2002b). ISO 14738:2002 Safety of Machinery: Anthropometric Requirements for the Design of Workstations at Machinery. Geneva: International Standards Organisation.

ISO. (2003a). ISO/CD 11228-3:2003 Ergonomics: Manual Handling Part 3: Handling of Low Loads at High Frequency. Geneva: International Organisation for Standardisation.

ISO. (2003b). ISO 15535:2003 General Requirements for Establishing Anthropometric Databases. Geneva: International Standards Organisation.

ISO. (2004). ISO 15537:2004 Principles for Selecting and Using Test Persons for Testing Anthropometric Aspects of Industrial Products and Designs. Geneva: International Standards Organisation.

Jaschinski, W. and Heuer, H. (2004). Vision and eyes. In N. J. Delleman, C. M. Haslegrave, and D. B. Chaffin (Eds.), *Working Postures and Movements.* Boca Raton, FL: CRC Press, 73–86.

Jaschinski, W., Heuer, H., and Kylian, H. (1998). Preferred position of visual displays relative to the eyes: a field study of visual strain and individual differences. *Ergonomics,* 41(7), 1034–1049.

Jaschinski, W., Heuer, H., and Kylian, H. (1999). A procedure to determine the individually comfortable position of visual displays relative to the eyes. *Ergonomics,* 42(4), 535–549.

Jonai, H., Villanueva, M. B. G., Takata, A., Sotoyama, M., and Saito, S. (2002). Effects of the liquid crystal display tilt angle of a notebook computer on posture, muscle activities and somatic complaints. *International Journal of Industrial Ergonomics,* 29, 219–229.

Jones, D. A. and McConell, A. (1997). Changes in muscle function with age and their consequences for activity and work capacity. In *Proceedings of the Health and Safety Executive Research Meeting on Ageing, Work and Health, London, 4 April 1997.*

Jones, J. C. (1963). Fitting trials: a method of fitting equipment dimensions to variation in the activities, comfort requirements and body sizes of users. *Architects Journal,* 6 February, 321–325.

Jones, J. R. and Hodgson, J. T. (1998). *Self Reported Work Related Illness in 1995: Results from a Household Survey.* Sudbury, U.K.: HSE Books.

Jones, M. A. (1986). *Textbook on Torts*, 6th ed. London: Blackstone.

Jones, P. R. M. and Rioux, M. (1997). Three-dimensional surface anthropometry: applications to the human body. *Optics and Lasers in Engineering,* 28, 89–117.

Jurgens, H. W., Aune, I. A., and Pieper, U. (1990). International Data on Anthropometry. Occupational and Health Series No. 65. Geneva: International Labour Office.

Kanis, H. (1997). Variation in results of measurement repetition of human characteristics and activities. *Applied Ergonomics,* 28(3), 155–163.

Kanis, H. and Steenbekkers, L. P. A. (1995). On variability in human characteristics. In S. A. Robertson (Ed.), *Contemporary Ergonomics 1995.* London: Taylor and Francis, 512–517.

Kapandji, I. A. (1974). *The Physiology of the Joints.* Edinburgh: Churchill Livingstone.

Kaplan, B. A. (1954). Environment and human plasticity. *American Anthropology,* 56, 780–800.

Karlqvist, L., Bernmark, E., Ekenvall, L., Hagberg, M., Isaksson, A., and Rostö, T. (1999). Computer mouse and track-ball operation: similarities and differences in posture, muscular load and perceived exertion. *International Journal of Industrial Ergonomics,* 23(7), 1261–1267.

Karlqvist, L., Hagberg, M., and Selin, K. (1994). Variation in upper limb posture and movement during word processing with and without mouse use. *Ergonomics,* 37(3), 157–169.

Kee, D. and Karwowski, W. (2001). The boundaries for joint angles of isocomfort for sitting and standing males based on perceived comfort of static joint postures. *Ergonomics,* 44(6), 614–648.

Kee, D. and Karwowski, W. (2004). Joint angles of isocomfort for female subjects based on the psychophysical scaling of static standing postures. *Ergonomics,* 47(4), 427–445.

Keegan, J. J. (1953). Alterations of the lumbar curve related to posture and seating. *Journal of Bone and Joint Surgery,* 35A, 589–603.

Kember, P., Ainsworth, L., and Brightman, P. (1981). *A Hand Anthropometric Survey of British Workers.* Ergonomics Laboratory, Cranfield Institute of Technology, Cranfield.

Kennedy, K. W. (1964). Reach capability of the USAF population: Phase I, the outer boundaries of grasping reach envelopes for the shirt sleeved, seated operator. Report AMRL-TDR-64-59. Wright Patterson Airforce Base, OH.

Kinoshita, H., Kawai, S., and Ikuta, K. (1995). Contributions and co-ordination of individual fingers in multiple finger prehension. *Ergonomics,* 38(6), 1212–1230.

Kira, A. (1976). *The Bathroom.* Harmondsworth, U.K.: Penguin.

Kirvesoja, H., Väyrynen, S., and Häikiö, A. (2000). Three evaluations of task-surface heights in elderly people's homes. *Applied Ergonomics,* 31(2), 109–119.

Kirwan, B. and Ainsworth, L. K. (Eds.). (1992). *A Guide to Task Analysis.* London: Taylor & Francis.

Klafs, C. E. and Lyon, M. J. (1978). *The Female Athlete: A Coach's Guide to Conditioning and Training.* St. Louis, MO: C. V. Mosby.

Knight, I. (1984). *The Heights and Weights of Adults in Great Britain.* London: HMSO Books.

Knowlton, R. G. and Gilbert, J. C. (1983). Ulnar deviation and short-term strength reductions as affected by a curve-handled ripping hammer and a conventional claw hammer. *Ergonomics,* 26(2), 173–179.

Koblianski, E. and Arensburg, B. (1977). Changes in morphology of human populations due to migration and selection. *Annals of Human Biology,* 4, 57–71.

Konz, S. (1986). Bent hammer handles. *Human Factors,* 28(3), 317–323.

Kovatz, F., Boszormenyi-Nagy, G., Nagy, G. G., and Ordig, L. (1988). Morphometry of the upright trunk during breathing. *SPIE Proceedings, Biostereometrics '88,* 1030, 255–262.

Kroemer, K. H. E. and Hill, S. G. (1986). Preferred line of sight angle. *Ergonomics,* 29, 1129–1134.

Lacey, A. (2004a) *Designing for Accessibility* (2004 Edition). London: Centre for Environments and RIBA Enterprises.

Lacey, A. (2004b). *Good Loo Guide.* London: Centre for Accessible Environments and RIBA Enterprises.

Lalueza Fox, C. and Frayer, D. W. (1997). Non-dietary marks in the anterior dentition of the Krapina Neanderthals. *International Journal of Osteoarchaeology,* 7, 133–149.

Le Carpentier, E. F. (1969). Easy chair dimensions for comfort. *Ergonomics*, 12, 328–337.

Lee, H. (2004). A new case of fatal pulmonary thromboembolism associated with prolonged sitting at computer in Korea. *Yonsei Medical Journal,* 45(2), 349–351.

Lewin, T. (1969). Anthropometric studies on Swedish industrial workers when standing and sitting. *Ergonomics,* 12, 883–902.

Li, G. and Buckle, P. (1999). Current techniques for assessing physical exposure to work-related musculoskeletal risks, with emphasis on posture-based methods. *Ergonomics,* 42(5), 674–695.

Li, G. and Haslegrave, C. M. (1999). Seated work posture for manual, visual and combined tasks. *Ergonomics,* 42(8), 1060–1086.

Li, G., Haslegrave, C. M., and Corlett, E. N. (1995). Factors affecting posture for machine sewing tasks: the need for changes in sewing machine design. *Applied Ergonomics,* 26(1), 35–46.

Liem, A. and Yan, H. (2004). Digital human models in work system design and simulation. In *Proceedings of the Digital Human Modeling for Design and Engineering Symposium, 15–17 June 2004, Rochester, MI.* Warrendale, PA: Society of Automotive Engineers. (on CD-ROM)

Lindgren, G. (1976). Height, weight and menarche in Swedish schoolchildren in relation to socio-economic and regional factors. *Annals of Human Biology,* 3, 510–528.

Little, K. B. (1965). Personal space. *Journal of Experimental Social Psychology,* 1, 237–247.

Lueder, R., Corlett, E. N., Danielson, C., Greenstein, G. C., Hsieh, J., Phillips, R., and Designworks/USA (1994). Does it matter that people are shaped differently, yet backrests are built the same? In R. Lueder and K. Noro (Eds.), *Hard Facts about Soft Machines.* London: Taylor & Francis, 205–217.

Lundervold, A. (1958). Electromyographic investigations during typewriting. *Ergonomics,* 1, 226–233.

MacFarlane, G. J. and Hunt, I. M. (2000). Role of mechanical and psychosocial factors in the onset of forearm pain: prospective population based study. *British Medical Journal,* 32, 676–679.

Magora, A. (1972). Investigations of the relation between low back pain and occupation. III. Physical requirements: sitting, standing and weight lifting. *Industrial Medicine,* 41, 5–9.

Magora, A. (1973a). Investigations of the relation between low back pain and occupation. IV. Physical requirements: bending, rotation, reaching and sudden maximal effort. *Scandinavian Journal of Rehabilitation Medicine,* 5, 186–190.

Magora, A. (1973b). Investigations of the relation between-low back pain and occupation. V. Psychological aspects. *Scandinavian Journal of Rehabilitation Medicine,* 5, 191–196.

Majoros, A. E. and Taylor, S. A. (1997). Work envelopes in equipment design. *Ergonomics in Design,* 5(1), 18–24.

Maki, B. E., Bartlett, S. A., and Fernie, G. R. (1984). Influence of stairway handrail height on the ability to generate stabilizing forces and moments. *Human Factors,* 26(6), 705–714.

Malina, R. M. and Zavaleta, A. N. (1976). Androgyny of physique in female track and field athletes. *Annals of Human Biology,* 3, 441–446.

Mandal, A. C. (1976). Work chair with tilted seat. *Ergonomics,* 19, 157–164.

Mandal, A. C. (1981). The seated man (Homo sedens): the seated work position, theory and practice. *Applied Ergonomics,* 12, 19–26.

Mandal, A. C. (1991). Investigation of the lumbar flexion of the seated man. *International Journal of Industrial Ergonomics,* 8(1), 75–87.

Marklin, R. W., Simoneau, G. G., and Monroe, J. F. (2000). Wrist and forearm posture from typing on split and vertically inclined computer keyboards. *Human Factors,* 41(4), 1559–1569.

Martin, B. J., Armstrong, T., Foulke, J. A., Natarajan, S., Klinenberg, E., Serina, E., and Rempel, D. (1996). Keyboard reaction force and finger flexor electromyograms during computer keyboard work. *Human Factors*, 38(4), 654–664.

Martin, F. and Niemitz, C. (2003). "Right-trunkers" and "left-trunkers": side preferences of trunk movements in wild Asian elephants (elephas maximus). *Journal of Comparative Psychology,* 117(4), 371–379.

Martin, J., Meltzer, H., and Eliot, D. (1988). *The Prevalence of Disability Among Adults.* London: HMSO Books.

Martin, J. I., Sabeh, R., Driver, L. L., Lowe, T. D., Hintze, R. W., and Peters, P. A. C. (1975). Anthropometry of law enforcement officers. NELC/TD 442, Naval Electronics Laboratory Center, San Diego, CA.

Martin, W. E. (1960). Children's body measurements for planning and equipping schools. Special Publication No. 4, U.S. Department of Health, Education and Welfare, Hyattsville, MD.

Mascie-Taylor, C. G. N. (1994). Statistical issues in anthropometry. In S. J. Ulijaszek and C. G. N. Mascie-Taylor (Eds.), *Anthropometry: The Individual and the Population.* Cambridge: Cambridge University Press, 56–77.

McClelland, I. L. and Ward, J. S. (1976). Ergonomics in relation to sanitary ware design. *Ergonomics,* 19, 465–478.

McClelland, I. L. and Ward, J. S. (1982). The ergonomics of toilet seats. *Human Factors,* 24, 713–725.

McConville, J. T. and Churchill, E. (1976). Statistical Concepts in Design. Technical Report AMRL-TR-76-29. Wright Patterson Air Force Base, OH: Aerospace Medical Research Laboratory.

McConville, J. T., Tebbetts, I., and Alexander, M. (1979). Guidelines for fit-testing and evaluation of USAF personal protective clothing and equipment. Technical Report AMRL-TR-79-2. Wright-Patterson Air Force Base, OH: Aerospace Medical Research Laboratory.

McCormick, E. J. (1970). *Human Factors Engineering.* New York: McGraw-Hill.

Medawar, P. B. (1944). Size, shape and age. In *Essays on Growth and Form Presented to D'Arcy Wentworth Thompson.* Oxford: Clarendon Press, 155–187.

Meindl, R. S., Zehner, G. F., and Hudson, J. A. (1993). A Multivariate Anthropometric Method for Crew Station Design (Statistical Techniques). Technical Report AL/CF - 1993-0054. Wright-Patterson Air Force Base, OH: Aerospace Medical Research Laboratory.

Melzack, R. and Wall, P. (1982). *The Challenge of Pain.* Harmondsworth, U.K.: Penguin.

Meredith, H. W. (1976). Findings from Asia, Australia, Europe and North America on secular change in mean height of children, youths and young adults. *American Journal of Physical Anthropology,* 44, 315–326.

Meunier, P. (2000). Use of body shape information in clothing size selection. In *Proceedings of the XIVth Triennial Congress of the International Ergonomics Association and 44th Annual Meeting of the Human Factors and Ergonomics Society,* Volume 6. Santa Monica, CA: Human Factors and Ergonomics Society, 715–718.

Miall, W. E., Ashcroft, M. T., Lovell, H. G., and Moore, F. (1967). A longitudinal study of the decline of adult height with age in two Welsh communities. *Human Biology,* 39, 445–454.

Miller, C. D. (1961). Stature and build of Hawaii-born youth of Japanese ancestry. *American Journal of Physical Anthropology,* 19, 159–171.

Mital, A. and Channaveeraiah, C. (1988). Peak volitional torques for wrenches and screwdrivers. *International Journal of Industrial Ergonomics,* 3(1), 41–64.

Mital, A., Fard, H. F., and Khaledi, H. (1987). A biomechanical evaluation of staircase riser heights and tread depths during stair climbing. *Clinical Biomechanics,* 2, 162–164.

Mokdad, A. H., Serdula, M. K., Dietz, W. H., Bowman, B. A., Marks, J. S., and Koplan, J. P. (1999). The spread of obesity epidemic in the United States, 1991-1998. *JAMA,* 282, 1519–1522.

Molenbroek, J. F. M. (1994). Op Maat Gemaakt:- Menselijke Maten voor het Ontwerpen en Beoordelen van Gebruiksgoederen. Doctoral thesis, Delft University Press, Delft, The Netherlands. (in Dutch)

Montoye, H. J. and Lamphier, D. E. (1977). Grip and arm strengths in males and females. *Research Quarterly of the American Association for Health, Physical Education and Recreation,* 48, 109–120.

Morris, J. N., Heady, J. A., and Raffle, P. A. B. (1956). Physique of London busmen: epidemiology of uniforms. *Lancet,* 15 September, 569–70.

Nachemson, A. and Elfstrom, G. (1970). Intra-vital dynamic pressure measurements in lumbar discs. *Scandinavian Journal of Rehabilitation Medicine,* Supplement 1.

Nakashima, K. and Sato, H. (1999). Personal distance against a mobile robot. *Japanese Journal of Ergonomics,* 35(2), 87–95.

Napier, J. R. (1956). The prehensile movements of the human hand. *Journal of Bone and Joint Surgery,* 38B, 902–13.

National Research Council (1999). Work-Related Musculoskeletal Disorders: Report, Workshop Summary and Workshop Papers. Washington, DC: National Academy Press.

Nicholson, A. S., Parnell, J. W., and Davis, P. R. (1985). Bed design for back pain sufferers. In I. D. Brown, R. Goldsmith, K. Coombes, and M. A. Sinclair (Eds.), *Ergonomics International 85.* London: Taylor and Francis, 7–9.

Nicholson, A. S. and Ridd, J. E. (1988). *Health, Safety and Ergonomics.* London: Butterworth-Heineman.

NIOSH. (1981). *Work Practices Guide for Manual Lifting.* Cincinatti, OH: National Institute for Occupational Safety and Health.

Noble, J. (1982). *Activity and Spaces: Dimensional Data for Housing Design.* London: Architectural Press.

Norfolk, D. (1993). Bedding design: a professional appraisal. *British Osteopathic Journal,* 12, 13–16.

Norkin, C. C. and White, D. J. (1995). *Measurement of Joint Motion: A Guide to Goniometry,* 2nd ed. Philadelphia: F. A. Davis.

Norris, B. J. and Wilson, J. R. (1995). The design and safety of swimming pool covers. *International Journal of Consumer Safety,* 1(3), 163–175.

Norris, B. and Wilson, J. R. (1995). *Childata. The Handbook of Child Measurements and Capabilities: Data for Design Safety.* London: Department of Trade and Industry.

Nowak, E. (1989). Workspace for disabled people. *Ergonomics,* 32(9), 1077–1088.

NRC (National Research Council). (1989). *Diet and Health: Implications for Reducing Chronic Disease Risk.* Washington, DC: National Academy Press.

Oborne, D. J. and Heath, T. O. (1979). The role of social space requirements in ergonomics. *Applied Ergonomics,* 10, 99–103.

O'Herlihy, E. and Gaughran, W. (2003). Identification of working height ranges for wheelchair users in workshop environments. In P. T. McCabe (Ed.), *Contemporary Ergonomics 2003.* London: Taylor and Francis, 567–572.

Oliver, R., Gyi, D., Porter, M., Marshall, R., and Case, K. (2001). A survey of the design needs of older and disabled people. In M. A. Hanson (Ed.), *Contemporary Ergonomics 2001,* London: Taylor & Francis, 365–370.

Onishi, N., Sakai, K., and Kogi, K. (1982). Arm and shoulder muscles load in various keyboard operating postures in women. *Journal of Human Ergology,* 11, 89–97.

OPCS. (1981). Adult Heights and Weights Survey. OPCS Monitor ref. 5581/1, London: Office of Population Census and Surveys.

Op de Beeck, R. and Hermans, V. (2000). Research on Work-Related Low Back Disorders. Report for the European Agency for Safety and Health at Work. Luxembourg: Statistical Office of the European Communities.

Open Ergonomics Ltd. (2000). *PeopleSize.* Loughborough: Open Ergonomics.

Oxenburgh, M. (1984). Musculoskeletal injuries occurring in world processor operators. In *Proceedings of the 21st Annual Conference of the Ergonomics Society of Australia and New Zealand,* Narrabundah: Ergonomics Society of Australia, 137–143.

Päivinen, M., Haapalainen, M., and Mattila, M. (1999/2000). Ergonomic design criteria for pruning shears. *Occupational Ergonomics,* 2(3), 163–177.

Patrias, K. (2000). Visible Human Project, Current Bibliographies in Medicine 2000-5. Bethesda, MD: US Department of Health and Human Services, Public Health Service, National Institutes of Health.

Paul, J. A., Frings-Dresen, M. H. W., Sall, H. J. A., and Rozendal, R. H. (1995). Pregnant women and working surface height and working surface areas for standing manual work. *Applied Ergonomics*, 26(2), 129–133.

Peebles, L. and Norris, B. (1998). ADULTDATA. The Handbook of Adult Anthropometric and Strength Measurements: Data for Design Safety. London: Department of Trade and Industry.

Pheasant, S. T. (1982). A technique for estimating anthropometric data from the parameters of the distribution of stature. *Ergonomics,* 25, 981–992.

Pheasant, S. T. (1983). Sex differences in strength: some observations on their variability. *Applied Ergonomics,* 14, 205–211.

Pheasant, S. T. (1984). Human proportions: sex, age and ethnic differences. In E. D. Megaw (Ed.), *Contemporary Ergonomics 1984.* London: Taylor & Francis, 142–147.

Pheasant, S. T. (1987). *Ergonomics: Standards and Guidelines for Designers.* PP 7317. London: British Standards Institution.

Pheasant, S. T. (1988a). The Zeebrugge-Harrisburg syndrome. *New Scientist,* 21 January, 55–58.

Pheasant, S. T. (1988b). User-centred design — an ergonomist's viewpoint. In A. S. Nicholson and J. E. Ridd (Eds.), *Health, Safety and Ergonomics.* London: Butterworths, 73–96.

Pheasant, S. T. (199la). *Ergonomics, Work and Health.* London: Macmillan.

Pheasant, S. T. (1991b). *Anthropometrics: An Introduction,* 2nd ed. PP 7310, London: British Standards Institution.

Pheasant, S. T. (1992). Does RSI exist? *Journal of Occupational Medicine,* 42, 167–168.

Pheasant, S. T. (1994a). Musculoskeletal injury at work: natural history and risk factors. in B. Richardson and A. Eastlake (Eds), *Physiotherapy in Occupational Health.* Oxford: Butterworth Heinemann, 146–170.

Pheasant, S. T. (1994b). Repetitive strain injury: towards a clarification of the points at issue. *Journal of Personal Injury Litigation,* September, 223–230.

Pheasant, S. T. (1995). A foreseeable risk of injury. In S. A. Robertson (Ed.), *Contemporary Ergonomics 1995.* London: Taylor & Francis, 2–13.

Pheasant, S. T. and O'Neill, D. (1975). Performance in gripping and turning. *Applied Ergonomics,* 6, 205–208.

Pheasant, S. T. and Scriven, J. G. (1983). Sex differences in strength: some implications for the design of hand tools. In K. Coombes (Ed.), *Proceedings of the Ergonomics Society's Conference 1983.* London: Taylor & Francis, 9–13.

Pheasant, S. T. and Stubbs, D. (1992a). Back pain in nurses: epidemiology and risk assessment. *Applied Ergonomics,* 23, 226–232.

Pheasant, S. T. and Stubbs, D. (1992b). *Lifting and Handling: An Ergonomic Approach.* London: National Back Pain Association.

Porter, M. L. (2000). The anthropometry of the fingers of children. In *Proceedings of the XIVth Triennial Congress of the International Ergonomics Association and 44th Annual Meeting of the Human Factors and Ergonomics Society, San Diego, 20 July – 4 August 2000,* Volume 6. Santa Monica, CA: Human Factors and Ergonomics Society, 27–30.

Psihogios, J. P., Sommerich, C. M., Mirka, G. A., and Moon, S. D. (2001). A field evaluation of monitor placement effects in VDT users. *Applied Ergonomics,* 32(4), 313–325.

Putz-Anderson, V. (1988). *Cumulative Trauma Disorders.* London: Taylor & Francis.

Quintner, J. (1991). The RSI syndrome in historical perspective. *International Disability Studies,* 13, 99–104.

Radl, G. W. (1980). Experimental investigations for optimal presentation mode and colour of symbols on the CRT-screen. In E. Grandjean and E. Vigliani (Eds.), *Ergonomic Aspects of Visual Display Terminals,* London: Taylor & Francis, 271–276.

Radwin, R. G., Oh, S., Jensen, T. R., and Webster, J. G. (1992). External finger forces in submaximal five-finger static pinch prehension. *Ergonomics,* 35(3), 275–288.

Rebiffé, R. (1966). An ergonomic study of the arrangement of the driving position in motor cars. In *Proceedings of the Institution of Mechanical Engineers,* 181, Part 3D 1966-67, 43–50.

Rebiffé, R., Guillien, J., and Pasquet, P. (1983). *Enquête Anthropométrique sur les Conducteurs Françaises.* Paris: Laboratoire de Physiologie et de Biomêchanique de l'Association Peugeout-Renault. (in French)

Rebiffé, R., Zayana, O., and Tarrière, C. (1969). Détermination des zones optimales pour l'emplacement des commandes manuelles dans l'espace de travail. *Ergonomics,* 12, 913–924. (in French)

Reilly, T., Tyrrell, A., and Troup, J. D. G. (1984). Circadian variation in human stature. *Chronobiology International,* 1(2), 121–126.

Reinecke, S. M., Hazard, R. G., and Coleman, K. (1994). Continuous passive motion in seating: a new strategy against low back pain. *Journal of Spinal Disorders,* 7, 29–35.

Rempel, D., Serina, E., Klinenberg, E., Martin, B. J., Armstrong, T. J., Foulke, J. A., and Natarajan, S. (1997). The effect of keyboard keyswitch make force on applied force and finger flexor muscle activity. *Ergonomics,* 40(8), 800–808.

Reynolds, H. M. (1978). The inertial properties of the body and its segments. In Webb Associates, *Anthropometric Source Book, Volume I: Anthropometry for Designers.* NASA Reference Publication No. 1024. Houston: U.S. National Aeronautics and Space Administration, Lyndon B. Johnson Space Center, IV.1-IV.76.

Reynolds, H. and Allgood, M. (1975). Functional strength of commercial airline stewardessses. FAA-AM-75-13, Department of Transportation, FAA, Washington, D.C.

Ridd, J. E. (1985). Spatial restraints and intra-abdominal pressure. *Ergonomics,* 28, 149–166.

Rioux, M. and Bruckart, J. (1997). Data collection. In K. M. Robinette, M. W. Vannier, M. Rioux, and P. R. M. Jones (Eds.), *3-D Surface Anthropometry: Review of Technologies.* AGARD Advisory Report 329. Neuilly-sur-Seine, France: Advisory Group for Aerospace Research and Development.

Roberts, D. F. (1973). *Climate and Human Variability.* An Addison-Wesley Module in Anthropology, No. 34. Reading, MA: Addison-Wesley.

Roberts, D. F. (1975). Population differences in dimensions, their genetic basis and their relevance to practical problems of design In A. Chapanis (Ed.), *Ethnic Variables in Human Factors Engineering,* Baltimore, MD: Johns Hopkins University Press, 11–29.

Robertson, D. and Bedford, S. (1999). The use of workspace modelling in the design of an offshore asset. In M. A. Hanson, E. J. Lovesey, and S. A. Robertson, (Eds.), *Contemporary Ergonomics 1999.* London: Taylor and Francis, 53–57.

Robinette, K. M. (2000). CAESAR measures up. *Ergonomics in Design,* 8(3), 17–23.

Robinette, K., Blackwell, S., Daanen, H., Fleming Boehmer, M., Brill, T., Hoeferlin, D., and Burnsides, D. (2002). *Civilian American and European Surface Anthropometry Resource (CAESAR), Final Report, Volume I: Summary.* AFRL-HE-WP-TR-2002-0169. Wright-Patterson AFB, OH: United States Air Force Research Laboratory, Human Effectiveness Directorate, Crew System Interface Division.

Robinette, K., Daanen, H., and Zehner, G. (2004). 3D anthropometry. In N. J. Delleman, C. M. Haslegrave, and D. B. Chaffin (Eds.), *Working Postures and Movements.* Boca Raton, FL: CRC Press, 29–49.

Robinette, K. M. and McConville, J. T. (1981). An Alternative to Percentile Models. SAE Technical Paper 810217. Warrendale, PA: Society of Automotive Engineers.

Roche, A. F. (1979). Secular trends in stature, weight and maturation. *Monographs of the Society for Research in Child Development,* Serial no. 179, 44(3-4), 3–27.

Roche, A. F and Davila, G. H. (1972). Late adolescent growth in stature. *Pediatrics,* 50, 874–880.

Roebuck, J. A. Jr. (1995). *Anthropometric Methods: Designing to Fit the Human Body.* Santa Monica, CA: Human Factors and Ergonomics Society.

Roebuck, J. A. Jr., Kroemer, K. H. E., and Thomson, W. G. (1975). *Engineering Anthropometry Methods.* New York: Wiley.

Roebuck, J. A. Jr. and Levendahl, B. N. (1961). Aircraft ground emergency exit design considerations. *Human Factors,* 3, 174–209.

Rona, R. J. (1981). Genetic and environmental factors in the control of growth in childhood. *British Medical Bulletin,* 37(3), 265–272.

Rona, R. J. and Altman, D. G. (1977). National study of health and growth: standards of attained height, weight and triceps skinfold in English children 5 to 11 years old. *Annals of Human Biology,* 4, 501–523.

Rona, R. J., Swan, A. V., and Altman, D. G. (1978). Social factors and health of primary schoolchildren in England and Scotland. *Journal of Epidemiology and Community Health,* 32, 147–154.

Ryan, G. A. and Bampton, M. (1988). Comparison of data processing operators with and without upper limb symptoms. *Community Health Studies,* 12, 63–68.

Savinainen, M., Nygard, C.-H., and Ilmarinen, J. (2004). A 16-year follow-up study of physical capacity in relation to perceived workload among aging employees. *Ergonomics,* 47(10) 1087–1102.

Schlichting, P., Hoilund-Carlsen, P., and Quade, F. (1981). Comparison of self-reported height and weight with controlled height and weight in women and men. International. *Journal of Obesity,* 5, 67–76.

Schulze, L. J. H., Koppa, R. J., Congleton, J. J., and Whiteley, J. D. (1991). Effect of pneumatic screwdrivers and workstations on operator body posture. *International Journal of Industrial Ergonomics*, 8, 17–31.

Seitz, T. and Bubb, H. (1999). Measuring of Human Anthropometry, Posture and Motion. SAE Technical Paper 1999-01-1913. Proceedings of Digital Human Modeling for Design and Engineering Symposium, 1999, The Hague. Warrendale, PA: Society of Automotive Engineers.

Sen, R. N. (1984). Application of ergonomics to industrially developing countries. *Ergonomics,* 27(10),1033–1050.

Sengupta, A. K. and Das, B. (2000). Maximum reach envelope for the seated and standing male and female for industrial workstation design. *Ergonomics,* 43(9), 1390–1404.

Sengupta, A. K. and Das, B. (2004). Determination of worker physiological cost in workspace reach envelopes. *Ergonomics,* 47(3), 330–342.

Shackel, B., Chidsey, K. D., and Shipley, P. (1969). The assessment of chair comfort. *Ergonomics,* 12(2), 269–306.

Shapiro, H. (1939). *Migration and Environment.* London: Oxford University Press.

Shih, Y.-C., Wang, M.-J., and Chang, C.-H. (1995). Evaluating the factors affecting MVTE in valve operation. In S. A. Robertson (Ed.), *Contemporary Ergonomics 1995*. London: Taylor and Francis, 524–529.

Shivers, L., Mirka, G. A., and Kaber, D. B. (2002). Effect of grip span on lateral pinch grip strength. *Human Factors*, 44(4), 569–577.

Silverstein, S. J., Fine, L. J., and Armstrong, T. J. (1986). Hand-wrist cumulative trauma disorders in industry. *British Journal of Industrial Medicine,* 43, 779–784.

Silverstein, S. J., Fine, L. J., and Armstrong, T. J. (1987). Occupational factors and carpal tunnel syndrome. *American Journal of Industrial Medicine,* 11, 343–358.

Simoneau, G. G. and Marklin, R. W. (2001). Effect of computer keyboard slope and height on wrist extension angle. *Human Factors,* 43(2), 287–298.

Simpson, E. and Buckle, P. (1999). The effect of telephone headsets on working posture and musculoskeletal symptoms: an intervention study. In M. A. Hanson, E. J. Lovesey, and S. A. Robertson (Eds.), *Contemporary Ergonomics 1999.* London: Taylor and Francis, 301–305.

Simpson, K. (1940). Shelter deaths from pulmonary embolism. *Lancet,* I, 744.

Smith, M. J. and Carayon, P. (1996). Work organization, stress, and cumulative trauma disorders. In S. D. Moon and S. L. Sauter (Eds.), *Beyond Biomechanics: Psychosocial Aspects of Musculoskeletal Disorders in Office Work.* London: Taylor & Francis, 23–42.

Smith, S. and Norris, B. (2001). Childata: Assessment of the Validity of Data. Report from the Product Safety Testing Group, Institute for Occupational Ergonomics, University of Nottingham for the Consumer Affairs Directorate, Department of Trade and Industry, London.

Smith, S. and Norris, B. J. (2004). Changes in the body size of UK and US children over the past three decades. *Ergonomics,* 47(11), 1195–1207.

Smith, S., Norris, B., and Peebles, L. (2000). OLDER ADULTDATA. *The Handbook of Measurements and Capabilities of the Older Adult: Data for Design Safety.* London: Department of Trade and Industry.

Snyder, R. G., Schneider, L. W., Owings, C. L., Reynolds, H. M., Golomb, D. H., and Schork, M. A. (1977). Anthropometry of Infants, Children and Youths to Age 18 for Product Safety Design. Report No. DB-270 277. Bethesda, MD: Consumer Product Safety Committee, U.S. Department of Commerce.

Sommer, R. (1969). *Personal Space: The Behavioural Basis of Design.* Englewood Cliffs, NJ: Prentice Hall.

Sommerich, C. M., Starr, H., Smith, C. A., and Shivers, C. (2002). Effects of notebook computer configuration and task on user biomechanics, productivity, and comfort. *International Journal of Industrial Ergonomics,* 30, 7–31.

Spence, S. (1990). Psychopathology amongst acute and chronic patients with occupationally related upper limb pain versus accident injuries of the upper limbs. *Australian Psychologist,* 25, 293–305.

Sperling, L., Dahlman, S., Wikström, L., Kilbom, Å., and Kadefors, R. (1993). A cube model for the classification of work with hand tools and the formulation of functional requirements. *Applied Ergonomics,* 24(3), 212–220.

Squires, P. C. (1956). The Shape of the Normal Working Area. Report No. 275. New London, CT: U.S. Navy Department, Bureau of Medicine and Surgery, Medical Research Laboratories.

Steenbekkers, L. P. A. (1998). Ranges of movements of joints. In L. P. A. Steenbekkers and C. E. M. van Beijsterveldt (Eds.), *Design-Relevant Characteristics of Ageing Users,* Delft, The Netherlands: Delft University Press, 60–68.

Steenbekkers, L. P. A. and Dirken, J. M. (1998). Conclusion and discussion. In L. P. A. Steenbekkers and C. E. M. van Beijsterveldt (Eds.), *Design-Relevant Characteristics of Ageing Users.* Delft, The Netherlands: Delft University Press, 199–207.

Steenbekkers, L. P. A. and van Beijsterveldt, C. E. M. (Eds.). (1998). *Design-Relevant Characteristics of Ageing Users.* Delft, The Netherlands: Delft University Press.

Stenlund, B., Goldie, I., Hagberg, M., Hogstedt, C., and Marions, O. (1992). Radiographic osteoarthrosis in the acromioclavicular joint resulting from manual work or exposure to vibration. *British Journal of Industrial Medicine,* 49(8), 588–593.

Stoudt, H. W. (1981). The anthropometry of the elderly. *Human Factors,* 23(1), 29–37.

Stoudt, H. W., Damon, A., and McFarland, R. A. (1970). Skinfolds, Body Girths, Biacromial Diameter and Selected Anthropometric Indices of Adults. National Centre for Health Statistics, Series 11, No. 35. Hyattsville, MD: U.S. Department of Health and Human Services, National Center for Health Statistics.

Stoudt, H. W., Damon, A. McFarland, R., and Roberts, J. (1965). Weight, Height and Selected Body Dimensions of Adults, United States 1961-1962. National Centre for Health Statistics, Series 11, No. 8. Hyattsville, MD: U.S. Department of Health and Human Services, National Center for Health Statistics.

Straker, L., Jones, K. J., and Miller, J. A. (1997). Comparison of the postures assumed when using laptop computers and desktop computers. *Applied Ergonomics,* 28(4), 263–268.

Straker, L., Pollock, C., Frosh, A., Aarås, A., and Dainoff, M. (2000). An ergonomic field comparison of a traditional computer mouse and a vertical computer mouse in uninjured office workers. In *Proceedings of the XIVth Triennial Congress of the International Ergonomics Association and 44th Meeting of the Human Factors and Ergonomics Society, 27 July - 6 August 2000, San Diego.* Santa Monica, CA: Human Factors and Ergonomics Society, Volume 6, 356–359.

Stranden, E. (2000). Dynamic leg volume changes when sitting in a locked and free floating tilt office chair. *Ergonomics,* 43(3), 421–433.

Strasser, H., Wang, B., and Hoffmann, A. (1994). Electromyographic and subjective assessment of masons' trowels equipped with different handles. In F. Aghazadeh (Ed.), *Advances in Industrial Ergonomics and Safety VI.* London: Taylor & Francis, 553–560.

Tanner, J. M. (1962). *Growth at Adolescence.* Oxford: Blackwell.

Tanner, J. M. (1978). *Foetus into Man.* London: Open Books.

Tanner, J. M., Hayashi, T., Preece, M. A., and Cameron, N. (1982). Increase in length of leg relative to trunk in Japanese children and adults from 1957-1977: a comparison with British and with Japanese Americans. *Annals of Human Biology,* 9, 411–423.

Tanner, J. M. and Whitehouse, R. H. (1976). Clinical longitudinal standards for height, weight, height velocity, weight velocity and stages of puberty. *Archives of Diseases in Childhood,* 51, 170–179.

Tanner, J. M., Whitehouse, R. H., and Takaishi, M. (1966). Standards from birth to maturity for height, weight, height velocity and weight velocity: British children, 1965. Part I. *Archives of Diseases of Childhood,* 41, 454–471; Part 2. *Archives of Diseases of Childhood,* 41, 613–635.

Taylor, J. H. (1973). Vision. In J. T. Parker and V. R. West (Eds.), *Bioastronautics Data Book,* 2nd ed. NASA Publication SP-3006. Washington, DC: National Aeronautics and Space Administration, 611–665.

Thompson, A. R., Plewes, L. W., and Shaw, E. G. (1951). Peritendinitis and simple tenosynovitis: a clinical study of 544 cases in industry. *British Journal of Industrial Medicine,* 9, 150–160.

Thompson, D. and Booth, R. T. (1982). The collection and application of anthropometric data for domestic and industrial standards. In R. Easterby, K. H. C. Kroemer, and D. B. Chaffin (Eds), *Anthropometry and Biomechanics: Theory and Applications.* New York: Plenum.

Thomson, A. M. (1959). Maternal stature and reproductive efficiency. *Eugenics Review,* 51, 157–162.

Tichauer, E. R. (1966), Some aspects of stress on forearm and hand in industry, *Journal of Occupational Medicine*, 8, 63–71.

Tichauer, E. R. (1975). *Occupational Biomechanics: The Anatomical Basis of Workplace Design.* Rehabilitation Monograph No. 51. New York: Institute of Rehabilitation Medicine, New York University Medical Centre.

Tichauer, E. R. (1978). *The Biomechanical Basis of Ergonomics.* New York: Wiley.

Tildesley, M. F. (1950). The relative usefulness of various characters on the living for racial comparison. *Man,* 50, 14–18.

Travell, J. (1967). Mechanical headache. *Headache,* 7, 23–29.

Travell, J. E. and Simons, D. E. (1983). *Myofascial Pain and Dysfunction: The Trigger Point Manual.* Baltimore, MD: Williams & Wilkins.

Treaster, D. E. and Marras, W. S. (2000). An assessment of alternate keyboards using finger motion, wrist motion and tendon travel. *Clinical Biomechanics,* 15, 499–503.

Troiano, R. P., Flegal, K. M., Kuczmarski, R. J., Campbell, S. M., and Johnson, C. L. (1995). Overweight prevalence and trends for children and adolescents. The National Health and Nutrition Examination Surveys, 1963 to 1991. *Archives of Pediatric Adolescent Medicine,* 149, 1085–1091.

Trotter, M. and Gleser, G. (1951). The effect of ageing upon stature. *American Journal of Physical Anthropology,* 9, 311–324.

Troy, J. and Guerin, J. (2004). Human swept volumes. *Proceedings of the Digital Human Modeling for Design and Engineering Symposium, 15-17 June 2004, Rochester, MI.* Warrendale, PA: Society of Automotive Engineers. (on CD-ROM)

Tuomi, K., Ilmarinen, J., Eskelinen, L., Järvinen, E., Toikkanen, J., and Klockars, M. (1991). Prevalence and incidence rates of diseases and work ability in different categories of municipal occupations. *Scandinavian Journal of Work, Environment and Health,* 17(Suppl. 1), 67–74.

Tutt, D. and Adler, D. (Eds). (1979). *New Metric Handbook.* London: Architectural Press.

Van Cott, H. P. and Kinkade, R. G. (Eds). (1972). *Human Engineering Guide to Equipment Design.* Washington, DC: U.S. Department of Defense.

Van Deursen, D. L., Van Deursen, L. L. J. M., Snijders, C. J., and Goossens, R. H. M. (2000). Effect of continuous rotary seat pan movements on physiological oedema of the lower extremities during prolonged sitting. *International Journal of Industrial Ergonomics,* 26(5), 521–526.

Vercruyssen, M. and Simonton, K. (1994). Effects of posture on mental performance: we think faster on our feet than on our seat. In: R. Lueder and K. Noro (Eds.), *Hard Facts about Soft Machines: The Ergonomics of Seating.* London: Taylor & Francis, 119–131.

Viitasalo, J. T., Era, P., Leskinen, A. L., and Heikkinen, E. (1985). Muscular strength profiles and anthropometry in random samples of men aged 31-35, 51-55 and 71-75 years. *Ergonomics,* 28, 1563–1574.

Vincent, L. M. (1979). *Competing with the Sylph: Dancers in Pursuit of the Ideal Body Form.* Kansas City: Andrews & McMeel.

Voorbij, A. I. M. and Steenbekkers, L. P. A. (1998). Step height. In L. P. A. Steenbekkers and C. E. M. van Beijsterveldt (Eds.), *Design-Relevant Characteristics of Ageing Users.* Delft, The Netherlands: Delft University Press, 85–91.

Voorbij, A. I. M. and Steenbekkers, L. P. A. (2001). The composition of a graph on the decline of total body strength with age based on pushing, pulling, twisting and gripping force. *Applied Ergonomics,* 32(3), 287–292.

Wang, Y., Das, B., and Sengupta, A. K. (1999). Normal horizontal working area: the concept of inner boundary. *Ergonomics,* 42(4), 638–646.

Ward, J. S. (1971). Ergonomic techniques in the determination of optimum work surface heights. *Applied Ergonomics,* 2, 171–177.

Ward, J. S. and Beadling, W. M. (1970). Optimum dimensions for domestic staircases. *Architects Journal,* 151, 513–520.

Ward, J. S. and Kirk, N. S. (1970). The relation between some anthropometric dimensions and preferred working surface heights in the kitchen. *Ergonomics,* 6, 783–797.

Wardell, L. and Mrozowski, P. (2001). Comparing the use of different designs of computer input devices in a real life working environment. In M. A. Hanson (Ed.), *Contemporary Ergonomics 2001.* London: Taylor & Francis, 99–104.

Warren, C. G. and Valois, T. (1991). Functional Categories of Persons with Disabilities and Operational Dimensions for Designing Accessible Aircraft Lavatories. Report prepared for sponsors. Seattle, WA: Paralyzed Veterans of America, National Easter Seal Society, Multiple Sclerosis Society and United Cerebral Palsy Association.

Wartenberg, C., Dukic, T., Falck, A. C., and Hallbeck, S. The effect of assembly tolerance on performance of a tape application task: a pilot study. *Industrial Journal of Ergonomics*, 33, 4, 369–379.

Waters, T. R., Putz-Anderson, V., Garg, A., and Fine, L. J. (1993). Revised NIOSH equation for the design and evaluation of manual lilting tasks. *Ergonomics,* 36, 749–776.

Webb Associates. (1978). *Anthropometric Source Book.* NASA Reference Publication No. 1024. Lyndon B. Johnson Space Center, Houston: U.S. National Aeronautics and Space Administration.

Wells, L. H. (1963). Stature in earlier races of mankind. In D. Bothwell and E. Higgs (Eds.), *Science in Archaeology.* London: Thames & Hudson.

Wells, R. (2004). Elbow, forearm, and wrist. In N. J. Delleman, C. M. Haslegrave, and D. B. Chaffin (Eds.), *Working Postures and Movements.* Boca Raton, FL: CRC Press, 297–310.

Westerstahl, M., Barnekow-Bergkvist, M., Hedberg, G., and Jansson, E. (2003). Secular trends in body dimensions and physical fitness among adolescents in Sweden from 1974 to 1995. *Scandinavian Journal of Medicine and Science in Sports,* 13, 128–137.

Weston, H. C. (1953). Visual fatigue with special reference to lighting. In W. F. Floyd and A. T. Welford (Eds.), *Symposium on Fatigue.* London: H. K. Lewis, 117–135.

WHO (World Health Organization). (1993). Aging and Working Capacity. WHO Technical Report 835. Geneva: World Health Organization.

WHO (World Health Organization). (1998a). *The World Health Report 1998. Life in the 21st Century: A Vision for All.* Geneva: World Health Organization.

WHO (World Health Organization). (1998b). Obesity: Preventing and Managing the Global Epidemic. Report of a WHO Consultation on Obesity, Geneva, 3-5 June 1997. Report No. WHO/NUT/NCD/98.1. Geneva: World Health Organization.

WHO (World Health Organization). (2001). *International Classification of Functioning, Disability and Health.* Geneva: World Health Organization.

Wick, J. and Drury, C. G. (1986). Postural change due to adaptations of a sewing workstation. In N. Corlett, J. Wilson, and I Manenica (Eds.), *The Ergonomics of Working Postures.* London: Taylor & Francis, 375–379.

Wigley, R. D. (1990). Repetitive strain syndrome: fact not fiction. *New Zealand Medical Journal,* 28 February, 75–76.

Wilmore, J. H. (1976). *Athletic Training and Physical Fitness: Physiological Principles and Practices of the Conditioning Process.* Boston: Allyn & Bacon.

Wilson, J. R. and Corlett, N. (2005). *Evaluation of Human Work,* 3rd ed. London: Taylor & Francis.

Wilson, S. (2001). The ergonomic ramifications of laptop computer use. In M. A. Hanson (Ed.), *Contemporary Ergonomics 2001.* London: Taylor & Francis, 93–98.

Wolstad, J. C., Mcmulkin, M. L., and Bussi, C. A. (1995). Forces applied to large hand wheels. *Applied Ergonomics,* 26(1), 55–60.

Woods, V., Hastings, S., Buckle, P., and Haslam, R. (2002). Ergonomics of Using a Mouse or Non-Keyboard Input Device. HSE Research Report 405. Sudbury, U.K.: HSE Books.

Woodson, W. E. (1981). *Human Factors Design Handbook.* New York: McGraw-Hill.

Wright, P. H., Ashford, N. J., and Stammer, R. Jr. (1998). *Transportation Engineering: Planning and Design.* New York: Wiley.

Xiao, H., Hua, D-H., and Liu, W. (1998). A research on head-face dimensions of Chinese adults. In *Proceedings of the 5th Pan-Pacific Conference on Occupational Ergonomics, 21-23 July 1998, Kitakyushu,* 60–63.

Yamana, N., Okabe, K., Nanako, C., Zenitani, Y., and Saita, T. (1984). The body form of pregnant women in monthly transitions. *Japanese Journal of Ergonomics,* 2, 171–178. (In Japanese)

Yanagisawa, S. and Kondo, S. (1973). Modernization of physical features of the Japanese with special reference to leg length and head form. *Journal of Human Ergology,* 2, 97–108.

Zecevic, A., Miller, D. I., and Harburn, K. (2000). An evaluation of three computer keyboards. *Ergonomics,* 43(1), 55–72.

Zehner, G. F., Meindl, R. S., and Hudson, J. A. (1993). A Multivariate Anthropometric Method for Crew Station Design. Technical Report AL-TR - 1992-0164. Wright-Patterson Air Force Base, OH: Aerospace Medical Research Laboratory.

Zhang, L., Helander, M. G., and Drury, C. G. (1996). Identifying factors of comfort and discomfort in sitting. *Human Factors,* 38(3), 377–389.

Zhang, X., Chaffin, D. B, and Thompson, D. (1997). Development of Dynamic Simulation Models of Seated Reaching Motions while Driving. SAE Technical Paper 970589. Warrendale, PA: Society of Automotive Engineers.

Index

A

B

C

D

E

F

G

H

I

J

K

L

M

N

O

P

R

S

T

U

V

W

Z